Microbolometers

Woodhead Publishing Series in Electronic and Optical Materials

Microbolometers

Fundamentals, Materials, and Recent Developments

Nuggehalli Ravindra
New Jersey Institute of Technology, Newark, New Jersey, United States

ELSEVIER
WP
WOODHEAD PUBLISHING
An imprint of Elsevier

Woodhead Publishing is an imprint of Elsevier
The Officers' Mess Business Centre, Royston Road, Duxford, CB22 4QH, United Kingdom
50 Hampshire Street, 5th Floor, Cambridge, MA 02139, United States
The Boulevard, Langford Lane, Kidlington, OX5 1GB, United Kingdom

Library of Congress Cataloging-in-Publication Data
A catalog record for this book is available from the Library of Congress

British Library Cataloguing-in-Publication Data
A catalogue record for this book is available from the British Library

ISBN: 978-0-08-102812-4 (print)
ISBN: 978-0-08-102813-1 (online)

For information on all Woodhead publications
visit our website at https://www.elsevier.com/books-and-journals

Publisher: Matthew Deans
Acquisitions Editor: Kayla Dos Santos
Editorial Project Manager: Andrea Gallego Ortiz
Production Project Manager: Kamesh Ramajogi
Cover Designer: Mark Rogers

Typeset by STRAIVE, India

Contents

Preface

The beginnings of the author's thinking along the lines of infrared detectors and temperature sensors had taken shape at the Indian Institute of Technology, Roorkee (then University of Roorkee), India, during the years 1977 through 1981 [1–3]. The author recalls the contributions and support of the late Professors V.K. Srivastava and Sushil Auluck at Roorkee and the late Dr. Trevor Simpson Moss [4–6] of the Royal Signals & Radar Establishment, Malvern, United Kingdom.

At NJIT, DARPA's funding of the Project in the 1990s, "Multiwavelength Imaging Pyrometry for Semiconductor Process Monitoring & Control," [7] and US DOD-DURIP's funding of the Project, "Spectral Emissometry for Semiconductor Process Monitoring & Control," [8] laid the foundation for the 2D mapping of temperature of semiconductors via noncontact temperature measurement techniques. These research activities—coupled with Professor William N. Carr's interests in vanadium oxides [9,10]; Professor Walter F. Kosonocky's contributions to Infra-Red Charge Coupled Devices (IRCCDs) based on platinum silicide Schottky Barrier Detectors (SBDs) [11,12]; and the author and his research group's interests in HgCdTe [13], vanadium oxides [14,15], silicides [16], thin films [17], and superconducting microbolometers [18]—were the inspiration for writing this book entitled *Microbolometers: Fundamentals, Materials, and Recent Developments.*

DARPA's commitment to the development of uncooled microbolometer has been a catalyst for the growth of uncooled microbolometers. The pioneering contributions of Honeywell and Texas Instruments continue to be the basis for novel approaches for the manufacture of microbolometers by several companies throughout the world. Recent applications such as autonomous vehicles, drones, rechargeable batteries/energy storage, and 3D printing are just a few examples of the use of microbolometers for infrared imaging/2D temperature mapping.

Next year, 2022, marks the 200th death anniversary of Frederick William Herschel [19]. Herschel's discovery of infrared light in 1800 has revolutionized various sectors of the global economy, including aeronautics, astronomy, civil infrastructure, energy production and monitoring, health and medical sciences, manufacturing, military, optical communications, safety, space technology, and transportation. It is also the 30th anniversary of the declassification of the Honeywell's Microbolometer technology by the United States government.

The 2021 Nobel Prize in Physiology or Medicine has been awarded to David Julius and Ardem Patapoutian for their discoveries of receptors for temperature and touch [**20**]. The need for noncontact temperature measurements will continue, each and every day—this is particularly the case today in light of Covid-19 and its variants.

Nuggehalli M. Ravindra
November 9th 2021
Newark, New Jersey

Online appendices

If, after reading this book, readers are interested in delving deeper into the historical details or patent information describing the evolution of infrared detectors, bolometers, and microbolometers, they could refer to further reading materials that are available on these subjects in the form of online Appendices. Online Appendices are available for download in a Word document format.

Appendix A: Honeywell approach

A brief historical perspective on infrared radiation and the evolution in infrared detectors and bolometers is presented. Examples of patents relating to Honeywell's portfolio on uncooled microbolometers are described. A summary of the chronological order of patents relating to uncooled microbolometers, which have been issued to Honeywell, is presented.

Appendix A is available for download online at https://doi.org/10.1016/B978-0-08-102812-4.09983-6

Appendix B: Texas Instruments approach

A brief summary of the Texas Instruments technology relating to uncooled microbolometers is presented. The evolution in the technology of infrared radiation detection with the use of pyroelectric material, such as barium strontium titanate, and semiconductor, such as amorphous silicon, as thermosensing materials, are discussed. Some of the key patents that have been issued to Texas Instruments on microbolometer technology are described.

Appendix B is available for download online at https://doi.org/10.1016/B978-0-08-102812-4.09983-6

Appendix C: Patents

A list of some of the major patents and inventions related to infrared detectors, including bolometers and microbolometers, is provided. The patent title, patent number, and publication date are shown, sorted by the date of publication. Some of the major industry players in the manufacture of microbolometers are discussed.

Appendix C is available for download online at https://doi.org/10.1016/B978-0-08-102812-4.09983-6

References

[1] N.M. Ravindra, V.K. Srivastava, Properties of PbS, PbSe and PbTe, Phys. Status Solidi A 58 (1) (1980) 311–316.

[2] N.M. Ravindra, V.K. Srivastava, Properties of liquid PbS, PbSe and PbTe, Infrared Phys. 20 (6) (1980) 399–418.
[3] N.M. Ravindra, S. Auluck, V.K. Srivastava, Temperature dependence of the energy gap of PbS, PbSe and PbTe, Phys. Status Solidi A 52 (1979) K151–K155.
[4] T.S. Moss, C. Hilsum, Handbook on Semiconductors, Elsevier, 1993.
[5] T.S. Moss, G.J. Burrell, B. Ellis, Semiconductor Opto-Electronics, Butterworths, 1973.
[6] T.S. Moss, Optical Properties of Semiconductors, Butterworths, 1959.
[7] W.F. Kosonocky, M.B. Kaplinsky, N.J. McCaffrey, E.S.H. Hou, C.N. Manikopoulos, N. M. Ravindra, S. Belikov, J. Li, V. Patel, Multiwavelength imaging pyrometer, in: SPIE's International Symposium on Optical Engineering and Photonics in Aerospace Sensing, 1994, Orlando, FL, United States, Proceedings Volume 2225, Infrared Detectors and Focal Plane Arrays III, 1994. https://doi.org/10.1117/12.179714.
[8] N.M. Ravindra, B. Sopori, O.H. Gokce, S.X. Cheng, A. Shenoy, L. Jin, S. Abedrabbo, W. Chen, Y. Zhang, Emissivity measurements and modeling of silicon-related materials—an overview, Int. J. Thermophys. 22 (5) (2001) 1593–1611.
[9] L. Jiang, W.N. Carr, Vanadium dioxide thin films for thermo-optical switching applications, in: MRS Online Proceedings Library (OPL), Volume 785: Symposium D—Materials and Devices for Smart Systems, 2003, p. D10.4. https://doi.org/10.1557/PROC-785-D10.4.
[10] W. Carr, D. Setiadi, Micromachined Pyro-Optical Structure, Patent No. US 6,770,882 B2, 2004.
[11] W.F. Kosonocky, H.G. Erhardt, G. Meray, F.V. Shallcross, H. Elabd, M.J. Cantella, J. Klein, L.H. Skolnik, B.R. Capone, R.W. Taylor, Advances in platinum silicide Schottky-Barrier IR-CCD image sensors, in: 1980 Technical Symposium East, 1980, Washington, D.C., United States, Proceedings Volume 0225, Infrared Image Sensor Technology, 1980. https://doi.org/10.1117/12.958705.
[12] W.F. Kosonocky, H. Elabd, H.G. Erhardt, F.V. Shallcross, T. Villani, G. Meray, 64x128-Element high-performance PtSi IR-CCD imager sensor, in: IEDM, IEEE, 1981.
[13] F.-M. Tong, Y. Haoxin, Y. Xiuzhen, N.M. Ravindra, HgCdTe photodiodes—a device study, Infrared Phys. 33 (6) (1992) 511–522.
[14] C. Lamsal, N.M. Ravindra, Optical properties of vanadium oxides—an analysis, J. Mater. Sci. 48 (2013) 6341–6351.
[15] C. Lamsal, N.M. Ravindra, Simulation of spectral emissivity of vanadium oxides (VO_x)-based microbolometer structures, Emerg. Mater. Res. 3 (4) (2014) 194–202.
[16] M.P. Lepselter, A.T. Fiory, N.M. Ravindra, Platinum and rhodium silicide—germanide optoelectronics, J. Electron. Mater. 37 (4) (2008) 403–416.
[17] V.R. Mehta, S. Shet, N.M. Ravindra, A.T. Fiory, M.P. Lepselter, Silicon-integrated uncooled infrared detectors: perspectives on thin films and microstructures, J. Electron. Mater. 34 (5) (2005) 484–490.
[18] F.M. Tong, N.M. Ravindra, L. Ganapathi, S. Giles, R. Rao, High T_c $YBa_2Cu_3O_{7-\delta}$ superconducting transition-edge microbolometers, Infrared Phys. Technol. 36 (7) (1995) 1053–1058.
[19] A. Rogalski, History of infrared detectors, Opto-Electron. Rev. 20 (3) (2012) 279–308.
[20] Press release: The Nobel Prize in Physiology or Medicine 2021, https://www.nobelprize.org/prizes/medicine/2021/press-release/, Accessed 5 October 2021.

Acknowledgments

The author is thankful to Dr. Asahel Bañobre and Dr. Sita Rajyalaxmi Marthi for their contributions during the initial phase of the write-up of the book. Lawrence Onyango, Intern from the Bergen County Technical High School, is acknowledged for his help with the chapter on patents. Over the years, the author has collaborated extensively with Dr. Anthony T. Fiory and the late Martin P. Lepselter in several areas of research relating to infrared imagers and temperature measurements. Dr. Chiranjivi Lamsal at SUNY-Plattsburgh worked on vanadium oxides for his doctoral studies at NJIT.

The author is extremely grateful to several colleagues, companies, journals, and institutions for help with the use of the figures: Charles Hanson; Chris North from Cardiff University; FLIR, www.teledyneflir.com; *Metrology and Measurement Systems*; Pierre Talbot, Business Development Manager at INO; Nick Gromicko, www.Gromicko.com, TEMATYS; MIT Lincoln Laboratory at Lexington, Massachusetts; GIRMET Company; *Defense Science Journal*; Teledyne Imaging; Sandrine Leroy, Director, Public Relations, www.yole.fr; James W. Beletic, PhD, President, Teledyne Imaging Sensors; Dr. A. Rogalski at the Military University of Technology in Poland; SPIE—The International Society for Optics and Photonics; and IEEE—Institute of Electrical and Electronics Engineers and Springer.

The writing of this book has been a long journey. The author thanks the team at Elsevier—Miss Andrea Gallego Ortiz, Miss Kayla Dos Santos, Miss Swapna Praveen, Mr. Kamesh Ramajogi, and Mr. Vignesh Tamilselvvan—for their constant help, guidance, and support.

Dedication

To my sister, Math. G.Y. Premalatha, and my brother-in-law, the late Professor G.N. Yoganarasimhan, Professor of Water Resources Engineering and Management for showing me the direction to science, creativity, and the passion for acquisition and sharing of knowledge and a purpose in life.

To the late Professor Walter F. Kosonocky for his friendship, collaboration, and teamwork—it has been 25 years since Walter passed away in 1996.

To the late Provost Gary Thomas for his unequivocal trust, confidence, and support.

To my students for keeping me intellectually occupied.

To the healthcare workers and mail carriers for their commitment to the communities they serve.

About the Author

N. M. Ravindra (Ravi) is Professor of Physics at the New Jersey Institute of Technology (NJIT). He was the Chair of the Physics Department (2009–2013) and Director, Interdisciplinary Program in Materials Science and Engineering at NJIT (2009–2016). Ravi is the Founding Editor of Emerging Materials Research. He serves on the editorial board of several international journals and book series that are dedicated to materials science and engineering.

Ravi and his research team have published over 350 papers in international journals, books, and conference proceedings; his team has several pending and two issued patents; he has organized over 30 international conferences; and he has given over 75 lectures in international meetings.

Historical perspective

1

1.1 Infrared radiation

The year 2020 marks 200 years since the formation of the Royal Astronomical Society (RAS) [1] with Sir William Herschel as its first president [1]. Since its inception, RAS has been actively involved in enhancing the study of Astronomy, Geophysics, Solar-System, and the associated branches of science. Herschel's discovery of infrared radiation in 1800 was the result of his interest in investigating the amount of heat that was being transmitted through the various colored filters which Herschel utilized to observe the Sun [2, 3].

A schematic of the electromagnetic spectrum is presented in Fig. 1.1 [4]. From the perspective of design and implementation of thermal sensors for a variety of applications, the spectral range between 1 and 50 μm is of interest to the scientific community [5], although a more general definition of infrared is the electromagnetic radiation in the wavelength range of 780 nm to 1 mm [6].

The ability to image distant objects in the sky, in the infrared range of wavelengths, is hampered by the infrared absorption by water vapor (5.5–7 μm) [8] and CO_2 (2.7, 4.3 and 15 μm) [9]. This is illustrated in Fig. 1.2, which shows the spectral irradiation from the sun before and after it passes through the atmosphere. In general, the range of a thermal imaging camera is a function of the lens, cooled or uncooled detector, sensitivity, object size, its temperature as well as that of the background [10]. Weather conditions such as humidity, fog, and rain can have an impact on the ability for an infrared camera to perform depending on the wavelength of operation of the infrared camera [10, 11]; additionally, droplets of water scatter the infrared signal. However, the infrared wavelength can penetrate through smoke (particle size $\sim$0.5 μm) due to the smaller particle size of smoke compared to that of the wavelength of IR radiation; thus, infrared imagers can help to fight fires. In general, the wavelength of choice for utility and usefulness of infrared imaging depends on the atmospheric conditions as well as the environment between the target and the imager [12]. The useful range of wavelengths for infrared imaging in the earth's atmosphere is illustrated in Fig. 1.3.

1.2 Infrared detector

The range of the electromagnetic spectrum, 0.2–1000 μm, is considered to be the optical radiation; it includes visible radiation (0.4–0.75 μm) and infrared radiation (0.75–1000 μm). The infrared radiation spectrum is not visible to the naked eye; the human eye can just detect wavelengths in the visible range. Infrared radiation is the second most intense source of radiation on earth. This is the reason for the efforts, during the last century, toward the study and development of infrared radiation detectors.

Microbolometers. https://doi.org/10.1016/B978-0-08-102812-4.00011-5

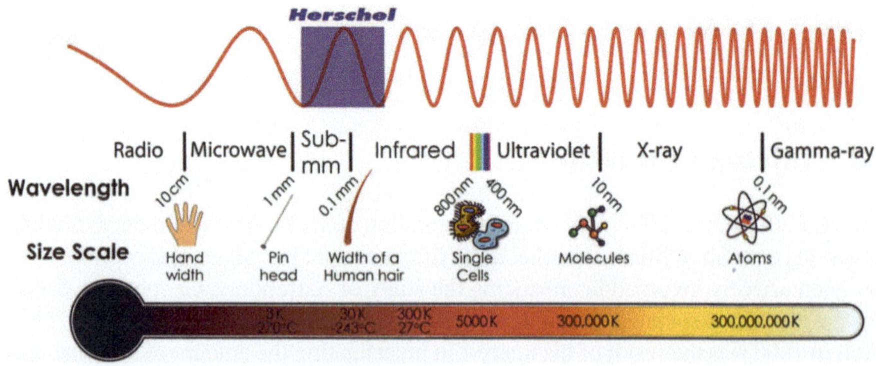

Fig. 1.1 The electromagnetic spectrum [4].
Reproduced with permission from Cardiff University/Chris North. http://herschel.cf.ac.uk/science/infrared

An infrared detector is an optoelectronic device that reacts to infrared radiation, transducing the infrared radiation into an electrical signal. Electromagnetic radiation interacts with matter in various ways that determine the physical mechanism involved in the detection process and can be classified as photon effect, thermal effect, and wave interaction effect. This radiation-matter interaction depends on the infrared absorber material and the infrared spectral range. Infrared detectors are broadly classified as quantum (optoelectronic) detectors, thermal detectors, such as thermoelectric devices, and pyroelectric, thermoresistive, and microelectronic mechanical devices (MEMS).

Infrared detectors are broadly classified into two types: photon detectors and thermal detectors [13].

Thermal detectors respond to the heating effect caused by the absorbed photon radiation that causes changes in the temperature of the detector and its electrical properties. The detection process of thermal detectors is divided into two steps. First, the radiation

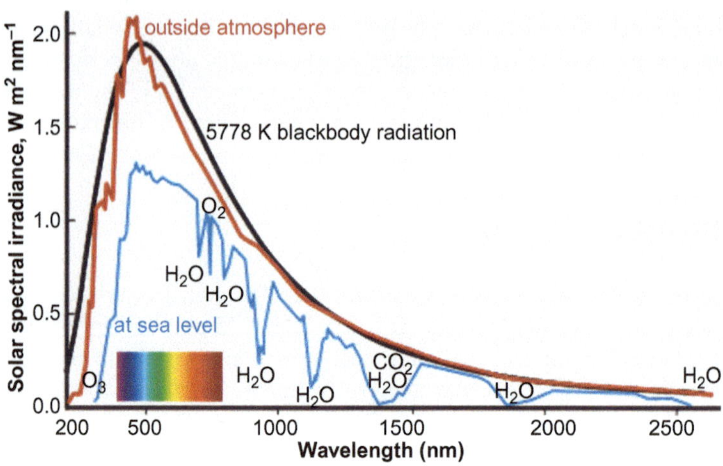

Fig. 1.2 Light from the Sun, before and after passing through the Earth's atmosphere and Water bands responsible for some of the difference [7].

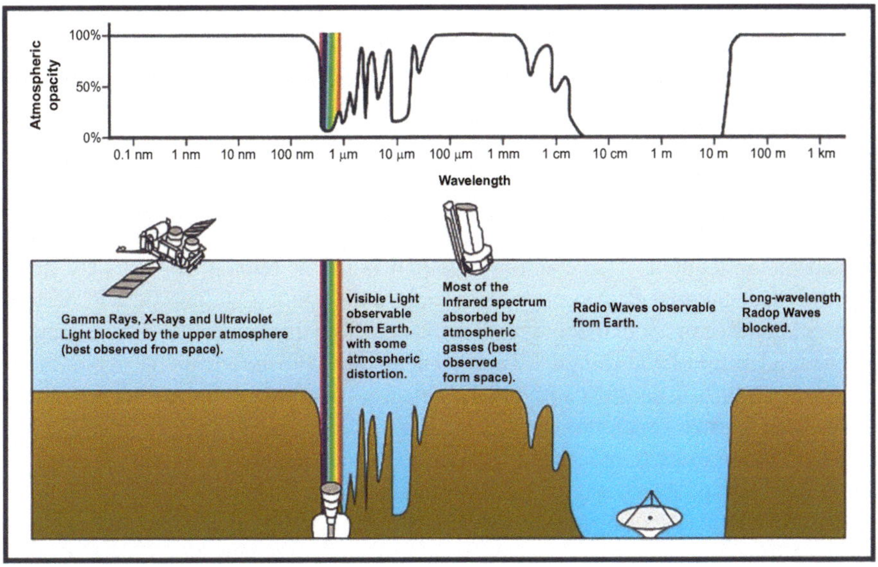

Fig. 1.3 Illustration of the spectral characteristics of the earth's atmosphere [8]. Credit NASA. Atmospheric Absorption & Transmission, GSP 216 Introduction to remote sensing, http://gsp.humboldt.edu/OLM/Courses/GSP_216_Online/lesson2-1/atmosphere.html.

is absorbed generating phonons and causes changes in the lattice temperature of the absorber material. Second, the change in temperature induces changes in a measurable parameter of the detector active element. The response of the thermal detector is proportional to the energy absorbed [13]. Thermal detectors, most used by the infrared industry, are bolometers, thermopiles, and pyroelectrics. Bolometers are detectors in which the absorption of radiation causes changes in the electrical resistance of its thermal sensing materials (for example, VO_x, a-Si—at room temperature). They exhibit responsivity that does not depend on their area. Bolometers are of several types, including carbon, metals, semiconductors, superconductors [14], and thermistors [15]. Thermoelectric detectors (thermopiles) are formed by the junction of two materials (Bi/Sb, Al/Poly-Si) with high Seebeck coefficients. They exhibit low responsivity. Pyroelectric detectors are made using dielectric materials ($LiTaO_3$, $Ba_xSr_{1-x}TiO_3$) that modify their polarization under the influence of changes in temperature and need a chopper system in order to generate a modulation in their polarization. They exhibit high responsivity.

Photon detectors are devices that detect optical radiation by direct interaction of individual photons with active charge carriers within the atomic lattice of the detector material. The transition of the active charge carriers from the valence band to the conduction band depends on the energy band gap of the material and causes changes in the physical parameters, such as resistance, capacitance, voltage, or current [16]. The photodetector response is proportional to the number of photons absorbed [17]. Among the many photon detectors, the photoconductive and the photovoltaic types of devices are the most used. Photoconductive detectors use the increase in electrical conductivity resulting from an increase in the number of free carriers generated when photons are absorbed (generation of current), whereas in photovoltaic detectors, current is generated as a result of the absorption of photons due to a voltage difference across a p-n junction (generation of electron-hole pair across the junction and a voltage).

A wide variety of photon detectors and thermal detectors have been developed by the infrared detection/imaging industry; each detection mechanism has advantages and disadvantages that determine the most appropriate approach depending on the infrared spectral range for the detection, application, and measurement conditions.

The time response of the thermal detector is a two-step process; it is slower than the corresponding time response of the photon detector, in which the absorption of a photon and the generation of an electron-hole pair is faster. Photon detectors are more sensitive than thermal detectors, but they can only operate at specific and narrow wavelength range in comparison with thermal detectors that can respond continuously over a broad range of wavelengths. Thermal detectors have the advantage to operate at room temperature, in contrast with photon detectors that need to be cooled, as a direct result of the thermal generation-recombination that increases the noise effect. In photon detectors, in order to reduce the thermally induced generation of charge carriers, the temperature of the detector must be maintained at cryogenic levels of 77 K or below [18], by using cooling systems such as Stirling cooling and Peltier cooling. This increases the cost, complexity, weight, size, and power consumption of the infrared detection imaging systems [19]. The low-cost and light weight/portability are the main reasons why uncooled thermal detectors are considered as an alternative to replace cooled photon detectors. The mechanism of infrared detectors is illustrated in Fig. 1.4. The absorber layer reacts to the incident infrared radiation by absorbing the radiation. In a bolometer, the transducer converts the absorbed infrared radiation into a change in resistance. In a photon detector, as a transducer, the absorbed photon leads to the formation of electron-hole pairs. In a pyroelectric detector, the increased

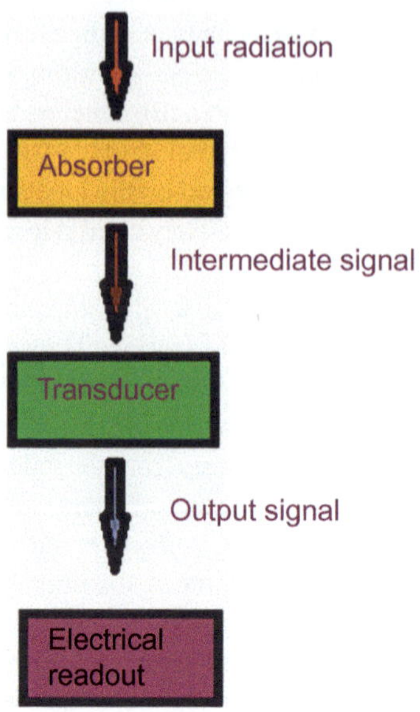

Fig. 1.4 Basic mechanism of operation of infrared detector [13].

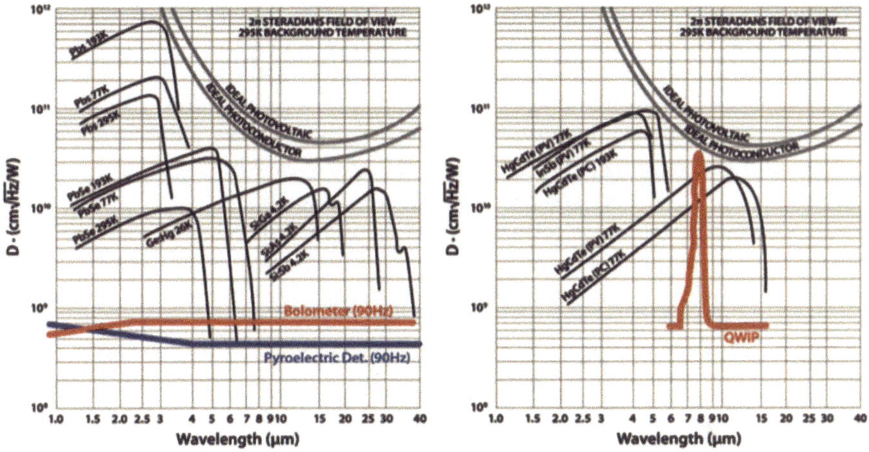

Fig. 1.5 Detectivity (D*) curves for different detector materials with typical operating (Kelvin) temperatures [20].

absorption of photons in the absorber layer leads to enhanced polarization. The electrical readout converts the output into an image.

The choice of materials for infrared detection is determined by the wavelength of operation of the detector/imager. This is illustrated in Fig. 1.5. As can be seen in Fig. 1.5, the detectivity of bolometers and pyroelectric detectors is generally independent of wavelength.

1.3 Microbolometer

In general, a bolometer is considered a thermal infrared detector or a sensitive thermometer. It functions on the basic principle of change in electrical resistance with temperature. In order for a bolometer to function, it is critical that the bolometer is thermally isolated from its support structure/substrate. A microbolometer represents an array of thermal sensors. When infrared radiation strikes the detector material, it causes the detector material to heat, thus changing its electrical resistance. The change in resistance is measured and used to create an image. Microbolometers usually do not require external cooling.

A microbolometer is a tiny vanadium oxide (VO_x) or amorphous silicon (a-Si) resistor, with a large temperature coefficient of resistance, on a silicon element with a large surface area, low heat capacity, and good thermal isolation. Infrared radiation from a source strikes the vanadium oxide or amorphous silicon film and changes its electrical resistance [21].

1.4 Fabrication process

A low-cost microbolometer is fabricated using standard CMOS process. It is followed by a post-CMOS process. Three metal interconnect layers are made by the CMOS process. The thicknesses of metal-2 and metal-3 layers are 0.55 and 0.8 μm, respectively.

In a standard microbolometer, thermally isolated absorber is supported by legs that are formed using a sacrificial layer technology. In this case, metal-2 may be used as surface sacrificial layer and metal-3 may be considered a thermistor. The readout integrated circuits (ROIC) are always designed to be located below the thermally isolated absorber. This facilitates in decreasing the area of a chip. The CMOS process is a key factor in determining the pixel size and fill factor by limiting metal-3 width to a minimum and also the etch-window dimensions between legs. Finer CMOS technologies enable increment in the performance of aluminum microbolometers.

Fig. 1.6 shows a schematic of the fabrication process for the low-cost microbolometer. Fig. 1.6(a) shows that the etch-windows are etched during the

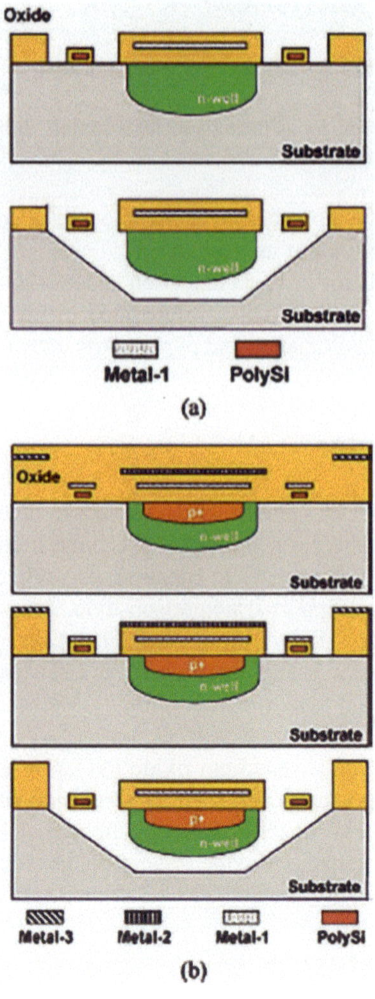

Fig. 1.6 The postprocessing steps of (a) resistive type and (b) diode type n-well microbolometers [22].

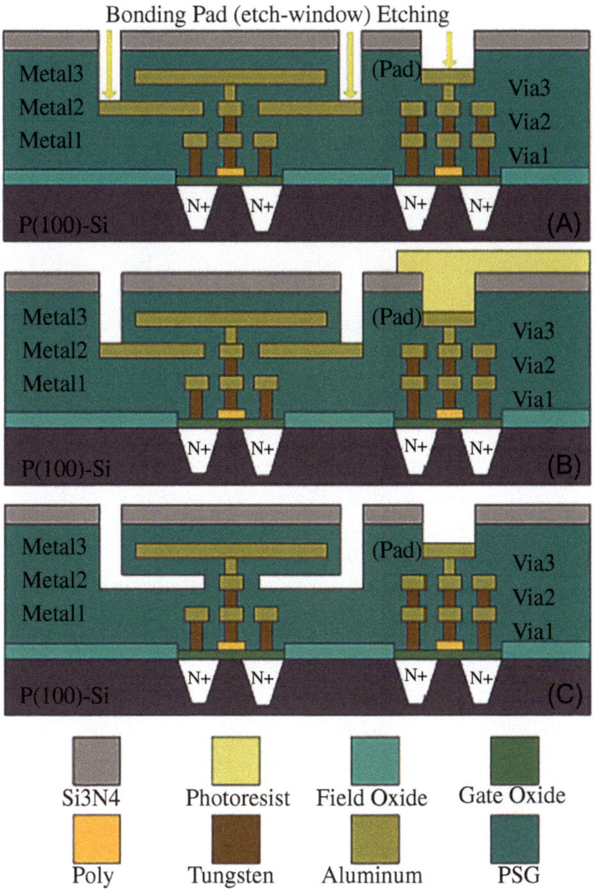

Fig. 1.7 Fabrication process of a microbolometer [23].

bonding-pad etching process in a standard CMOS technology so that the metal-2 layer (the sacrificial layer) is exposed. A photoresist layer is added to prevent the pads from being etched, as shown in Fig. 1.6(b). The pads are situated approximately 200 μm away from the etched window so that the photoresist layer can easily be spread manually, without adding lithography steps. Later, the exposed metal-2 part is removed by a solution comprising a mixture of 80% phosphoric acid, 5% nitric acid, 5% acetic acid, and 10% water [22, 23].

Fabrication process of a microbolometer is shown in Fig. 1.7.

In Table 1.1, the major milestones in the development of infrared detectors are summarized.

Figs. 1.8, 1.9, and 1.10 illustrate the history of the development of infrared detectors, industrial configuration evolution of uncooled infrared detector manufacturers in the United States, and historical trends in infrared sensor technologies, respectively.

Table 1.1 Major milestones in the development of infrared detectors [21].

Year	Event
1800	Discovery of Infrared by W. Herschel
1821	Discovery of Thermoelectric effect by T. J. Seebeck
1830	Thermal element for thermal radiation measurement by L. Nobili
1833	Thermopile by L. Nobili and M. Melloni
1834	Discovery of Peltier effect by J. C. Peltier
1835	Formulation of hypothesis for similarities in light and EM radiation by A. M. Ampere
1839	Solar absorption spectrum and role of water vapor by M. Melloni
1840	Discovery of three atmospheric windows by J. Herschel
1857	Integration of thermoelectric effects by W. Thomson
1859	Relation between absorption and emission by G. Kirchhoff
1864	Theory of EM radiation by J. C. Maxwell
1873	Discovery of photoconductive effect in Selenium by W. Smith
1876	Discovery of photovoltaic effect in Selenium by W. G. Adams and A. E. Day
1879	Establishing relation between radiation intensity and temperature of blackbody by J. Stefan
1880	Study of absorption characteristics of the atmosphere using a Pt bolometer by S. P. Langley
1883	Study of transmission characteristics of infrared transparent materials by M. Melloni
1884	Derivation of Stefan law by L. Boltzmann
1887	Photoelectric effect in ultra violet light by H. Hertz
1890	J. Elster and H. Geitel constructed a photoemissive detector of alkali-metal cathode
1894, 1900	Derivation of blackbody radiation law by J. W. Rayleigh and W. Wein
1900	Discovery of quantum properties of light by M. Planck
1903	Temperature measurements of stars and planets using spectrometry and infrared radiometry by W. Coblentz
1905	Establishment of photoelectric theory by A. Einstein
1911	First television image tube made based on the principles of cathode ray tubes by R. Rosling
1914	Application of bolometers for remote exploration of people and aircrafts
1917	First infrared photoconductor from Th and S developed by T. W. Case
1923	Establishment of theory of dry rectifiers by W. Schottky
1928	Proposal of electro-optical converter by G. Holst, J. H. De Boer, M. C. Teves, and C. F. Veenemans
1929	A converter tube with a photocathode sensitive in the NIR made by L. R. Kohler
1930	Infrared direction finders based on PbS quantum detectors in the wavelength range of 1.5–3.0 μm for military applications
1934	Introduction of first IR image converter
1939	Development of first IR display unit in US
1941	Observation of photovoltaic effect in a Si PN Junction by R. S. Ohl

Table 1.1 Continued

Year	Event
1942	First infrared sensitive film by G. Eastman
1947	Pneumatically acting, high-detectivity radiation detector by M. J. E. Golay
1954	First imaging cameras based on thermopiles and bolometers
1955	Initiation of production of infrared seeker heads for infrared guided rockets in US
1957	Discovery of HgCdTe as infrared detector material by W. D. Lawson, S. Nelson and A. S. Young
1961	Discovery of extrinsic Ge:Hg and its application in the first LWIR FLIR systems
1965	Mass production of infrared cameras for civilian applications in Sweden
1970	Discovery of charge-coupled devices (CCDs) by W. S. Boyle and G. E. Smith and start of production of infrared sensor arrays
1975	First in bump hybrid infrared focal plane arrays
1977	Discovery of InAs/GaSb superlattices by G. A. Sai Halasz, R. Tsu, and L. Esaki
1980	Development and production of second generation systems
1985	Development and mass production of cameras fitted with Schottky diode FPAs
1990	Development and production of quantum well infrared photodetector (QWIP)
1995	Initial production of IR cameras with uncooled focal plane arrays (FPAs)
2000	Development of third generation infrared systems

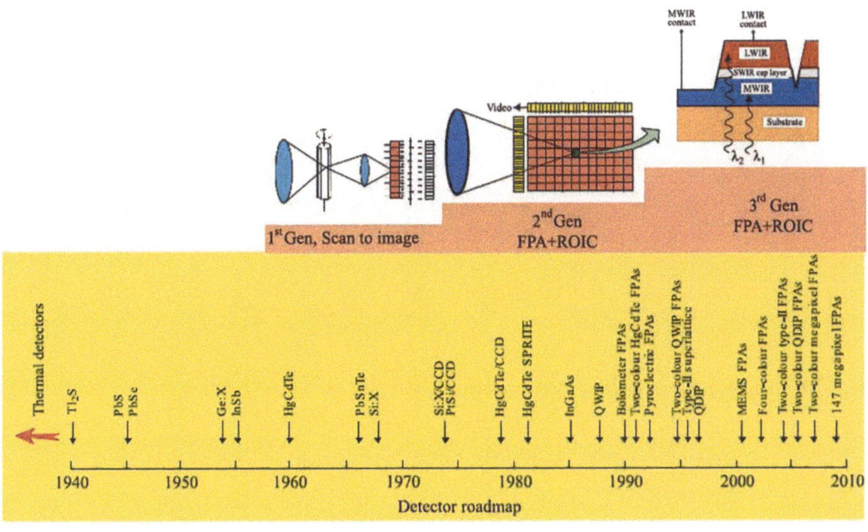

Fig. 1.8 History of the development of infrared detectors [21].

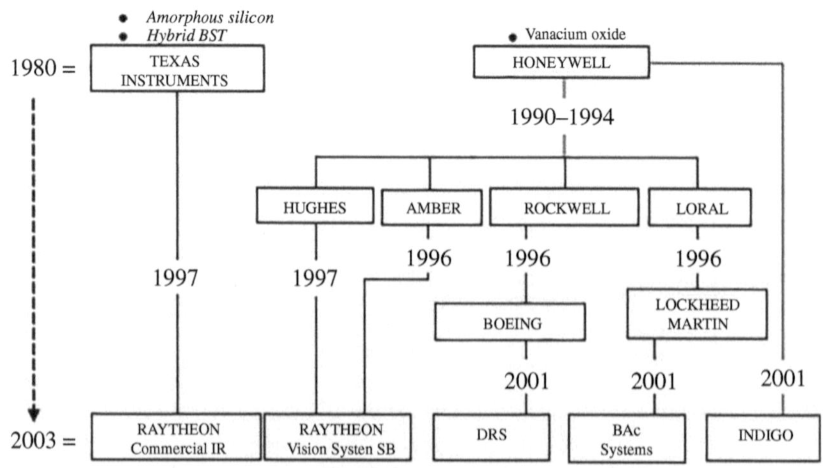

Fig. 1.9 Industrial configuration evolution of uncooled infrared detector manufacturers in the United States [24].

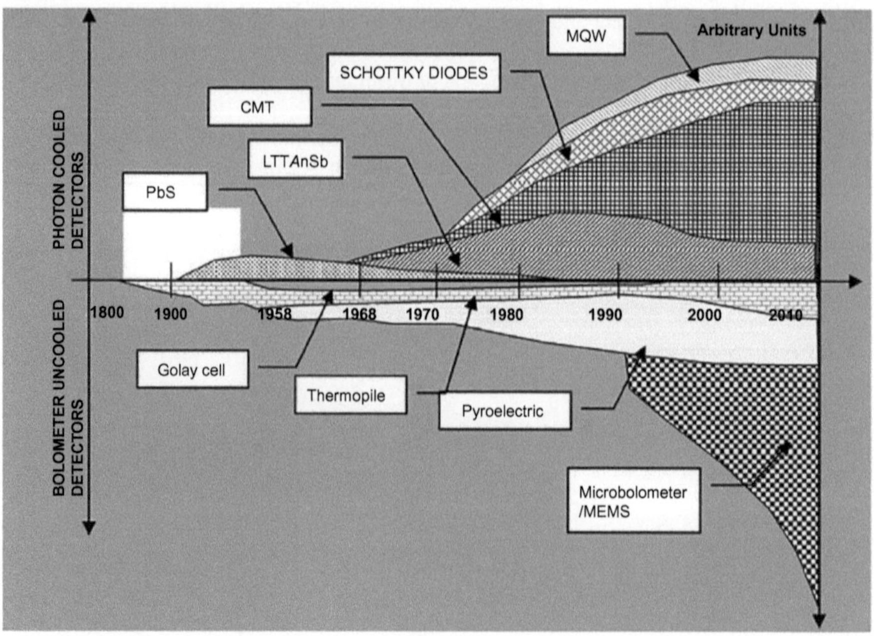

Fig. 1.10 Historical evolution in IR sensor technologies [25].

Fig. 1.11 illustrates the trend in array sizes for various technologies over the past 40 years and some forecasting for the future.

While the early evolution of bolometers was due to the dedicated efforts by astronomers, much of the later development was the result of its significant applications in the defense/military sector. Today, materials such as vanadium oxide and amorphous

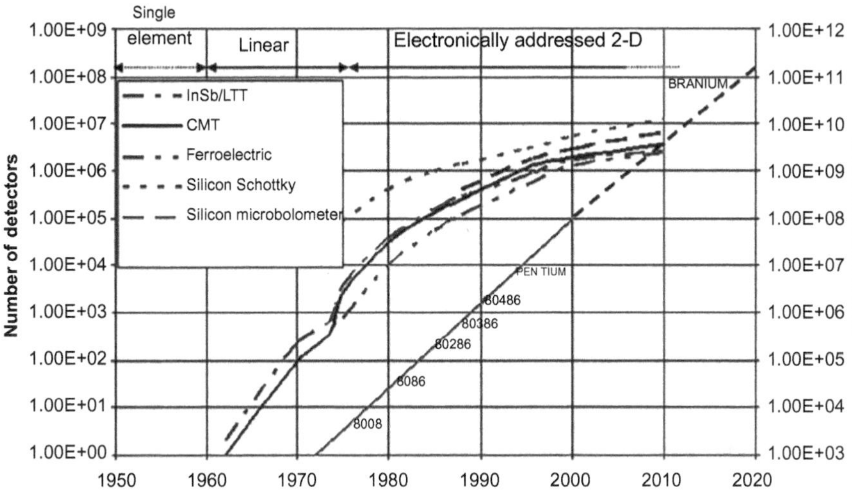

Fig. 1.11 Number of detectors/pixels in infrared detector arrays versus time (Moore's law for comparison) [25].

silicon have become standards in the manufacture of uncooled microbolometers. In recent years, microbolometers, based on two-dimensional materials, such as graphene [26–28], transition metal dichalcogenides, such as MoS_2 and WS_2 [29, 30], are being developed.

References

[1] https://ras.ac.uk/about-the-ras, Accessed 11 November 2019.
[2] https://www.britannica.com/science/electromagnetic-radiation/Microwaves#ref307323, Accessed 11 November 2019.
[3] http://coolcosmos.ipac.caltech.edu/cosmic_classroom/classroom_activities/herschel_bio.html, Accessed 11 November 2019.
[4] http://herschel.cf.ac.uk/science/infrared, Accessed 11 November 2019.
[5] https://www.hgh-infrared.com/FAQ/Infrared-Testing/What-are-the-wavelength-boundaries-of-infrared-radiation, Accessed 11 November 2019.
[6] https://www.icnirp.org/en/frequencies/infrared/index.html, Accessed 11 November 2019.
[7] http://www1.lsbu.ac.uk/water/water_vibrational_spectrum.html#d, Accessed 11 November 2019.
[8] gsp.humboldt, Atmospheric Absorption & Transmission, GSP 216 Introduction to Remote Sensing, 2019. http://gsp.humboldt.edu/OLM/Courses/GSP_216_Online/lesson2-1/atmosphere.html. (Accessed 13 November 2019).
[9] nov79, CO_2 Absorption Spectrum, 2019. https://nov79.com/gbwm/ntyg.html. (Accessed 13 November 2019).
[10] FLIR, Seeing Through Fog and Rain with a Thermal Imaging Camera, FLIR Technical Note, FLIR Commercial Vision Systems B.V., Breda, Netherlands, 2021.

[11] C. Hill, R.L. Jones, Absorption of solar radiation by water vapor in clear and cloudy skies: implications for anomalous absorption, J. Geophys. Res. 105 (D7) (2000) 9421–9428.
[12] N.K. Dhar, R. Dat, A.K. Sood, Advances in infrared detector array technology, chapter 7, in: Optoelectronics—Advanced Materials and Devices, INTECH, 2013, https://doi.org/10.5772/51665.
[13] P.W. Kruse, The Photon Detection Process in Optical and Infrared Detectors, second ed., Springer-Verlag, New York Heidelberg Berlin, 1980. ISBN:0–387–10176-4.
[14] F.M. Tong, N.M. Ravindra, L. Ganapathi, S. Giles, R. Rao, High T_c $YBa_2Cu_3O_{7-\delta}$ superconducting transition-edge microbolometers, Infrared Phys. Technol. 36 (1995) 1053–1058.
[15] W.L. Wolfe, P.W. Kruse, Thermal detectors, chapter 19, in: M. Bass (Ed.), Handbook of Optics, McGraw Hill, Inc, 1995.
[16] A. Rogalski, Infrared detectors: an overview, Infrared Phys. Technol. 43 (2002) 187–210.
[17] A. Rogalski, Infrared Photon Detectors, SPIE The International Society of Optical Engineering, SPIE Optical Engineering Press, 1995. ISBN: 978–1420076714.
[18] R.B. Darling, S. Inawaga, Structure, properties and MEMS and microelectronic applications of vanadium oxides, Sadhana 34 (2009) 531–542.
[19] G. Hyseni, et al., Infrared thermal detectors parameters: semiconductor bolometers versus pyroelectrics, WSEAS Trans. Circuits Syst. 9 (2010) 238–241.
[20] Thermographic imaging selecting an IR camera with the right detector, Photonics (2008).
[21] A. Rogalski, History of infrared detectors, Opto-Electron. Rev. 20 (3) (2012) 279–308.
[22] S. Eminoglu, D.S. Tezcan, M.Y. Tanriklu, et al., Low-cost uncooled infrared detectors in CMOS process, Sensors Actuators A 109 (2003) 102–113.
[23] N. Shen, J. Yu, Z. Tang, An uncooled infrared microbolometer array for low-cost applications, IEEE Photon. Technol. Lett. 27 (12) (2015) 1247–1249.
[24] J.L. Tissot, IR detection with uncooled sensors, Infrared Phys. Technol. 46 (2004) 147–153.
[25] C. Corsi, History highlights and future trends of infrared sensors, J. Mod. Opt. 57 (18) (2010) 1663–1686.
[26] A. Blaikie, D. Miller, B.J. Aleman, A fast and sensitive room-temperature graphene nanomechanical bolometer, Nat. Commun. 10 (2019) 4726, https://doi.org/10.1038/s41467-019-12562-2.
[27] X. Du, D.E. Prober, H. Vora, C.B. Mckitterick, Graphene-based bolometers, Graphene 2D Mater. 1 (2014) 1–22.
[28] D.K. Efetov, et al., Fast thermal relaxation in cavity-coupled graphene bolometers with a Johnson noise read-out, Nat. Nanotechnol. 13 (2018) 797–801.
[29] J. Wang, J. Han, X. Chen, X. Wang, Design strategies for two-dimensional material photodetectors to enhance device performance, InfoMat (2019) 33–53, https://doi.org/10.1002/inf2.12004.
[30] A. Goswami, et al., Effect of interface on mid-infrared photothermal response of MoS_2 thin film grown by pulsed laser deposition, Nano Res. 10 (10) (2017) 3571–3584.

Introduction

2

2.1 Electromagnetic (EM) radiation

Electromagnetic (EM) radiation is classified into various spectral regions, viz., X-rays, ultraviolet rays, visible, as function of its fundamental characteristics: energy, frequency, or wavelength.

In 1800, Sir William Hershel discovered infrared radiation and wrote "...There are rays coming from the sun...invested with a high power of heating bodies, but with none of illuminating objects..." [1]. This component of the electromagnetic spectrum attracted the attention of physicists during the second half of the nineteenth century, and experimental and theoretical approaches were developed in efforts to explain the energy distribution of what Hershel called "invisible rays." In the study of thermal radiation, the work of scientists such as Kirchhoff, Boltzmann, Wien, Rayleigh, and Planck led to the introduction of the term Blackbody and the evolution of the application of the laws of classical physics and quantum physics [1].

The electromagnetic spectrum (Fig. 2.1) is the arrangement of all known frequencies and their relation to wavelengths of the known electromagnetic radiations. The electromagnetic radiation ranges are grouped depending on the nature of the source (heat, radio waves, sunlight, and X-rays) and are transmitted in the form of waves (electromagnetic waves) or particles (photons). The optical laws of reflection, refraction, interference, polarization, and refraction are the same for all the electromagnetic radiation [1]. Electromagnetic waves are described by physical properties such as frequency ν (Hz), wavelength λ (m), and photon energy E_{ph} (J). The photon energy is expressed as

$$E_{ph} = hv = \frac{hc}{\lambda}, \tag{2.1}$$

where h (6.626×10^{-34} J.s) is Planck's constant, ν is the frequency (Hz), λ is the wavelength (m), and c ($2.998 \times 10^{8}\ \mathrm{ms}^{-1}$) is the speed of light in vacuum [3].

At temperatures above absolute zero Kelvin (−273°C), the molecules and atoms oscillate and will radiate electromagnetic waves; this radiation is known as thermal radiation [4]. Thermal energy can be transferred by conduction and convection in the presence of a material medium and by radiation in the absence of any medium. The thermal radiation emitted by an object is called infrared radiation (IR) and it depends on its surface conditions and temperature [5].

Microbolometers. https://doi.org/10.1016/B978-0-08-102812-4.00017-6

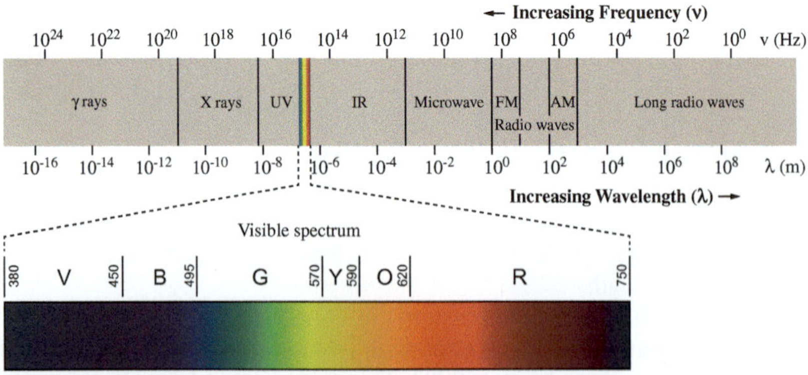

Fig. 2.1 Electromagnetic spectrum with visible light highlighted [2]. https://en.wikipedia.org/wiki/Electromagnetic_radiation.

2.2 Infrared radiation

The electromagnetic infrared region extends from 0.75 μm to 1000 μm and is classified into four regions: 0.75–3 μm (near IR), 3–6 μm (intermediate IR), 6–15 μm (far IR), and 15–1000 μm (extreme IR). Infrared radiation, or simply infrared or IR, is electromagnetic radiation with wavelengths longer than those of visible light, and is therefore invisible. IR radiation is the component of the EM spectrum that people encounter most in everyday life, although much of it goes unnoticed. It is invisible to human eyes, but people can feel it as heat.

Infrared extends from the nominal red edge of the visible spectrum at 700 nm (frequency 430 THz) to 1,000,000 nm (300 GHz).

Infrared can be subdivided into multiple spectral regions, or bands, based on wavelength; however, there is no uniform definition of exact boundaries of each of the bands. Infrared is commonly separated into near-, mid-, and far-infrared. It can also be divided into five categories: near-wavelength, short-wavelength, mid-wavelength, long-wavelength, and far-infrared.

The near-IR band contains the range of wavelengths closest to the red end of the visible light spectrum. It is generally considered to consist of wavelengths measuring from 750 nm to 1300 nm—or 0.75 to 1.3 μm. Its frequency ranges from about 230 THz to 400 THz. This group consists of the longest wavelengths and shortest frequencies, and it produces the least heat. The intermediate IR band, also called the mid-IR band, covers wavelengths ranging from 1300 nm to 3000 nm—or 1.3 to 3 μm. The frequencies range from 100 THz to 230 THz. Wavelengths in the far-IR band, which are closest to microwaves, extend from 3000 nm to 1 mm—or 3 to 1000 μm. Frequencies range from 0.3 THz to 100 THz. This group consists of the shortest wavelengths and longest frequencies, and it produces most heat.

Infrared light is invisible to the human eye, although longer infrared waves can be sensed as heat. It does, however, share some characteristics with visible light, namely, infrared light can be focused, reflected, and polarized. Almost half of the energy arriving to the Earth from the sun comes as infrared radiation.

2.3 Applications

A variety of applications utilize infrared range of wavelengths. Among the most well-known are heat sensors, thermal imaging, and night vision equipment. In communications and networking, infrared radiation is used in wired and wireless operations. Remote control components and systems use near-infrared light, transmitted with light-emitting diodes (LEDs), to send focused signals to home-entertainment devices, such as televisions. Infrared light is also used in fiber optic cables to transmit data.

Infrared is used extensively in astronomy to observe objects in space that cannot be detected in whole or part by the human eye, including molecular clouds, stars, planets, and active galaxies. Astronomers have found that infrared radiation is especially useful when attempting to probe areas of our universe that are surrounded by clouds of gas and dust. Because of the longer wavelength of the infrared, it can pass through these clouds and reveal details that are otherwise invisible by observing with other types of radiation. Especially interesting areas are stars, formation of planets, and the cores of galaxies where it is believed that huge black holes might reside. Infrared imaging has numerous applications.

Some of the applications of infrared component technology and infrared imaging are summarized in Table 2.1.

Table 2.1 Examples of applications of infrared component technology and infrared imaging.

Applications of infrared component technology	Applications of infrared imaging
Amplifiers	Astronomy
Coatings	Autonomous vehicles
Communications	Chemical imaging
Control systems	Conditional monitoring
Detectors/imagers, detector/imager arrays	Drones—Agricultural mapping, fighting forest fires; situational awareness
Filters	Failure analysis
Lasers	Materials research
Lenses	Medical imaging
Light-emitting diodes	Meteorology—Weather satellites
Microscopes	Military applications
Modulators	Moisture inspection
Photonics	Night vision—Imaging objects in the dark
Repeaters	Nondestructive testing
Sources	Photography—Thermography
Transmitters	Preventive maintenance in structures
Waveguides	Process monitoring and control
	Pyrometry
	Thermal management
	Veterinary thermal imaging
	Volcanology
	Warfare—Heat seeking missiles

2.4 Microbolometers

Bolometers are thermal sensors that operate on the basis of the changes in electrical resistance with temperature. A microbolometer consists of an array of pixels, each pixel being made up of several layers. Historically, thermal detectors always required cooling systems. The cooling systems sometimes include cryogenic cooling systems such as liquid helium or liquid nitrogen coolers or thermoelectric coolers (based on Peltier effect). The addition of cooling systems makes the microbolometers heavy and expensive.

Since the invention of the bolometer in 1878 by Samuel Pierpont Langley [6], there has been a continued evolution in its development. The US Department of Defense (DOD)/Defense Advanced Research Projects Agency (DARPA) has been a leader and a strong supporter of the microbolometer technology due to its significant military-related applications. DARPA continues to invest on the research and development of low-cost, high-performance infrared detector technology using wafer-scale fabrication techniques [7].

The fundamental principles that govern the operation of microbolometers can be explained by Radiometry and Blackbody radiation theory. Radiometry is the branch of optical physics that focuses on the study of the electromagnetic spectrum covering from the ultraviolet region to the infrared region. Radiometric measurements characterize and quantify the spectral parameters of the optical radiation sources and the distribution of their emitted optical radiation.

An example of a room-temperature microbolometer is illustrated in Fig. 2.2 [8]. In collaboration with the US DOD, Honeywell has been a pioneer in the research, development, and manufacture of uncooled vanadium oxide based microbolometer technology. This technology continues to dominate over other uncooled microbolometer technologies, such as amorphous silicon [9] or pyroelectric sensors [10,11]. The uncooled amorphous silicon microbolometer and barium strontium titanate (BST) ferroelectric infrared detectors were originally developed by Texas

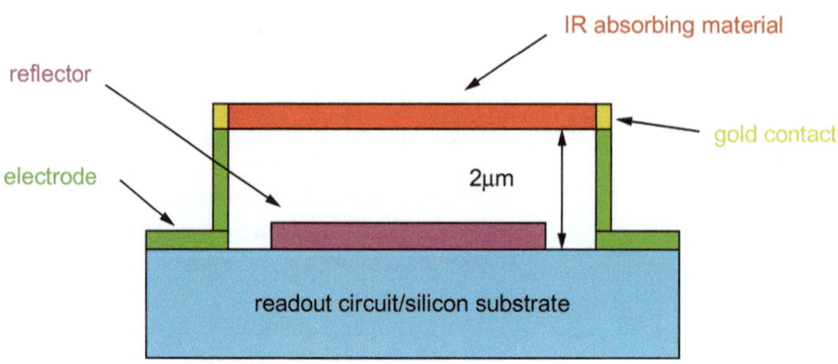

Fig. 2.2 Cross-sectional view of a microbolometer [8].
From https://en.wikipedia.org/wiki/Microbolometer.

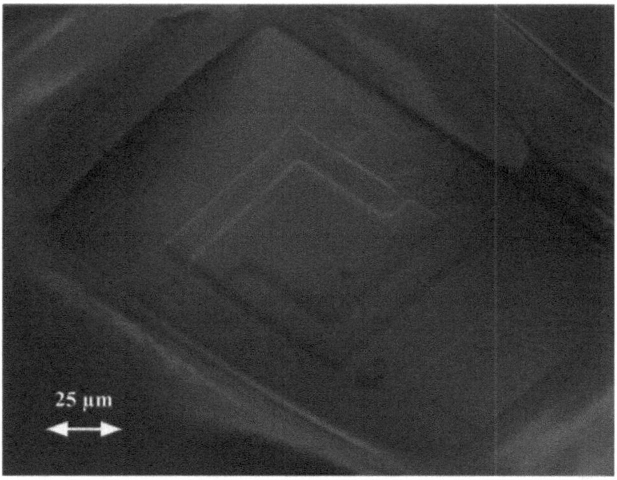

Fig. 2.3 A micrograph shows the typical micromachined structures in a microbolometer focal-plane array (FPA). The pixels are suspended above the readout integrated circuit, and the long skinny legs provide thermal isolation for the pixels [12].
From Device-level vacuum packaged uncooled microbolometer on a polyimide substrate; Moinuddin Ahmed, Donald P. Butler, Zeynep Celik-Butler; Infrared Phys. Technol. 79(2016) 50–57.

Instruments. A micrograph of the amorphous silicon microbolometer is shown in Fig. 2.3 [12].

2.5 Radiometry and black body radiation

Optical radiant emission is investigated in terms of the ideal concept of a Blackbody model. A Blackbody is a hypothetical body that absorbs and emits all possible incident electromagnetic radiation, regardless of its frequency or angle of incidence, while it is in thermal equilibrium [13]. A Blackbody, in thermal equilibrium, emits an electromagnetic radiation spectrum that is determined by its temperature and not by its body composition or its shape [14].

Planck's law considers that the energy is quantized at the atomic level and describes the radiant properties of a Blackbody in a continuum of wavelength, as a function of wavelength and source temperature [15], and is expressed as

$$L_{e,\lambda}(\lambda, T) = \frac{2hc^2}{\lambda^5\left(e^{\frac{hc}{\lambda kT}} - 1\right)}, \tag{2.2}$$

where $L_{e,\lambda}$ ($\mathrm{Wcm^{-2}\mu m^{-1}\,sr^{-1}}$) is the spectral radiance emittance, h (6.626×10^{-34} J s) is the Planck's constant, c ($2.998 \times 10^8\ \mathrm{ms^{-1}}$) is the speed of light

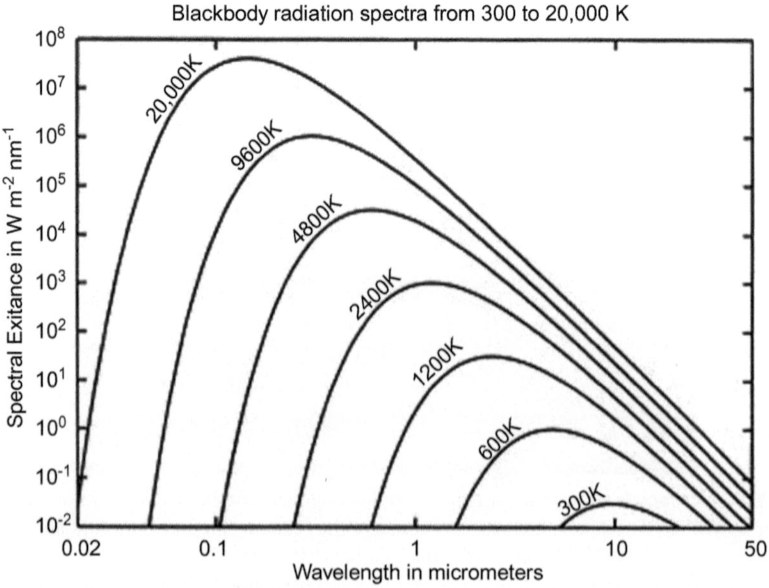

Fig. 2.4 Calculated spectral exitance/emittance spectra for black bodies at various temperatures from 300 to 20,000 K [16].

in vacuum, λ is the wavelength (m), k_B (1.38×10^{-23} J K^{-1}) is the Boltzmann's constant, and T (K) is the absolute temperature of the optical source. The calculated spectral exitance/emittance spectra of black bodies is plotted in Fig. 2.4 [16].

A Blackbody source is considered to be a perfect Lambertian radiator, in which the radiance is the same in all the directions and is independent of the viewing angle. The relationship between the spectral radiance emittance and the radiant exitance M_e, ($\mathrm{Wcm^{-2}\ \mu m^{-1}}$) for a planar Lambertian source can be written as

$$M_{e,\lambda}(\lambda, T) = \pi L_{e,\lambda}. \tag{2.3}$$

The total spectral radiance emittance L_e ($\mathrm{Wcm^{-2}\ sr^{-1}}$) and the total spectral radiant exitance M_e ($\mathrm{Wcm^{-2}}$) of an optical source, at a specific temperature, are integrated over a finite spectral range [15], and are expressed by Eqs. (2.4) and (2.5):

$$L_e = \int_{\lambda_1}^{\lambda_2} L_{e,\lambda}(\lambda, T), \tag{2.4}$$

$$M_e = \int_{\lambda_1}^{\lambda_2} M_{e,\lambda}(\lambda, T) d\lambda = \pi \lambda_e. \tag{2.5}$$

In Eq. (2.5), integration is performed over all wavelengths ($\lambda_1 = 0$ to $\lambda_2 = \infty$) and can be modified as in Eq. (2.6) and is called as Stefan-Boltzmann law. It states that the

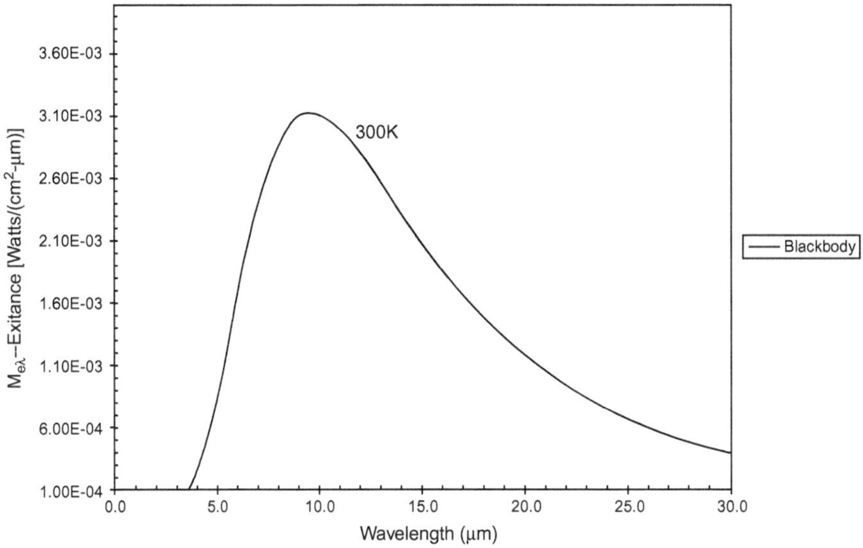

Fig. 2.5 Total radiant exitance from a blackbody at 300 K is the area under the spectral exitance curve [15].

relationship between the total radiant exitance and the temperature of the Blackbody is as shown in Fig. 2.5, where σ (5.67×10^{-12} $\mathrm{Wcm^{-2}\,K^{-4}}$) is called the Stefan-Boltzmann constant [15]

$$M_e = \int_{\lambda_1}^{\lambda_2} M_{e,\lambda}(\lambda, T)d\lambda = \sigma T^4. \tag{2.6}$$

Differentiating Planck's law and solving for the wavelength at the maximum spectral exitance yields Wien's displacement law. It quantifies the decreasing wavelength of peak exitance with an increase in object temperature and is expressed by the following equation:

$$\lambda_{max} = \frac{2898}{T}, \tag{2.7}$$

where λ_{max} (μm) is the wavelength at maximum spectral exitance, emitted by an object at a particular temperature T [15]. A Blackbody source at $T = 300$ K (room temperature) has a peak exitance at $\lambda_{max} = 9.66$ μm.

The Blackbody energy spectrum model is an ideal concept, but real sources behave differently. The radiation spectrum of a real body and an ideal Blackbody is related by a factor called emissivity, that is, the ratio of the radiant exitance of the real source and the radiant exitance of a Blackbody at the same temperature. Emissivity is a dimensionless number that is ≤ 1, and it is equal to 1 for a Blackbody or equal to 0 for a nonradiant source. A real source is called as a selective radiator if its emissivity depends on wavelength, and emissivity is less than unity, or gray-body, if its emissivity is constant and is independent of wavelength. Emissivity depends on the type of material, material surface, wavelength, and temperature of the material [15]; it is expressed as

$$\varepsilon(\lambda, T) = \frac{M_{e,\lambda}(\lambda, T) Source}{M_{e,\lambda}(\lambda, T) Blackbody} \tag{2.8}$$

When radiant energy, ϕ (W), is incident on a surface that is in thermal equilibrium with its surroundings, in order for the energy to be conserved, a fraction of the energy may be absorbed, a fraction of the energy may be reflected, and a fraction of the energy may be transmitted [15],

$$\phi_{incident} = \phi_{absorbed} + \phi_{reflected} + \phi_{transmitted} \tag{2.9}$$

Dividing Eq. (2.9) by the incident radiation $\phi_{incident}$, Eq. (2.10) is obtained:

$$1 = a\lambda + r\lambda, T + t\lambda, T \tag{2.10}$$

where a is the absorptance, r is the reflectance and t is the transmittance. By definition, a Blackbody absorbs all the incident radiation; therefore, $a=1$ and $r=t=0$.

Kirchoff's law states that if a body is in thermal equilibrium with its surrounding environment and is at a temperature T, the integrated absorptance a is equal to the integrated emissivity ε, where the absorptance and emittance are function of wavelength λ and temperature T. Kirchhoff's law is valid if there is no significant temperature gradient or if the phonons and electrons are in thermal equilibrium. Kirchhoff's law, on a spectral basis, can be stated as [15]

$$\alpha_{\lambda,T} = \varepsilon_{\lambda,T}. \tag{2.11}$$

2.6 Optical imaging system

The detection of infrared radiation depends on the energy radiated by the source and the optical system that is used for detection. An optical lens is required to collect the incident radiation in order to be directed to the infrared radiation detector element. Fig. 2.6 shows the ray schematic of an imaging system: source-lens-image (detector) [15].

The lens aperture A_{lens} acts as a collector receiver and the amount of radiant energy (flux) ϕ (W) collected by the optical system can be expressed as

$$\phi = LA_{obj}\Omega_{lens} = LA_{lens}\Omega_{obj}, \tag{2.12}$$

where A_{obj}, A_{lens}, and A_{img} are the area (cm^2) of the object (radiation source), the area of the lens and the area of the radiation detector A_{det}, respectively; and

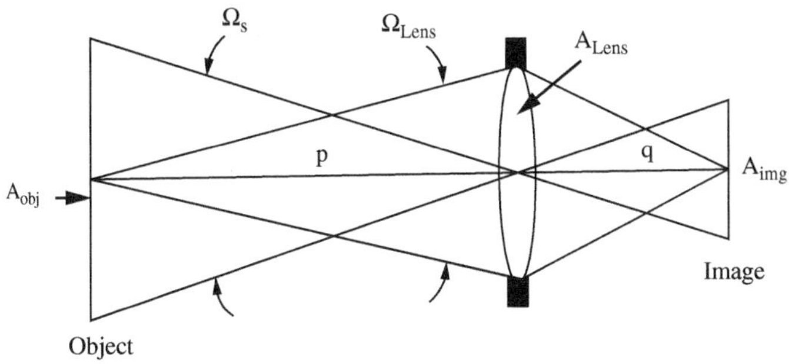

Fig. 2.6 Optical detection imaging system [15].

L (Wcm^{-2}μm^{-1}sr^{-1}) is the spectral radiance emittance. The solid angle of the lens and the solid angle of the image are

$$\Omega_{lens} = \frac{A_{lens}}{p^2}, \tag{2.13}$$

$$\Omega_{obj} = \frac{A_{det}}{q^2}, \tag{2.14}$$

where A_{det} is the area of the radiation detector, p (cm) is the distance between the object (radiation source) and the transmission lens, and q (cm) is the distance between the transmission lens and the radiation detector. Substituting the solid angles into Eq. (2.12), the total amount of flux can be expressed as

$$\phi = L\frac{A_{lens}A_{obj}}{p^2} = L\frac{A_{lens}A_{det}}{q^2}, \tag{2.15}$$

where the maximum image quality is obtained when $A_{det} = A_{obj}\left(\frac{q}{p}\right)^2$.

From Eq. (2.15), the optical radiation that reaches the detector area is called irradiance E_{det} (W cm^{-2}). It is the total flux collected by the lens divided by the radiation detector area,

$$E_{det} = \frac{\phi}{A_{det}} = L\frac{A_{lens}}{q^2}. \tag{2.16}$$

For distant objects (radiation sources) $p \to \infty$, and distance $q = f$, where f is the focal length of the lens; the focal number of the optic system is defined as

$$f/_{\#} = \frac{f}{D_{lens}}, \tag{2.17}$$

where D_{lens} is the diameter of the lens. The focal number is related to the dimensions of the diameter of the lens and the distance between the lens and the detection element.

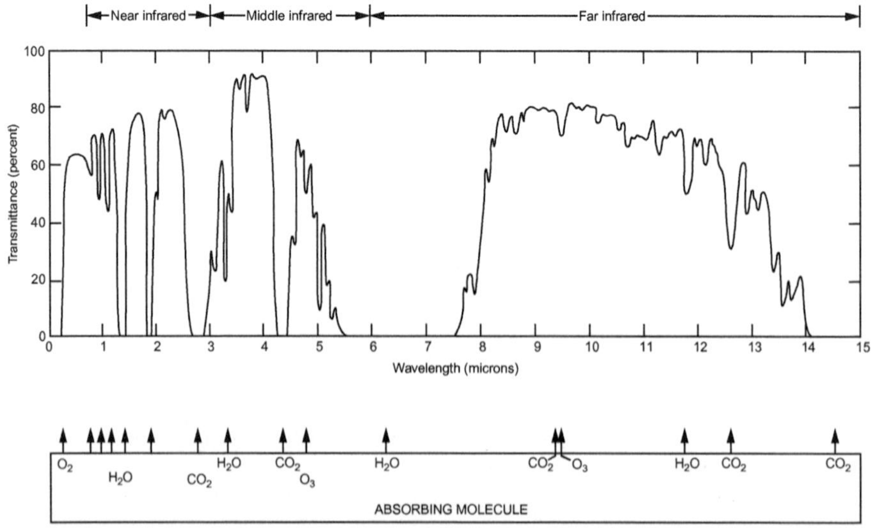

Fig. 2.7 Infrared radiation transmission in earth's atmosphere, over 6000 ft. horizontal path at sea level, containing 17 mm of precipitate water.
Source: Hudson Jr. R. D. Infrared System Engineering, John Wiley & Sons Inc., ISBN: 471-41850 (1969).

2.7 Infrared radiation transmission in Earth's atmosphere

Earth's atmosphere selectively absorbs and scatters infrared radiation by its composition of gaseous molecules. Fig. 2.7 shows the infrared transmission in the earth's atmosphere over 1.8 km as a function of wavelength (0–15 μm) [1]. The infrared radiation transmission in the atmosphere is restricted to three main windows in the following wavelength regions: 0.8–2.5 μm (near IR), 3–5 μm (intermediate IR), and 8–14 μm (far IR). The transmission bands in the atmosphere are due to the IR absorption and scattering by gas molecules such as carbon dioxide (CO_2), ozone (O_3), oxygen (O_2), and water (H_2O) [1].

The three atmospheric infrared transmission windows determine the design and optimization of infrared detectors for each specific infrared window and with a specific detection application [15]. The IR band, 8–14 μm (far IR), is the infrared spectral region of choice for uncooled infrared imaging detection systems due to the existence of an atmospheric window, and, at room-temperature (300 K), objects register maximum infrared radiation emissivity [17].

References

[1] R.D. Hudson Jr., Infrared System Engineering, John Wiley & Sons Inc., 1969. ISBN: 471-41850.

[2] https://en.wikipedia.org/wiki/Electromagnetic_radiation, Accessed 23 October 2021.

[3] G.H. Rieke, Detection of Light, second ed., Cambridge University Press, United Kingdom, 2003, ISBN: 0-521-01710-6.
[4] E. Hecht, Optics, third ed., Addison Wesley Longman, Inc., 1998, ISBN: 0-201-83887-7.
[5] S.L. Gupta, S. Gupta, Unified Physics: Thermodynamics and Physics, twenty-fourth ed., vol. 2, Jai Prakash Nath & Co., 2002.
[6] P.T. S., The bolometer, Nature 25 (1881) 14–16, https://doi.org/10.1038/025014a0.
[7] See for example, Defense Advanced Research Projects Agency > Program Information Wafer Scale Infrared Detectors (WIRED), Dr. Whitney Mason, https://www.darpa.mil/program/wafer-scale-infrared-detectors, Accessed 30 January 2020.
[8] https://en.wikipedia.org/wiki/Microbolometer, Accessed 23 October 2021.
[9] M. Moreno, R. Jiménez, A. Torres, R. Ambrosio, Microbolometers based on amorphous silicon-germanium films with embedded nanocrystals, IEEE Trans. Electron Devices 62 (7) (2015) 2120–2127.
[10] M.D. Aggarwal, A.K. Batra, P. Guggilla, M.E. Edwards, B.G. Penn, J.R. Currie, Pyroelectric Materials for Uncooled Infrared Detectors: Processing, Properties and Applications, *NASA/TM*—2010—*21673*, National Aeronautics and Space Administration, Marshall Flight Center, MSFC, Alabama, 2010.
[11] N. Neumann, V. Banta, Comparison of pyroelectric and thermopile detectors, in: *AMA Conferences 2013—Sensor 2013*, OPTO 2013, IRS^2, 2013, pp. 139–143, https://doi.org/10.5162/irs2013/iP12.
[12] M. Ahmed, D.P. Butler, Z. Celik-Butler, Device-level vacuum packaged uncooled microbolometer on a polyimide substrate, Infrared Physics & Technology 79 (2016) 50–57.
[13] R. Siegel, J. Howell, Thermal Radiation Heat, second ed., Hemisphere Publishing Corporation, 1972, ISBN: 0-07-057316-6.
[14] F.W. Sears, Mechanics, Wave Motion and Heat, Addison-Wesley Publishing Company, Inc., 1950, ISBN: 978-0201069051.
[15] E.L. Dereniak, G.D. Boreman, Infrared Detectors and Systems, John Wiley & Sons, Inc., 1996, ISBN: 0-471-12209-2.
[16] R. McCluney, Radiometry and Photometry, in Encyclopedia of Physical Science and Technology, Third Edition, Elsevier, 2003.
[17] C. Lamsal, N.M. Ravindra, Simulation of spectral emissivity of vanadium oxides (VO_x)-based microbolometer structures, Emerg. Mater. Res. 3 (2014) 194–202.

Figure of merit

3

3.1 Introduction

The operation of a bolometer is based on the change in the electrical resistance of a thermosensing material as a function of temperature. The optical radiation absorber and the thermosensing material are connected thermally and electrically through a support structure to the substrate that acts as a heat sink. The basic representation of the operation of the thermal detector is shown in Fig. 3.1. The thermal detector with thermal capacitance C_{th} is connected by a thermal link of thermal conductance G_{th} to a heat sink, assuming that there is no other path for heat loss [1].

In order to measure changes in its resistance, the bolometer needs to be connected to a DC circuit biased with a constant current source. The electrical circuit, commonly used to interface the bolometer into the reading system, is a voltage divider circuit configuration [2]. Fig. 3.2 shows the schematic of a voltage divider circuit. A bias voltage is applied across a load resistor and the bolometer, where V_{bias} (V) is the bias voltage applied to the circuit, R_b (Ω) is the bolometer resistance, and R_L (Ω) is the load resistance. The measurement readout electric circuit introduces Joule heating $I_b^2R_b$ (W) as a source of heat to be considered, where I_b is constant current (A).

3.2 Thermal detector heat-balance equation

The relationship between the incident radiation and heat loss of the thermal detector (bolometer), considering the Joule heat source, can be described by the heat-balance equation [3]

$$C_{th}\frac{d(\Delta T)}{dt}+G_{th}\Delta T=\varepsilon\Phi_i+{I_b}^2R_b \tag{3.1}$$

where C_{th} (JK^{-1}) is the thermal capacitance coupled via a thermal conductance G_{th} ($W\ K^{-1}$) to a heat sink, t (s) is the radiation absorption time, ΔT is the change in temperature between the bolometer and the heat sink due to the incident optical signal ϕ_i ($W\ cm^{-2}$), ε ($0<\varepsilon<1$) is the emissivity (absorptance) of the IR sensitive film, I_b (A) is the circuit biased current, and R_b (Ω) is electrical resistance of the bolometer active element.

The thermal capacitance (heat capacity) of the thermal bolometer is given by

$$C_{th}=\sum\nolimits_i c_i\rho_i d_i A \tag{3.2}$$

where c ($J\ g^{-1}\ K^{-1}$) is the specific heat capacity of the material, ρ ($g\ m^{-3}$) is the density of the material, d (m) is the thickness of the bolometer, and A (m^2) is the area of the

Microbolometers. https://doi.org/10.1016/B978-0-08-102812-4.00004-8

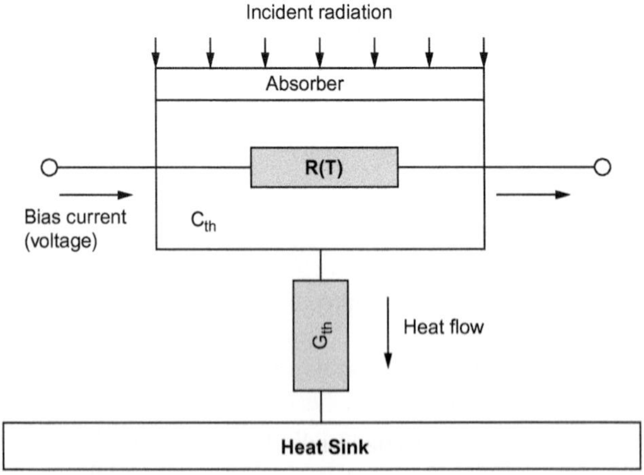

Fig. 3.1 Schematic diagram of the operation of a thermal detector [1].

Fig. 3.2 Schematic of electrical biasing circuit for bolometer.

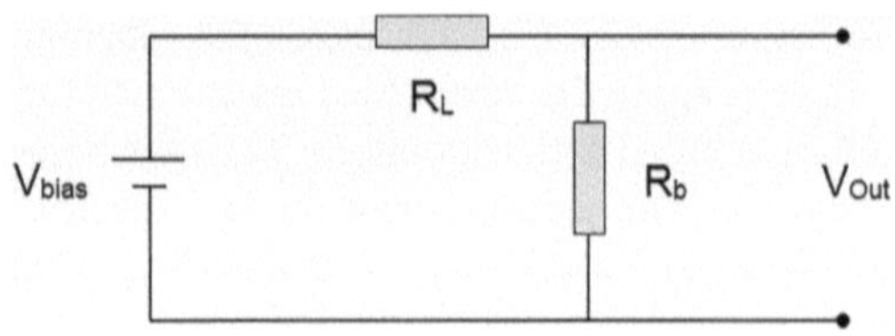

bolometer. The thermal detector is the combination of a series of thin film layers and the subscript i denotes these layers.

Assuming that the absorbed incident optical radiation is a modulated function $\phi_i = \phi_{0e}^{iwt}$, where ϕ_0 is the amplitude of the sinusoidal radiation and $\omega = 2\pi f$ (rad s^{-1}) is the angular frequency of modulation, the heat-balance equation can be rewritten as

$$C_{th}\frac{d(\Delta T)}{dt} + G_{th}\Delta T = \varepsilon\Phi_0 e^{i\omega t} + {I_b}^2 R_b \quad (3.3)$$

The thermal conductance G_{th} of the bolometer is the combination of the conductance by convection G_{conv}, the conductance by radiation G_r, and the conductance by conduction G_{cond} throughout the link to the heat sink. Considering that the bolometer is completely isolated from the environment under vacuum conditions, the heat loss of thermal conductance by convection G_{conv} is negligible; therefore, the thermal conductance can be expressed by.

$$G_{th}\Delta = G_{cond} + G_r \quad (3.4)$$

The thermal conductance by conduction G_{cond} is the heat conductance through the link to the heat sink and is written as

$$G_{cond} = \sum_{\mathrm{i}} \frac{K_i A_i}{L_i} \quad (3.5)$$

where K (W m^{-1} K^{-1}) is the thermal conductivity of the link, A (m^2) is the area of the link, L (m) is the respective length of the link to the heat sink, and the subscript i denotes the thermal conductance and the dimensions of each thin film layer of the heat link.When the bolometer absorbs the incident optical radiation and its temperature increases by an amount ΔT, it will radiate a total flux that is estimated by the Stefan-Boltzmann total radiation law. The thermal conductance by radiation, where σ is the Stefan-Boltzmann constant [3], is given by

$$G_r = 4A\varepsilon\sigma T^3 \tag{3.6}$$

Considering all the heat losses in the thermal detector, the heat-balance Eq. (3.3), that describes the relationship between the absorbed optical radiation and the heat loss, can be rewritten as

$$C_{th}\frac{d(\Delta T)}{dt} + \left(G_{cond} + 4\varepsilon\sigma T^3\right)\Delta T = \varepsilon\Phi_0 e^{i\omega t} + I_b{}^2 R_b \tag{3.7}$$

The equation relating the thermal detector resistance change with temperature is given by

$$R_b(T)R_0(1+\alpha\Delta T) \tag{3.8}$$

where $R_b\,(T)$ is the bolometer resistance as a function of temperature, R_0 is the bolometer resistance at room-temperature, and ΔT (K) is the change in the bolometer temperature. TCR, the temperature coefficient of resistance α (%/K), is defined as.

$$\alpha = \frac{1}{R_0}\frac{dR_b}{dT} \tag{3.9}$$

where dR_b is the change in the bolometer resistance due to the change in bolometer temperature dT.

The response of the bolometer can be measured with a readout circuit interface that typically has a voltage divider configuration. The schematic of the voltage divider circuit is shown in Fig. 3.2. When a DC voltage source sends in a constant current I_b, the output voltage of the voltage divider circuit is given by

$$V_{out} = V_{bi}\left(\frac{R_b}{(R_b + R_L)}\right) \tag{3.10}$$

At a constant biased current, from Eq. (3.9), the bolometer change in output voltage is

$$\Delta V_b = I_b \Delta R_b = I_b R_0 \alpha \Delta T = V_{out}\alpha\Delta T \tag{3.11}$$

where ΔV_b (V) is the voltage change across the bolometer, as function of its temperature change ΔT and resistance change ΔR_b.

The Joule effect term $I_b^2 R_b$ can be expressed as

$$I_b{}^2 R_b = \frac{v_b^2 R_b}{(R_L + \mathrm{R}_b)^2} \tag{3.12}$$

At thermal equilibrium, when the bolometer is not exposed to optical radiation $\phi_0 = 0$ and a current bias is applied, then the steady-state solution to Eq. (3.7) is

$$G_0 \Delta T = I_b{}^2 R_b = \frac{v_b^2 R_b}{(R_L + \mathrm{R}_b)^2} \tag{3.13}$$

where G_0 is the thermal conductance of the material at temperature T.

At a constant current bias, the Joule heating depends on the resistance that is temperature dependent; therefore, the joule heating effect is temperature dependent in the dynamic heat Eq. (3.7) and solving its derivative with respect to temperature yields.

$$\frac{d}{dt}\left(I_b{}^2\right) = \frac{d}{dR_b}\left(I_b{}^2 R_b\right)\frac{dR_b}{dT} \tag{3.14}$$

By substituting Eqs. (3.9) and (3.13) into Eq. (3.14), and solving the derivative with respect to bolometer resistance, the following equations are obtained:

$$\frac{d}{dt}\left(I_b{}^2\right)\frac{dR_b}{dT} = \frac{d}{dR_b}\left(\frac{v_b^2 R_b}{(R_L + \mathrm{R}_b)^2}\right)\alpha R_b = \alpha\left[\frac{v_b^2 R_b}{(R_L + \mathrm{R}_b)^2}\right]\left(\frac{R_L - R_b}{R_L + R_b}\right) \tag{3.15}$$

From Eqs. (3.14) and (3.15), the following equation is obtained:

$$I_b{}^2 R_b = \alpha\left[\frac{v_b^2 R_b}{(R_L + \mathrm{R}_b)^2}\right]\left(\frac{R_L - R_b}{R_L + R_b}\right)\Big)\Delta T \tag{3.16}$$

For the thermal detector, the thermal heat-balance Eq. (3.7) can be rearranged as

$$C_{th}\frac{d(\Delta T)}{dt} + \left(G_{cond} + 4A\varepsilon\sigma T^3\right)\Delta T = \varepsilon \Phi_0 e^{i\omega t} + \alpha\left[\frac{v_b^2 R_b}{(R_L + \mathrm{R}_b)^2}\right]\left(\frac{R_L - R_b}{R_L + R_b}\right)\Delta T \tag{3.17}$$

$$C_{th}\frac{d(\Delta T)}{dt} + \left[\left(G_{cond} + 4A\varepsilon\sigma T^3\right)\Delta T - \alpha\left[\frac{v_b^2 R_b}{(R_L + \mathrm{R}_b)^2}\right]\left(\frac{R_L - R_b}{R_L + R_b}\right)\right]\Delta T$$

$$= \varepsilon \Phi_0 e^{i\omega t} \tag{3.18}$$

where the total thermal conductance of the bolometer depends on the thermal conductance by conductivity in the arms, the thermal conductance by radiation and the joule effect as a function of bolometer temperature change. The effective thermal conductance G_e is defined as

$$G_e = \left[G_{cond} + 4A\varepsilon\sigma T^3 - \alpha \left[\frac{v_b^2 R_b}{(R_L + R_b)^2} \right] \left(\frac{R_L - R_b}{R_L + R_b} \right) \right] \tag{3.19}$$

The dynamic thermal heat-balance of the thermal detector can be finally written as

$$C_{th} \frac{d(\Delta T)}{dt} + G_e \Delta T = \varepsilon \Phi_0 e^{i\omega t} \tag{3.20}$$

Eq. (3.20) is an inhomogeneous first-order differential equation and the change in temperature of the bolometer, due to the absorption of incident optical radiation flux, is the solution,

$$\Delta T = T_0 e^{-\left(\frac{G_e}{C_{th}}\right)t} + \frac{\varepsilon \Phi_0 e^{i\omega t}}{G_e + i\omega C_{th}} \tag{3.21}$$

The first term of the heat transfer solution is a transient term and as time increases, it decreases exponentially to zero. The second term represents the relationship between the incident radiation power and heat loss. The change in temperature of a thermal detector, due to absorption of incident optical radiation and Joule effect, is

$$\Delta T = \frac{\varepsilon \Phi_0}{\left(G_e + \omega^2 C_{th}^2\right)^{1/2}} \tag{3.22}$$

Considering a fixed infrared radiation power, the change in temperature of the device is dominated by the radiation absorption in the absorber material and the thermal conductivity of the support structure. Eq. (3.22) shows that, in order to achieve a large change in the bolometer temperature, the thermal conductance and the thermal capacity should be as small as possible, and a high emissivity of the optical radiation absorber material is required. It can also be observed that, as ω increases, the term $\omega^2 C_{th}^2$ increases and the temperature change will decrease.

Introducing the effective thermal time constant response τ_e(s) as the ratio of thermal capacity to the thermal conductance, it is defined as

$$\tau_e = \frac{C_{th}}{G_e} \tag{3.23}$$

The effective thermal time constant or integration time of the system is the time it takes for the detector output to reach a value of 70% of its final, steady state value.

The time response constant is given by Eq. (3.23) where the cut-off frequency f_c (−3 dB frequency) of the responsivity is improved by reducing the heat capacity and reducing the thermal conductance of the detector element.

Eq. (3.22) can be rewritten as

$$\Delta T = \frac{\varepsilon \Phi_0}{G_e\left(1+\omega^2\tau_e^2\right)^{1/2}} \tag{3.24}$$

The effective thermal time constant characterizes the radiation flux transition between the low and high frequency regions. At low frequencies, the detector temperature is inversely proportional to the thermal conductance of the heat link. At high frequencies, the detector temperature is inversely proportional to the thermal capacitance and falls off inversely with the frequency of the optical signal.

At low frequencies, $\omega^2\tau_e^2 \ll 1$; therefore,

$$\Delta T = \frac{\varepsilon \Phi_0}{G_e} \tag{3.25}$$

At high frequencies, $\omega^2\tau_e^2 \gg 1$; thus,

$$\Delta T = \frac{\varepsilon \Phi_0}{\omega C_{th}} \tag{3.26}$$

3.3 System noise

The detector output signal is disturbed by a variety of sources of noise, resulting in random fluctuations of the measured signal. The performance of an optical detector is determined by the noise, which is the combination of the noise generated by the detector, the noise generated by the readout electronic system, and the noise from the radiation background against which the signal is detected [4]. Noise sets the limit to the minimum optical radiation power to be detected and dictates the detector design parameters to be considered to achieve a high signal to noise detection system. The sources of noise are required to define the figures of merit of the detectors such as noise equivalent power *NEP* and normalized detectivity D^*.

The total noise of the bolometer is the combination of thermal fluctuation noise, Flicker noise (1/f), thermal background fluctuation noise, and Johnson noise [3]. The square of the total noise voltage of the detection system is

$$V_n^2 = V_j^2 + V_{th}^2 + V_b^2 + V_{1/f}^2 \tag{3.27}$$

The Johnson noise or Johnson-Nyquist noise V_j (V) is associated with electrical resistors and is caused by the thermal agitation of atoms and charge carriers within a conductor material. In order to minimize this, the detector resistance R, temperature T, and the measurement noise bandwidth Δf, related to the integration time, should be minimized. Johnson noise is defined by Eq. (3.28), where k_B is the Boltzmann constant and R is the detector resistance at room temperature [4].

$$V_j^2 = 4k_B \mathrm{TR}\Delta f \tag{3.28}$$

The thermal fluctuation noise V_{th} (V) is caused by the temperature fluctuations in the detector and is due to fluctuation in the rate at which heat is released from the detector to its surroundings with which it is thermally connected. This noise is inherent of thermal detectors and its main heat loss is through the thermal link connected to the heat sink. The thermal fluctuation voltage noise is derived from the heat-balance equation [4] and is expressed as

$$V_{th}^2 = \frac{4k_B T^2 \Delta f}{G_e\left(1+\omega^2\tau_e^2\right)} \frac{\Delta V}{\Delta T} \tag{3.29}$$

where ΔV is the change in the electrical output of the detector corresponding to the detector change of temperature ΔT.

The background noise V_b (V) is caused by the radiative heat exchange between the detector and the surrounding environment that is being observed [5].

$$V_b^2 = \frac{8k_B \varepsilon\sigma A\left(T_d^2 - T_b^2\right)}{G_e^2\left(1+\omega^2\tau_e^2\right)} \left(\frac{\Delta V}{\Delta T}\right)^2 \tag{3.30}$$

where A is the optical radiation absorber area of the detector, T_d is the detector temperature, and T_b is the temperature of the surrounding observed environment.

The Flicker noise (1/f) is a component caused by conductivity fluctuations [5] in all resistors and can be described by the empirical equation

$$V_{1/\mathrm{f}}{}^2 = k_{1/f} \frac{I^\delta}{f^\beta} (\Delta f) \tag{3.31}$$

where the coefficient $k_{1/f}$ is a factor of proportionality, and coefficients β and δ have values of about 1. These coefficients depend on the methods of material preparation and processing.

The Flicker noise (*1/f*) is attributed to surface factors (McWhorter model) and to bulk defects (Hoogen model). The McWhorter model implies fluctuations in the number of carriers [3], and the Hoogen model considers the carrier mobility [6, 7]. 1/f is a conductivity noise that, for low frequencies, tends to be the limiting factor, depending on the number of effective carriers and not the number of carriers. The number of effective carriers is reduced due to scattering with crystal lattice imperfections, non-Ohmic contacts, and grain boundaries [8].

3.4 Figures of merit

The detector performance can be described quantitatively and evaluated in terms of figures of merit. Figures of merit are parameters that relate the detector response to the absorbed optical radiation and characterize the performance of optical detectors. Some figures of merit allow to compare the performance between optical detectors, independent of their optical radiation absorption mechanism, while others are dependent on these mechanisms and detector geometry dimensions.

The performance of a bolometer is determined by figures of merit such as responsivity R_v, noise equivalent power *NEP*, and normalized detectivity D^*.

The responsivity R_v (*V/W*) of the detector anticipates the signal output for a specific optical radiation absorbed. Responsivity is defined as the ratio of the output signal ΔV_b to the input radiant power incident that is absorbed, ϕ_0, by the detector and is given by [9]

$$R_v = \frac{\Delta V_b}{\Phi_0} \tag{3.32}$$

When the detector is interfaced with a voltage divider configuration readout circuit as shown in Fig. 3.2, substituting Eqs. (3.10), (3.11) and (3.24) in Eq. (3.32), the responsivity can be rewritten as

$$R_v = \frac{\varepsilon\alpha}{G_e\left(1+\omega^2\tau_e^2\right)^{1/2}}\frac{V_{bias}R_b}{(R_b+R_L)} \tag{3.33}$$

The detector output signal must be above the detector noise level that determines the minimum absorbed optical radiation level the detector can discern; the noise equivalent power *NEP* (W) is the signal power needed to achieve a signal-to-noise ratio of 1, and it can be expressed in terms of responsivity as [9]

$$NEP = \frac{V_n}{R_v} \tag{3.34}$$

The normalized detectivity D^* (cmHz$^{1/2}$ W^{-1}) is the reciprocal of the noise equivalent power normalized to 1-cm^2 area and 1-Hz noise equivalent bandwidth and is defined as [9]

$$D^* = \frac{\sqrt{A_d\Delta f}}{NEP} \tag{3.35}$$

where A_d (cm^2) is the area of the detector (area of radiation absorption) and Δf (Hz) is the detector noise equivalent bandwidth related to the modulation frequency of the radiation source. The normalized detectivity defines the resolving power of the detector and allows to compare the detector performance with different sizes and technology.

The detector noise equivalent bandwidth is the frequency over which the noise is measured and is expressed as

$$\Delta f = \frac{1}{2\pi\tau_e} \tag{3.36}$$

The figures of merit are the parameters that allow the characterization of infrared detectors (single element) and their performance comparison with other infrared detectors, independent of the radiation absorption mechanism and the active area of absorption. The figures of merit and the various device parameters that are described above are applicable to array/s of detectors.

An optical infrared imaging system consists of an array of infrared detectors located at the focal plane (FPA) of an optical detection imaging system, integrated with a read-out integrated circuit (ROIC). The arrays can be one-dimensional (1D) or two-dimensional (2D), depending on the purpose and application of the imaging system/s. A significant figure of merit of the infrared imaging system is the noise equivalent temperature difference *NETD* (mK) [10], which is defined as

$$NETD = \frac{4\left(\frac{f}{\#}\right)^2 \sqrt{\Delta f}}{\tau_a \tau_0 \sqrt{A_d} D^* \left({}^{\Delta L}/_{\Delta T}\right)_{\lambda 1-\lambda 2}} \tag{3.37}$$

where $f/\#$ is the focal number of the optical detection imaging system, A_d is the detector area, τ_0 is the transmittance of the optics system, τ_a is the atmospheric transmittance, and $\left({}^{\Delta L}/_{\Delta T}\right)_{\lambda 1-\lambda 2}$ is the temperature contrast of total spectral radiance emittance at 300 K and represents the change in IR radiation power detected on the detector element per unit change in temperature of the source object within the spectral wavelength band under consideration.

The *NETD* parameter is referred as thermal contrast and measures the smallest temperature change that a thermal infrared detector is able to distinguish between the infrared radiation source and its background. *NETD* is typically expressed in milli-Kelvin (mK) and is inversely proportional to the detectivity. Commercial state-of-the-art uncooled infrared bolometer arrays have typically *NETD* values in the order of less than 100 mK.

The performance of the microbolometer is characterized by certain characteristic figure of merit, which includes but is not limited to responsivity, noise equivalent power (NEP), and detectivity. The purpose of a microbolometer is to convert an input light signal to the output electrical signal. The term responsivity (*R*) is used to describe the amplitude of the electrical signal with respect to the incident flux

$$R = \frac{V_{out}}{\varphi_{incident}} = \frac{V_{out}}{EA_d} [\mathrm{V/W}] \tag{3.38}$$

where V_{out} is the detector output voltage, $\varphi_{incident}$ is the incident power in Watts, E is the incident irradiance in Watt per square centimeter, and A_d is the detector active area in square centimeters. The unit of R is Volt per Watts.

The voltage responsivity may also be written as the RMS (root mean square) signal voltage output per unit radiant power incident upon the detector. The responsivity R is written as

$$R = \frac{V}{P_{incident}} = I_b \frac{dR}{dT}\frac{dT}{dP}\frac{P_{absorbed}}{P_{incident}} = I_b\, Z_{thermal} \frac{dR}{dT}\frac{1}{\sqrt{1+\omega^2\tau^2}}\eta \tag{3.39}$$

where I_b is the bias current, $\frac{dR}{dT}$ is the change in resistance of the bolometer for given change in Temperature, T and η is the coupling efficiency of incident wave, $Z_{thermal}$ is the thermal impedance, ω is the frequency and τ is the time constant.

If $\frac{dR}{dT}$ is constant over the considered temperature range, the temperature change in the device is proportional to change in dissipated power. One characteristic of a microbolometer is the linear relationship between resistance and dissipated power over the operating temperature range.

In order to increase the responsivity, bias current cannot be increased infinitely. It is limited mainly by Joule heating and detector instability.

The responsivity of a microbolometer is related to the voltage signal level and the incident flux measured on the detector. Large responsivity does not necessarily imply that the microbolometer can easily discern small optical signal changes. Hence, a more appropriate measure of the performance of a detector is how well it extracts a signal from the surrounding noise. This characteristic is referred as noise equivalent power (NEP). Noise equivalent power is defined as the root-mean-square value of sinusoidally modulated radiant power falling on the detector required to produce signal to noise ratio of one (S/N = 1). NEP is most frequently specified at bandwidth of 1 Hz and is expressed in units of Watts/Hz$^{1/2}$. NEP is related to the noise voltage (V_n) and responsivity (R) by.

$$\text{NEP} = \frac{V_n}{R} \tag{3.40}$$

The detectivity, or D^*, is the reciprocal of the NEP, normalized to detector area 1 cm^2, and 1 Hz electrical bandwidth.

$$D^* = \frac{A_{bolo}^{1/2} b^{1/2}}{NEP} \tag{3.41}$$

where D^* = detectivity in cm-Hz$^{1/2}$/watt, A_{bolo} = detector area in cm^2, and b = postdetection electrical bandwidth.

Spectral detectivity curves for a number of commercially available IR detectors are shown in Fig. 3.3. The major interest is in the performance of the detectors in the wavelengths of the two atmospheric windows, 3–5 μm (middle wavelength infrared or MWIR) and 8–14 μm (the LWIR region). Atmospheric transmission is the highest in the MWIR and LWIR bands, and the emissivity maximum of the objects at $T \approx 300$ K is at the wavelength $\lambda \approx 10$ μm. However, in recent years, applications in space have increased the interest to infrared imaging in longer wavelengths.

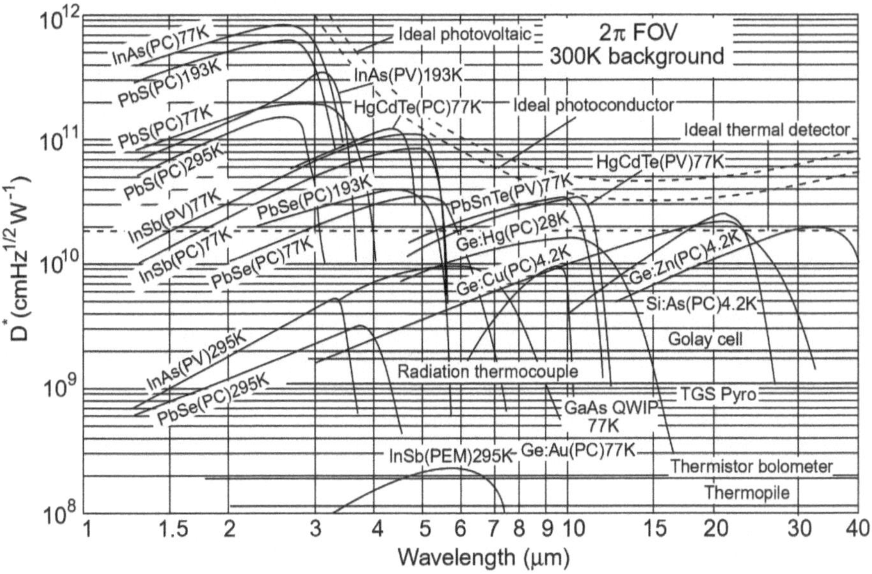

Fig. 3.3 Comparison of the D^* of various commercially available IR detectors when operated at the indicated temperature [11].
Reproduced with permission from Antoni Rogalski, Infrared detectors: status and trends, Progr. Quantum Electronics 27(2–3), 2003, 59–210, ISSN 0079–6727, https://doi.org/10.1016/S0079-6727(02)00024-1

Due to the low signal voltage from microbolometer devices, the most important figure of merit to consider during the design and optimization is responsivity. Maximizing the bias current and detector coupling efficiency results in optimization of responsivity. By converting into physical constants, the role of the material properties and structures of the microbolometers becomes much clearer.

$$R = I_b\, Z_{thermal}\, \alpha R_{bolo} \frac{1}{\sqrt{1+\omega^2\tau^2}}\eta \tag{3.42}$$

where α is the temperature coefficient of resistance (TCR).

R depends on bias current, thermal conductance, the temperature coefficient of resistance (TCR), and coupling efficiency. To maximize the responsivity, the thermal conductance of the bolometer must be minimized. Cross-sectional area must be decreased and the length of the thermal link must be increased. The responsivity of a microbolometer can be increased by the use of suspended bridges that prevent contact between the detector and the substrate. The temperature coefficient of resistance

(TCR) is critical in the selection of the material of the bolometer. The TCR may be written as.

$$\mathrm{TCR} = \alpha = \frac{1}{R_{bolo}} \frac{dR_{bolo}}{dT} \tag{3.43}$$

For low temperature applications, the superconducting bolometer [12] exploits the resistive transition at the superconducting threshold. The TCR is maximum at the superconducting threshold temperature. Depending on the material characteristics, some materials show a positive TCR. Some materials may also have a negative TCR. Negative TCR is more advantageous than positive TCR due to biasing in order to have a stable performing microbolometer. TCR of materials is a result of two factors, the carrier concentration that increases with increase in temperature and the carrier mobility generally decreasing with increase in temperature. The resulting sign of differential conductivity, as a factor of temperature, is proportional to the product of carrier concentration and carrier mobility in a given temperature range. A desirable microbolometer material has large TCR value with negative TCR at its operating temperature.

By increasing the bias current, a higher responsivity may be achieved. The maximum bias current is likely to be limited by one of the many factors which leads to failure of the microbolometer. Thermally induced current due to joule heating, high filament current density which leads to electromigration failure in the detector element, and detector instability are a few of the failure mechanisms.

The power dissipated in the microbolometer detector mainly comprises incoming radiation and joule heating from the bias current. However, the power dissipated in the detector is majorly caused by joule heating from the bias source. The maximum allowable detector bias current is given by.

$$I_{max} = \sqrt{\frac{\Delta T_{max}}{R_{bolo}\, Z_{thermal}}} \tag{3.44}$$

where ΔT_{max} is the maximum allowed temperature rise.

The relationship between the bias current and the responsivity is given by

$$R = \sqrt{\Delta T_{max}}\,\sqrt{R_{bolo}}\,\sqrt{Z_{thermal}}\,\frac{\alpha}{\sqrt{1+\omega^2\tau^2}}\eta \tag{3.45}$$

The usage of detector materials that can withstand high temperatures improves the responsivity of microbolometers. Materials that melt at high temperatures qualify as good candidates to be used in detectors. Increasing the lower ambient temperature increases the maximum temperature difference. This is also likely to improve NEP by reducing the thermally induced component of noise. Choosing high resistance detector automatically increases the responsivity. The maximum allowable current may be restricted to one of the supporting leg structures.

The thermal impedance is also dependent on the length to width ratio of the supporting structure of the detector. For any given detector material, the length to width ratio of the detector will determine the maximum responsivity of the device. A crucial factor that may limit the responsivity through bias current is the stability of the detector under constant bias conditions. If the device is biased at constant current, it results in positive or negative feedback, depending on whether the detector has a positive or negative temperature coefficient of resistance. The feedback is due to the absorption of incident radiation affecting the bias power with change in detector resistance. When the bias current is held constant, it increases the bias power, which further increases the resistance and as a result the bias power.

An absorption enhancement device can be fabricated by maximizing the coupling efficiency. By modulating the absorption of the incoming optical signal, the amount of absorbed power by the device can be modified and optimized. An approach is to alter the structure of the microbolometer to make the coupling efficiency a function of wavelength.

Noise sources may exist within the IR detector itself, as the detector interacts with its environment or can be introduced by the detector readout system. Micromachining enables batch fabrication of highly efficient transducers that convert very small heat fluxes or temperature differences into conveniently measured output signals. While it significantly reduces sizes and heat capacitances improve the sensitivity of thermal and calorimetric sensors, it makes them useful as IR detectors and is governed by the influence of the various noise sources. The noise characteristics of microscopic devices tend to impose fundamental limitations to the performance of thermal detectors. Two of such constraints are applicable to all types of thermal IR detectors and arise from the fact that every material absorbs heat. The constraints depend on its thermal mass and the degree of heat exchange with the environment. They undergo certain temperature fluctuations. These spontaneous temperature changes are mainly negligible in macroscopic devices but they make a compelling improvement to micro devices such as microbolometers.

The magnitude of thermal fluctuation noise increases with improved thermal isolation. It reaches maximum for ideally isolated detectors, in the case of purely radiative heat exchange. Better thermal isolation improves the performance of a thermal IR detector by increasing the responsivity of the detector.

Noise equivalent temperature difference (*NETD*) is described by the noise rating of the IR detector. It specifies the amount of radiation required to produce an output signal that is equivalent to the noise of the detector. Being able to detect minute difference in temperature plays important role in IR imaging signifying a better *NETD* value. It also enables a better performance.

3.5 Optical infrared imaging system

A determining figure of merit used by the imaging industry, to characterize and compare the performance of an optical infrared imaging system (array of infrared detectors) that could be used to calculate the performance of a single detector, is the noise

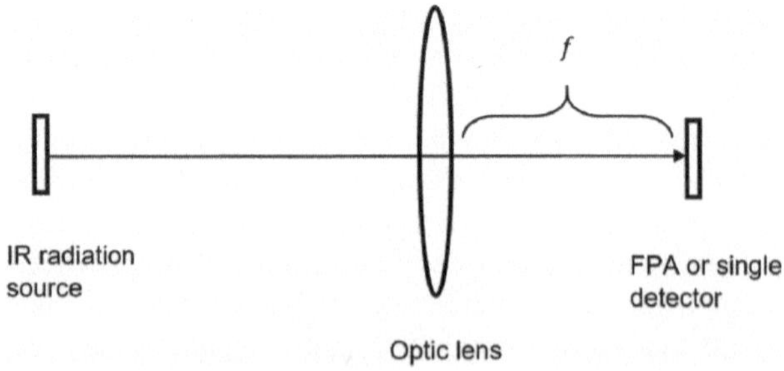

Fig. 3.4 Schematic diagram of optical system.

equivalent temperature difference *NETD* (Eq. 3.37); this parameter specifies the minimum detectable temperature difference by the detection system.

Manufacturers of IR cameras measure *NETD*, using different parameters, but the most important is the focal number $f/\#$ of the optical lens used during the measurements. The comparison of the detection performance of different imaging systems is based on the comparison of the *NETD* values with the same $f/\#$, that is usually normalized to 1. For an imaging system to have a considerable performance, the *NETD* should be less than 100 mK, and for a state-of-the-art system, it should be less than 50 mK. Another important parameter, in the imaging industry, is the frame rate at which the camera operates; standard focal plane array (FPA) systems operate in a range of 9–100 Hz and typically in the range of 30–60 Hz. The frame rate is the rate at which the FPA imaging system creates images and a common practice is to make the thermal time response of the detector (pixel) not to exceed one-third of the reciprocal of the frame rate. Eq. (3.46) describes the frame rate relationship with the detector thermal time constant

$$\text{Frame rate} = \frac{1}{3\tau_e} \tag{3.46}$$

Fig. 3.4 shows the schematic diagram of a standard optical system, where *f* is the focal length. Using optics (lens) with larger diameters allows to build more sensitive IR thermal imaging systems (cameras), but glass lens does not have a high IR radiation transmittance and optical lenses made of high IR transmittance germanium (Ge) are used by the IR industry. Germanium lenses are expensive and an increase in the diameter of the optical lens made of Ge results in an increase in the cost of the IR thermal camera; this is the reason why manufacturers use lenses with small diameters, or high focal number.

References

[1] A. Kosarev, et al., Thin film uncooled microbolometers based on plasma deposited materials, Can. J. Phys. 92 (2014) 570–575.

[2] J.W. Nilsson, Electric Circuits, Addison-Wesley Publishing Company, 1983, ISBN: 0-201-06238-0.

[3] A. Rogalski, Infrared Detectors, second ed., CRC Press, 2011, ISBN: 978-1-4200-7671-4.
[4] J.C. Doll, B.L. Pruitt, Piezoresistor Design and Applications, Springer, 2013. ISBN: 978-1-4614-8516-2.
[5] W. Kruse, et al., Elements of Infrared Technology: Generation, Transmission and Detection, Wiley, New York, 1962, ISBN: 0471508861.
[6] L.K.J. Vandamme, G. Trefan, 1/f noise in homogeneous and inhomogeneous media, IEEE Proc.—Circuits Devices Syst. 149 (2002) 3–12.
[7] F.N. Hooge, 1/f noise sources, IEEE Trans. Electron Devices 41 (1994) 1926–1935.
[8] F.N. Hooge, Discussion of recent experiments on 1/f noise, Physica 60 (1972) 130–144.
[9] R.D. Hudson Jr., Infrared System Engineering, John Wiley & Sons Inc., 1969. ISBN: 471-41850.
[10] J. Baars, M. Schulz, Performance characteristics, measurement procedures, and figures of merit for infrared planes arrays, Proc. SPIE 2470 (1995) 141–155.
[11] W. Vandermeiren, et al., in: F.J. Duarte (Ed.), Infrared Thermo-Electric Photodetectors, in Laser Pulse Phenomena and Applications, IntechOpen, 2010.
[12] F.M. Tong, et al., High T_c $YBa_2Cu_3O_{7-\delta}$ superconducting transition-edge microbolometers, Infrared Phys. Technol. 36 (1995) 1053–1058.

Infrared detector materials

4

4.1 Introduction

Temperature measurement, mapping and control is of enormous interest in a variety of applications in numerous sectors of the global economy. These sectors include the following: aerospace, astronomy, communication, defense, energy, environment, food, health, infrastructure, manufacturing, pharmaceuticals, space, telecommunication, and transportation. Temperature measurements are routinely performed using contact thermometers and noncontact thermometers in various biological and chemical systems. Contact thermometers use the fundamental property of thermal conduction by requiring physical contact with the object whose temperature needs to be measured. These include the thermal expansion/contraction-based liquid thermometers, thermoelectric effect-based thermocouples/thermopiles and resistance based-thermistors/resistance temperature detectors (RTDs).

From a historical perspective, the first known attempt at the measurement of temperature has been reported to be performed by Galen (130–210) in AD 170 [1]. This was followed by contributions to the thermoscope by Galileo Galilei (1564–1642), Santorio (1561–1636), Giuseppe Biancani (1566–1624), Robert Fludd (1574–1637), Cornelius Drebbel (1572–1633), Otto von Guericke (1602–1686), Ferdinando II de'Medici, the Grand Duke of Tuscany (1610–1670), and Francesco Sagredo (1571–1620) [2]. The addition of a temperature scale, known as the RØmer scale, to the thermoscope, was accomplished by Ole Christensen RØmer (1644–1710) three hundred and twenty years ago [3]. In 1665, Christiaan Huygens (1629–1695) proposed the use of melting point of ice and the boiling point of water as standards for measurements of temperature [3,4]. In 1694, ice and boiling water were proposed as limits on the universal scale by Carlo Renaldini (1615–1698) [5]. Daniel Farenheit (1686–1736) invented the modern mercury thermometer and the Farenheit scale in 1714 [6]. The Celsius or the Centigrade scale was invented by Anders Celsius (1701–1744) in 1742 [7]. Sir Thomas Clifford Albut (1836–1925) invented the Clinical Thermometer [8]. Sir William Thomson, Baron Kelvin of Largs, Lord Kelvin of Scotland (1824–1907) invented the Kelvin scale in 1848 [9].

2021 marks the 200th anniversary since the discovery of the Thermoelectric Effect by Thomas Johann Seebeck (1770–1831) in 1821 [10]. Today, thermoelectrics provide an excellent, all-solid state solution to recover waste heat, direct power generation and cooling without the need for any moving parts. The thermoelectric generator market is anticipated to grow to US $741 million by 2025 [11].

The history of IR detectors is summarized in Table 4.1 [12].

With the onslaught of the Covid-19 pandemic, the need for non-contact temperature measurements and monitoring has become essential across the world. The advancements in infrared detector technologies has led to numerous techniques for measurement and mapping of temperature in real-time in a variety of applications.

Microbolometers. https://doi.org/10.1016/B978-0-08-102812-4.00003-6

Table 4.1 History of IR detectors [12].

1800	IR radiation Sir W. Herschel
1821	Thermoelectric effect Seebeck
1829	Thermocouple G. Nobili
1833	Thermopile Macedonio Melloni
1836	Optical pyrometer Becquerel
1873	Photo-detection (Selenium) Smith
1884	IR radiation law Boltzmann
1902	Photoconductivity effect Bose
1917	Lead sulphide Case
1933	Lead sulphide (galena) Kutzsher
1940	TI2S Cashman
1942	Golay cell Golay—Queen Mary College
1948	Transistor Bardeen-Brattain-Shockley
1930s	PbS, PbSe, PbTe T. Moss RRSE
1959	HgCdTe W. Lawson, J. Putley
1960s	Ge: X, InSb
1969	CCD Boyle-Smith (Bell Labs)
1970s	PbSnTe/HgCdTe, Si: X Lincoln Labs, SBRCHughes, Honeywell, Rockwell, Mullard
1973	Common modules Night vision Lab
1975	IR Fly Eye Smart Sensors: C. Corsi, Elettronica SpA Si: X/CCD/PtSi/CCDHgCdTe/CCD RCA Princeton Lab
1978	W. F. Kosonocky, F. Shepherd D. Barbee-F. Milton-J. Steckel
1980s	HgCdTe SPRITE InGaAs QWIP T. Elliott RSE, F. Capasso L. Esaki, B. FLevine, M. Razeghi, L. J. Kozlowski
1990s	Pyroelectric FPAs/MicroBolometer FPAs/Multi-colour FPAs/Advanced FPAs RRSE-BAE,. R. A. Wood, J. L. Tissot, P. R. Norton, A. Rogalski, H. Zogg S. D. Gunapala, D. Z. Ting
2000	MEMS FPAs—Cantilever IR Nanotubes/Nanowires B. Coole, S. R. Hunter, X. Zhang, J. M. Xu, S. Huang, Y. Zhao, J. Xu, Maurer, G. Jiang, D. J. Zook

4.2 Photon detectors

Fig. 4.1 presents an illustration of the IR detector technology [13] along with the details of the materials and their corresponding range of operating wavelengths.

In general, infrared detectors can be classified into two types: (a) photon detectors and (b) thermal detectors. Photon detectors are made from semiconductors. The electronic transition from the valence band to the conduction band in a semiconductor occurs via the photoelectric effect or by thermal excitation. The light induced electronic transition in a semiconductor occurs under the condition when the photon energy, $h\nu > E_g$, the bandgap of the semiconductor, h is the Planck's constant and ν is the photon frequency. $\nu = (c/\lambda)$, where c is the velocity of light in vacuum, and

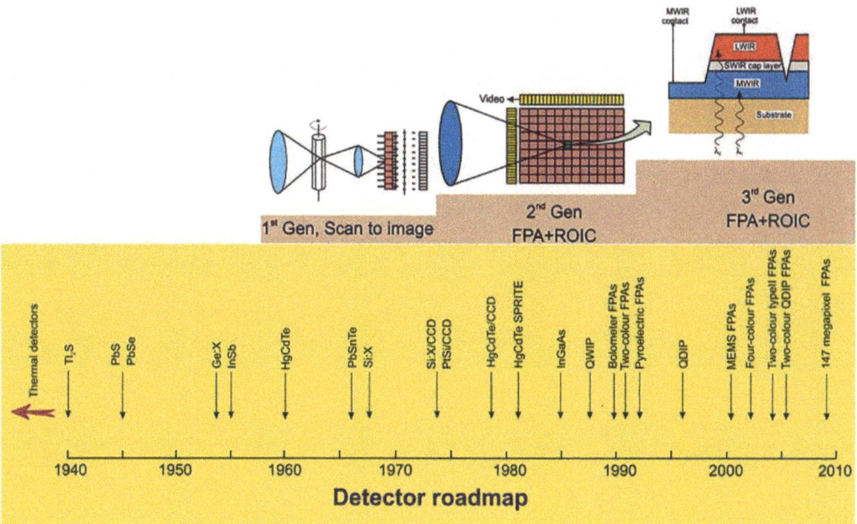

Fig. 4.1 Illustration of IR detector technology [13].
From Recent progress in infrared detector technologies; Infrared Phys. Technol. 54(3), 2011, 136–154.

λ is the wavelength of light. λ_{cutoff}, the cutoff wavelength, i.e., the longest wavelength of sensitivity of a detector is given by the following equation:

$$\lambda_{cutoff} = \left(hc/E_g\right) = \left(1.24/E_g\right); \lambda_{cutoff} \text{ in } \mu m; E_g \text{ in eV}, \tag{4.1}$$

$$\lambda_{cutoff} = \left(hc/E_g\right) = \left(1240/E_g\right); \lambda_{cutoff} \text{ in nm}; E_g \text{ in eV}. \tag{4.2}$$

Thermal excitation of electrons in an intrinsic (pure, undoped) semiconductor is given by

$$n_e = N \exp\left(-E_g/2kT\right), \tag{4.3}$$

where n_e is the number of electrons excited across the energy gap, i.e., the same number of holes in the valence band; N is the number of electrons available at the top of the valence band for excitation, k is the Boltzmann constant, and T is the temperature in Kelvin. This is the origin of dark current in semiconductor detectors and provides the rationale for cooling them [14].

$$T_{max} = \left(200K/\lambda_{cutoff}\right). \tag{4.4}$$

The constant evolution in infrared detector technology has been summarized in several books [15–23]. In his paper, "Effect of light on selenium during the passage of an electric current," Willoughby Smith (1828–1891) reported his discovery of photoconductivity in selenium in 1873 [24]. The emission of electrons as a result of material-light interactions due to Albert Einstein's photoelectric effect and Max Planck's Black Body Radiation laid the foundation for the modern photon detectors. Narrow bandgap semiconductors, based on lead chalcogenides, PbS, PbSe, and PbTe, were developed as infrared sensors in the 1940s and early 1950s. InSb extended the spectral range of PbS by providing sensitivity in the MWIR spectral range, i.e., 3–5 μm [25].

Bandgap engineering has been the prime enabler of tunable bandgaps in IR photon detectors. The discovery of variable band gap HgCdTe alloy in 1959 by Lawson et al. [26] has provided an unprecedented degree of freedom in infrared detector design. The bandgap energy tunability results in IR detector applications that span the short wavelength IR (SWIR: 1–3 μm), middle wavelength (MWIR: 3–5 μm), long wavelength (LWIR: 8–14 μm), and very long wavelength (VLWIR: 14–30 μm) ranges. HgCdTe technology development was and continues to be primarily for military applications [27]. The change in bandgap and the corresponding cut-off wavelength in $Hg_{1-x}Cd_xTe$, as function of the cadmium fraction x, is shown in Fig. 4.2 [14].

In Table 4.2, the energy gap of $Hg_xCd_{1-x}Te$, as function of composition X of Cd, is summarized. As can be seen in this table, as the cadmium content in $Hg_xCd_{1-x}Te$ increases, the energy gap increases and the corresponding cut-off wavelength of the detector decreases.

In general, the responsivity of a photodiode is defined as the ratio of the photocurrent output, I_p, to the radiant energy, P:

$$R_\lambda = (P/I_p) \tag{4.5}$$

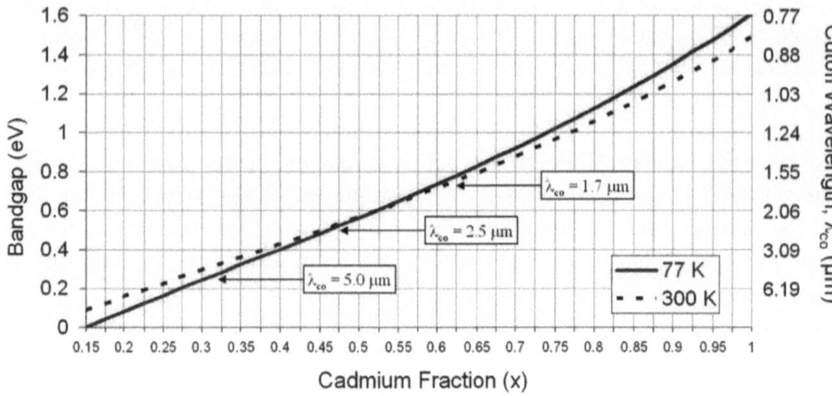

Fig. 4.2 Tunable bandgaps and cutoff wavelength in $Hg_{1-x}Cd_xTe$ [14]. Reproduced with permission from James Beletic, Teledyne Imaging Sensors.

Table 4.2 E_g of $Hg_{1-x}Cd_xTe$ as a function of x [14].

x	E_g (eV)	λ_c (μm)
0.196	.09	14
0.21	.12	10
0.295	.25	5
0.395	.41	3
0.55	.73	1.7
0.7	1.0	1.24

Reproduced with permission from James Beletic, Teledyne Imaging Sensors.

and is expressed in (Ampere/Watt). The radiant power, P, is the product of the photon flux, r_p and the photon energy, $h\nu$ [28]:

$$P = (r_p \times h\nu) \tag{4.6}$$

Fig. 4.3 [29] presents a summary of the types of IR detectors and the associated materials of interest.

In Table 4.3, a detailed summary of the mode of operation, schematic, operation and properties of photon detectors is presented [30].

Semiconductors of interest to photon detectors and their fundamental properties are illustrated in Fig. 4.4 [31].

Single photon detectors are of significant interest in various areas of applications such as in astronomy and astrophysics, fiber-optic communication, light detection and ranging, materials science, medical imaging, quantum encryption, and quantum information science. Examples of infrared single-photon detectors and their performance characteristics are summarized in Table 4.4 [32].

In Fig. 4.5, the temperatures and wavelengths of operation of various semiconductor IR photon detectors is presented [33].

In recent years, with the advent of 1D and 2D materials and structures, new material candidates have been proposed as IR photodetectors. A summary of the responsivity of infrared photodetectors based on various nanostructured materials, as function of wavelength, is presented in Fig. 4.6 [34]. Their performance characteristics is listed in Table 4.5 [34].

Examples of commercially available photon detectors from Horiba are summarized in Table 4.6 [53].

The generation-recombination noise due to the exchange of photons with the radiating background limits the operation of photon detectors. Photon detectors need to be cooled to reduce this noise.

The relative spectral response of a photon detector and a thermal detector for (a) constant incident radiant power and (b) photon flux [33] is illustrated in Fig. 4.7. As can be seen in the figure, the relative spectral response of a thermal detector per incident radiant energy is constant and is independent of the wavelength. The photon detector exhibits a peak that is characteristic of the material of the detector.

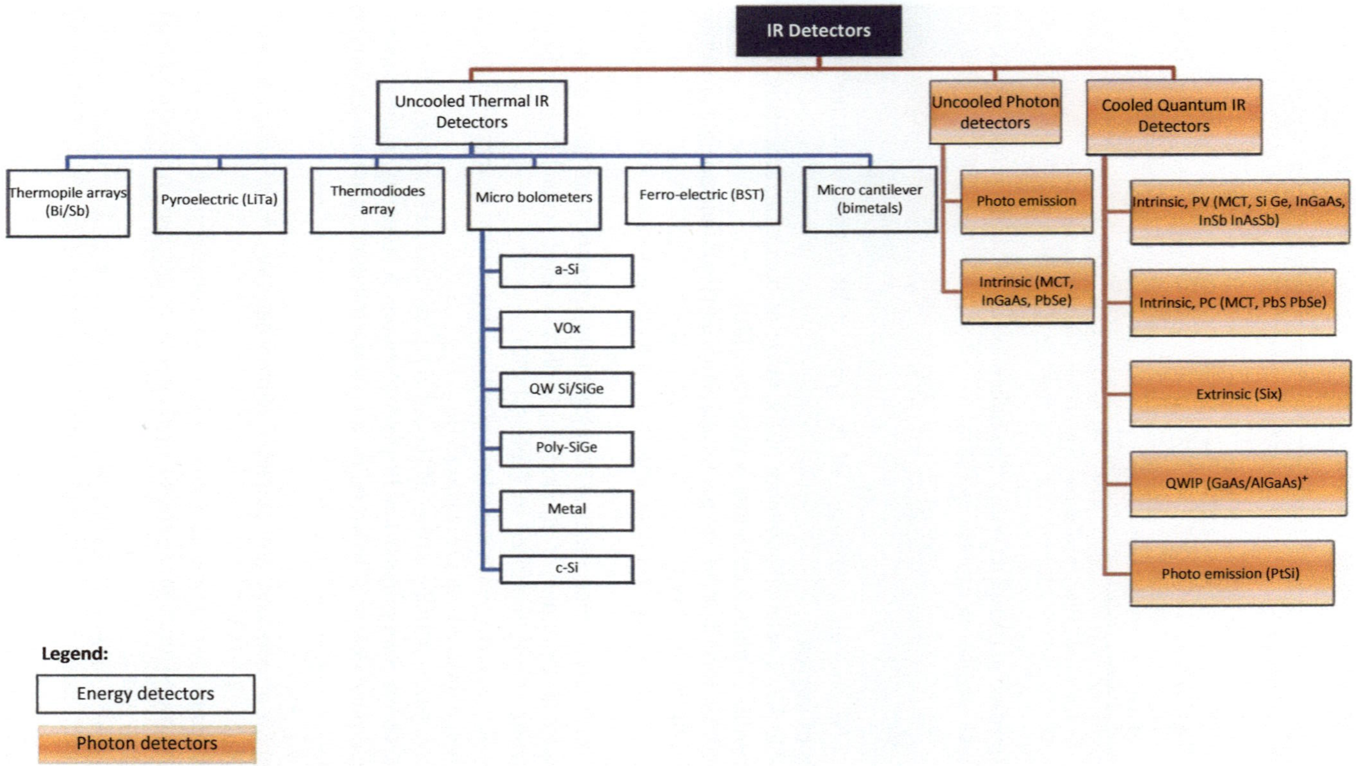

Fig. 4.3 IR detector classification [29].

Table 4.3 Summary of mode of operation, schematic, operation, and properties of photon detectors [30].

Mode of operation	Schematic of detector	Operation and properties
Photoconductor	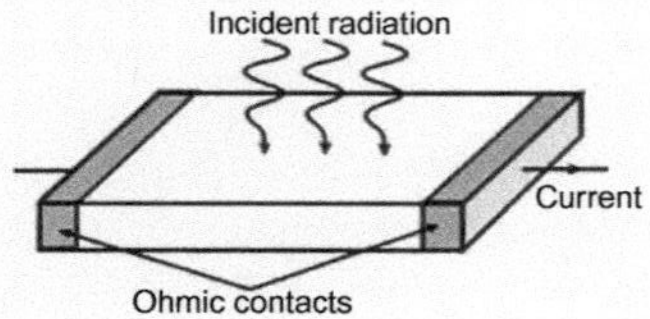	This is essentially a radiation-sensitive resistor, generally a semiconductor, either in thin film or bulk form. A photon may release an electron-hole pair or an impurity-bound charge carrier, thereby increasing the electrical conductivity. In almost all cases the change in conductivity is measured by means of electrodes attached to the sample. For low-resistance material, the photoconductor is usually operated in a constant current circuit. For high-resistance photoconductors, a constant voltage circuit is preferred and the signal is detected as a change in current in the bias circuit
Blocked impurity band detector	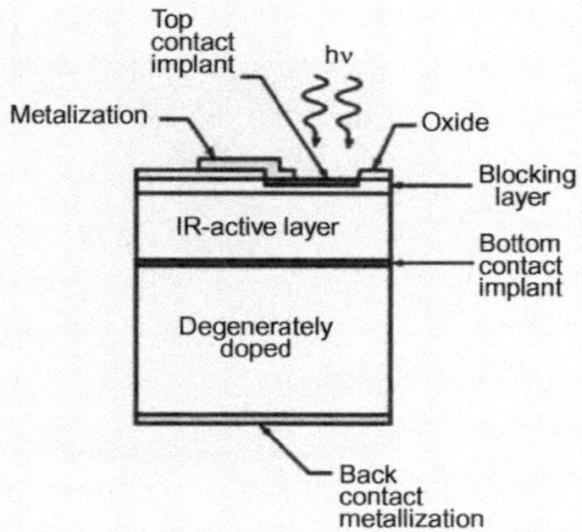	The active region of a BIB detector structure, usually based on epitaxially grown *n*-type material, is sandwiched between a higher-doped degenerate substrate electrode and an undoped blocking layer. Doping of the active layer is high enough for the onset of an impurity baud to display a high quantum efficiency for impurity ionization (in the case of Si:As BIB. the active layer is doped to $\approx 5\times10^{17}$ cm^{-3}). The device exhibits a diode-like characteristic, except that photoexcitation of electrons takes place between the donor impurity and the conduction band. The heavily doped *n*-type IR-active layer has a small concentration of negatively charged compensating acceptor impurities. In the absence of an applied bias, charge neutrality requires an equal concentration of ionized donors. Whereas the negative charges are fixed at acceptor sites, the positive charges associated with ionized donor sites (D^+ charges) are mobile and can propagate through the IR-active layer via the mechanism of hopping between occupied (D°) and vacant (D^+) neighboring sites. A positive bias to the transparent contact creates a field that drives the pre-existing D^+ charges toward the substrate, while the undoped blocking layer prevents the injection of new D^+ charges. A region depleted of D^+ charges is therefore created, with a width depending on the applied bias and on the compensating acceptor concentration

Continued

Table 4.3 Continued

Mode of operation	Schematic of detector	Operation and properties
p-n junction photodiode	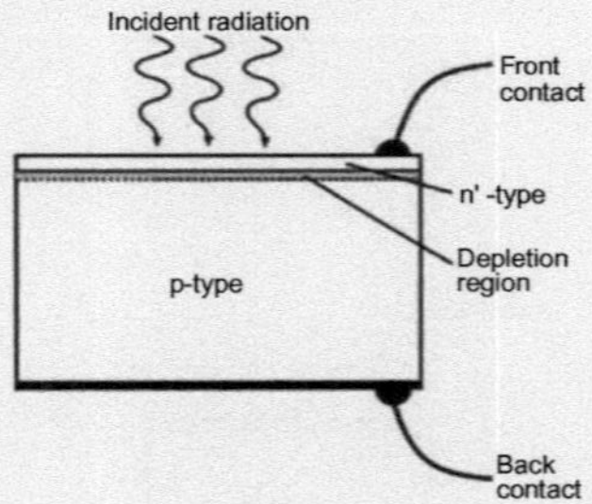	This is the most widely used photovoltaic detector but is rather rarely used as a THz detector. Photons with energy greater than the energy gap create electron-hole pairs in the material on both sides of the junction. By diffusion, the electrons and holes generated within a diffusion length from the junction reach the space-charge region, where they are separated by the strong electric field; minority carriers become majority carriers on the other side. This way a photocurrent is generated causing a change in voltage across the open-circuit cell or a current to flow in the short-circuited case. The limiting noise level of photodiodes can ideally be $\sqrt{2}$ times lower than that of the photoconductor, due to the absence of recombination noise. Response times are generally limited by device capacitance and detector-circuit resistance
nBn detector	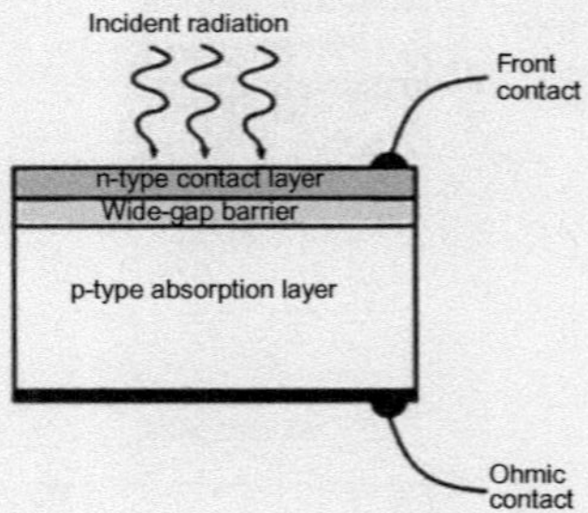	The nBn detector consists of a narrow-gap *n*-type absorber layer (AD), a thin wide-gap barrier layer (BL), and a narrow-gap *n*-type contact layer (CL). The thin wide-gap BL presents a large barrier in the conduction band that eliminates electron flow. Current through the nBn detector relics on transport of mobile holes through drift and diffusion in the BL between the two *n*-type narrow-gap regions. Effectively, the nBn detector is designed to reduce the dark current (generation-recombination current originating within the depletion layer) and noise without impeding the photocurrent (signal). In particular, the barrier serves to reduce the surface leakage current. The nBn detector operates as a unipolar unity-gain detector, and its design can be stated as a hybrid between photoconductor and photodiode

Metal-insulator-semiconductor (MIS) photodiode	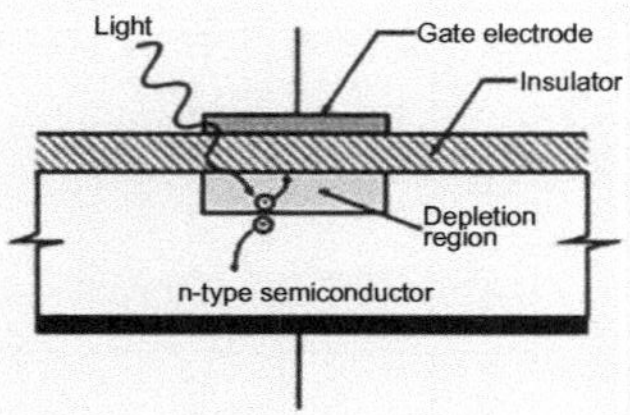	The MIS device consists of a metal gate separated from a semiconductor surface by an insulator (I). By applying a negative voltage to the metal electrode, electrons are repelled from the I-S interface, creating a depletion region. When incident photons create hole-electron pairs, the minority carriers drift away to the depletion region and the volume of the depletion region shrinks. The total amount of charge that a photogate can collect is defined as its well capacity. The total well capacity is decided by the gate bias, the insulator thickness, the area of the electrodes, and the background doping of the semiconductor. Numerous such photogates with proper clocking sequence form a CCD imaging array
Schottky barrier photodiode	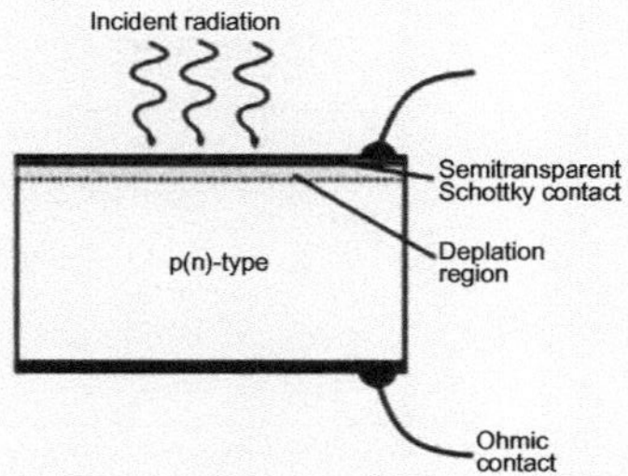	Schottky barrier photodiodes reveal some advantages over the p-n junction photodiode: fabrication simplicity (deposition of metal barrier on n(p)-semiconductor), absence of high-temperature diffusion processes, and high speed of response. Since it is a majority-carrier device, minority-carrier storage and removal problems do not exist, and therefore, higher bandwidths can be expected. The thermionic emission process in the Schottky barrier is much more efficient than the diffusion process, and therefore for a given built-in voltage, the saturation current in a Schottky diode is several orders of magnitude higher than in the p-n junction

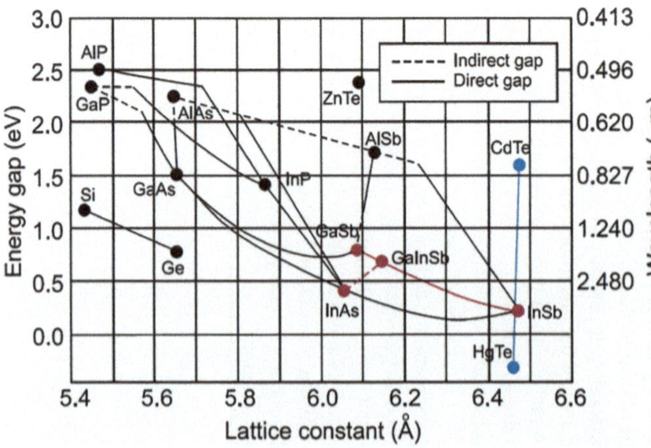

Fig. 4.4 Semiconductors of interest to photon detectors and their fundamental properties [31].

Table 4.4 Examples of infrared single-photon detectors and their performance characteristics [32].

Infrared single-photon detector comparison table				
	Semiconducting APDs		Superconducting SPDs	
Detectors	**Si**	**InGaAs**	**SSPD (NbN)**	**TES (W)**
Temperature (K)	300	200	2	0.1
Wavelength (μm)	0.25–1.1	1.1–1.8	0.4–5.6	0.1–5
Time resolution (ps)	48	~300	18	3×10^5
Quantum efficiency	25%@0.7 μm	14%@1.54 μm	10%@1.55 μm	>80%@1.5 μm
Apertures (μm)	200	30-80	SM fiber	SM fiber
Dark count rates (cps)	$<10^4$	$>3 \times 10^4$	<1	<0.001
Data rate	<3 MHz	<15 MHz	>2 GHz	20 kHz
Dynamic range(#*h*)	1–3000	NA	$>10^8$	50
Electrical quenching	Yes (0.3 ns)	Yes (0.3 ns)	No	No
Photon number resolution	Limited	No	No	Yes
Ruggedness	Very high	High	High	Low
Availability	Yes	Limited	Limited	Very limited

APD, Avalanche photodiode; *SSPD*, Superconducting single photon detector; *TES*, Transition edge sensors.

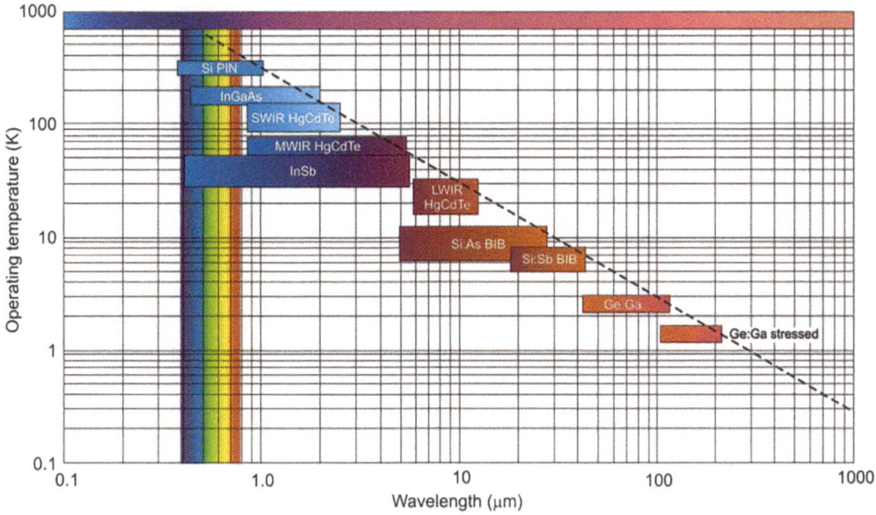

Fig. 4.5 Operating temperatures for low-background material systems with their spectral band of greatest sensitivity. The dashed line indicates the trend toward lower operating temperature for longer-wavelength detection [33].

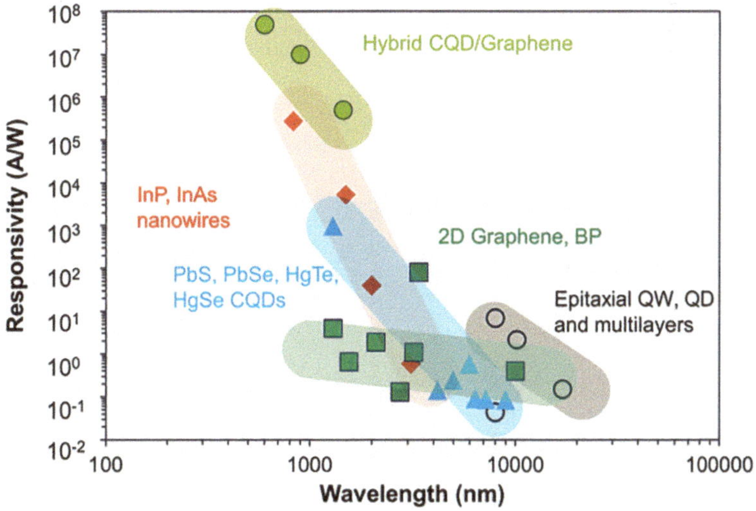

Fig. 4.6 A summary of the responsivity of infrared photodetectors based on various nanostructured materials, as function of wavelength [34].

Table 4.5 List of typical nanostructured materials and their performances in infrared photodetection [34].

Category	Detection bands	Material	Wavelength (μm)	Responsivity (A W^{-1})	Detectivity (Jonse)	Speed (s, Hz)	Ref.	Technical notes
Epitaxial QWs, QDs, and multilayers	SWIR to FIR	InGaAs/GaAs QDs	10.2	2.16	1.01×10^{11}	–	[35]	Strain relaxed
	Narrow band	InGaAs/InP QW	8	7	7.4×10^{10}	–	[36]	Plasmonic enhancement
		InGaAs/GaAs/AlGAs QDs	17	0.15	10^7	–	[37]	–
		GaAs/AlGaAs QCD	8	0.044	4.5×10^{11}	–	[38]	Photovoltaic mode
		CaSb/AlSb Barrier QCD	3.8	–	1.81×10^{11}	–	[39]	Electron barrier introduced
Solution casted QDs	SWIR to LWIR	PbS QD	1.3	10^3	1.8×10^{13}	18 Hz	[40]	–
		PbSe QD	1.15	–	1.6×10^{12}	10 kHz	[41]	Field effect tuned
		HgTe QD	3-5	0.15-0.25	2×10^9	1 KHz	[32]	–
		HgSe QD	4.2-9	0.145	–	–	[42]	Plasmonic enhancement
		HgSe QD	6	0.6	10^8	–	[43]	–
1D nanowires	NIR to SWIR	InP	0.83	2.8×10^5	9.1×10^{15}	139 ms	[44]	Field effect depletion
		InAs NW	1.5	5.3×10^3	–	–	[45]	Schottky contact
		InAs NW	2–3.113	40–0.6	2×10^{12}–10^{10}	≈0.1 ms	[46]	Negative photoresponse

2D materials	NIR to FIR	Graphene QD	0.532–10	1.25–0.4	–	–	[47]	Quantum dots
		Graphene	2.75	0.13	–	–	[48]	–
		Graphene	1.3–3.2	4–1.1	–	–	[49]	Double layer tunneling junction
		BP	1.55	0.657	–	3 GHz	[20]	Waveguide integrated
		BP	3.39	82	–	0.13 ms	[50]	Field effect tuned
Hybrid materials	SWIR	Graphene/PbS QD	0.895	10^7	–	0.3–1.7 s	[51]	Photogate effect
		Graphene/PbSe/TiO_2	1	–	5.7×10^{12}	750 ns	[52]	–
		Graphene/PbS	0.600–1.45	5×10^7–5×10^5	7×10^{13}	–	[23]	–

Table 4.6 Typical wavelength ranges of HORIBA scientific solid state detectors [53].

Solid state detector	Wavelength range (μm) [Photovoltaic (PV) or Photoconductive (PC)]
Si	0.2–1.1 [PV]
InGaAs	0.8–1.7 [PV]
InGaAs (LN_2)	0.8–1.55 [PV]
InGaAs (2.2 μm, LN_2)	1.1–2.2 [PV]
Ge	0.8–1.8 [PV]
Ge (LN_2)	0.8–1.55 [PV]
PbS	1–2.8 [PC]
PbSe	1–4.5 [PC]
InSb	2–5.5 [PC]
Hg_xCd_{1-x}	2–20 [PC]

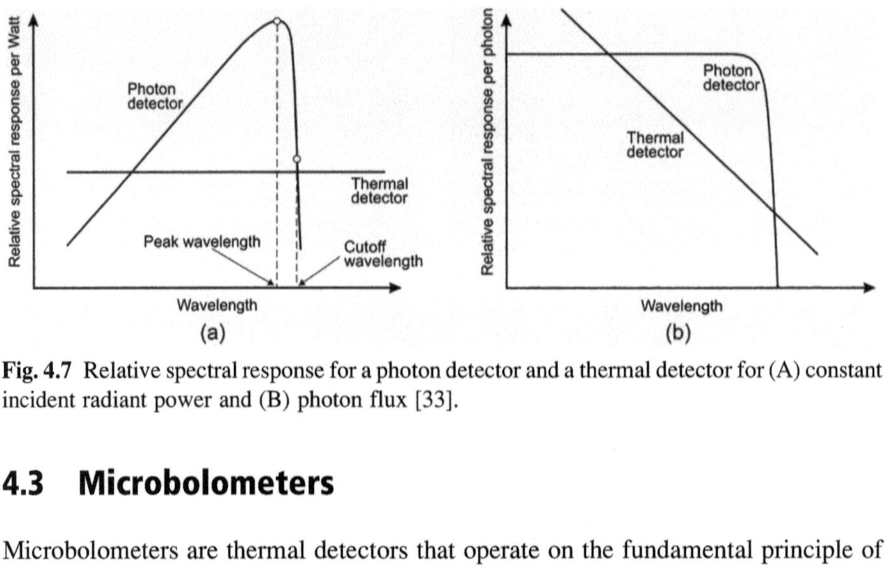

Fig. 4.7 Relative spectral response for a photon detector and a thermal detector for (A) constant incident radiant power and (B) photon flux [33].

4.3 Microbolometers

Microbolometers are thermal detectors that operate on the fundamental principle of the change in resistance with change in temperature of a material. Most important design parameters for microbolometers are as follows:

(i) Low thermal conductance between substrate and detector layer;
(ii) small heat capacity of detector layer;
(iii) thermosensitive material for sensing change in temperature of detector element implying low internal noise and high temperature coefficient (TCR).

Various materials that can be used in the manufacture of microbolometers are expected to have certain characteristics such as high absorption in the infrared range of wavelengths. Wide varieties of materials have been used as absorbers of optical infrared radiation since the 1930s [43,54–57] and have been extensively utilized as infrared absorber coatings. Traditionally, various surface modification techniques

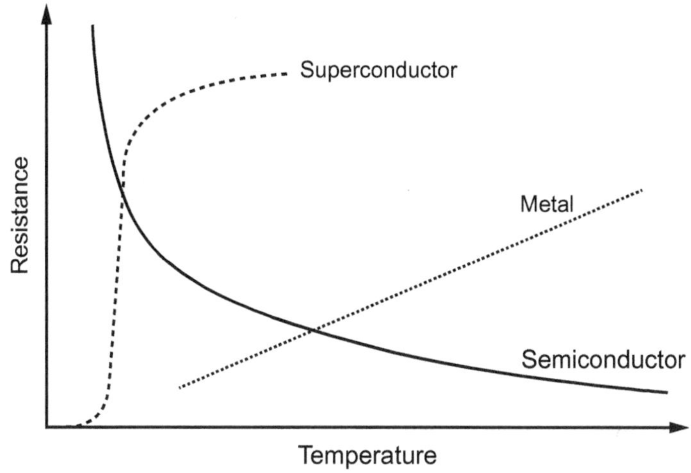

Fig. 4.8 Temperature dependence of three bolometer material types [33].

such as electroplating, chemical, and mechanical etching have been used to achieve highly porosity metal surfaces. These surfaces provide the most effective approach for maximizing the absorption of incident infrared radiation [36,58]. Modern surface modification techniques such as femtosecond laser-assisted-etching/processing have been implemented in the fabrication of black metals and semiconductors that exhibit an extremely low reflectance across the infrared wavelength spectra [59,60]. In order to obtain higher responsivity, materials that exhibit a phase transformation near the operating temperature are highly desirable.

An illustration of the temperature dependence of three bolometer material types—superconductor, metal, and semiconductor is presented in Fig. 4.8 [33]. In general, metals are characterized by a linear increase in resistance with temperature. Semiconductors exhibit an increase in electrical conductivity with temperature, or a decrease in electrical resistivity with temperature. Above a critical temperature, T_c, the electrical conductivity increases by several order of magnitude in superconductors.

A variety of materials such as polycrystalline vanadium oxide (VO_x) and amorphous silicon (a-Si) have been widely used as IR thermosensing materials [61–64]. These materials have the advantage of high temperature coefficient of resistance that leads to detectors with high sensitivity.

An example of the unit cell of VO_x-based microbolometer is presented in Fig. 4.9 [65]. Fig. 4.10 [65] shows an SEM image of a microbolometer focal plane array.

Materials of interest for the manufacture of microbolometer include vanadium oxide, amorphous silicon, titanium oxide, yttrium barium copper oxide, and Si/SiGe. Various vanadium oxide compounds are known to undergo insulator to metal phase transitions at certain temperatures, across which their optical and electrical properties are changed immensely. As a consequence of the drastic change in the optical and electrical properties, high thermo-optic coefficients (TOCs) as well as high

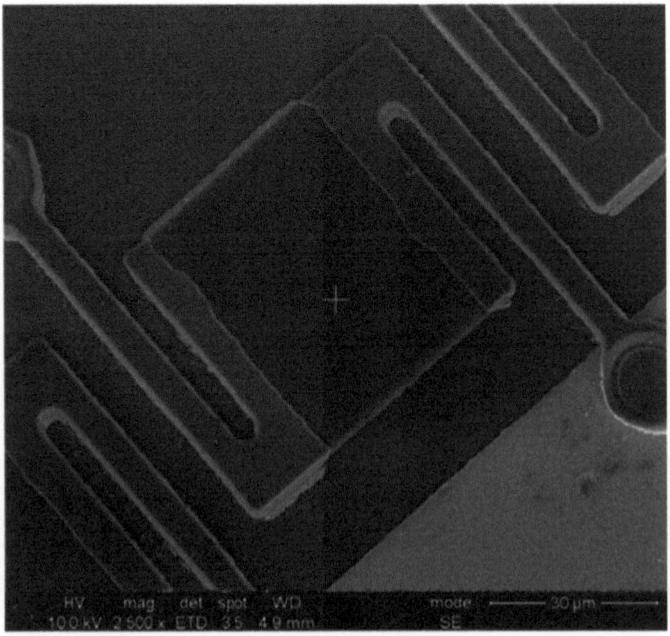

Fig. 4.9 Depiction of a VO_x microbolometer unit cell [65].
From B. Wang, J. Lai, H. Li, H. Hu, S. Chen, Nanostructured vanadium oxide thin film with high TCR at room temperature for microbolometer, Infrared Phys. Technol. 57 (2013) 8–13, ISSN 1350–4495, https://doi.org/10.1016/j.infrared.2012.10.006.

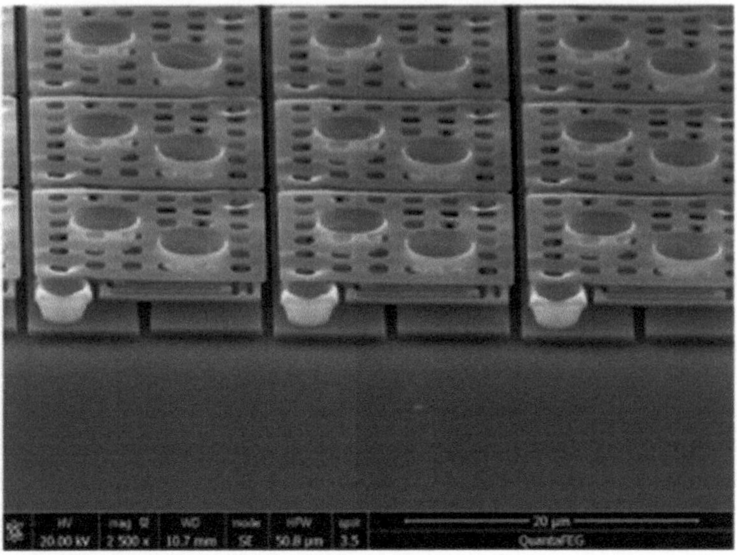

Fig. 4.10 SEM image of a microbolometer FPA [65].
From Z. Luo, H. Wang, N. An, Z. Zhu, A tunnel dielectric-based tunnel FET, IEEE Electron Device Lett. 36 (9) (2015) 966–968, https://doi.org/10.1109/LED.2015.2458932.

temperature coefficients of resistance (TCRs) are expected, a fact that has led vanadium oxide compounds to be employed in many modern solid state devices. The electrical properties across the insulator to metal phase transition and, in particular, in the semiconductor phase region where the TCRs of some vanadium oxide compounds are very high, have successfully led to the development of vanadium oxide microbolometers for thermal imaging applications. In Fig. 4.11 [66], a case study of the preparation and electrical properties of vanadium oxide films is presented.

The temperature coefficient of resistance, $\alpha(T)$, is given by the following equation [67]:

$$\alpha(T) = (1/R)\,[\mathrm{d}R/\mathrm{d}T] \approx \{E_a/kT^2\}, \tag{4.7}$$

where R is the resistance, T is the temperature, E_a is the activation energy, and k is the Boltzmann constant.

The details of the mode of operation, properties, and schematic of thermal detectors is summarized in Table 4.7 [30].

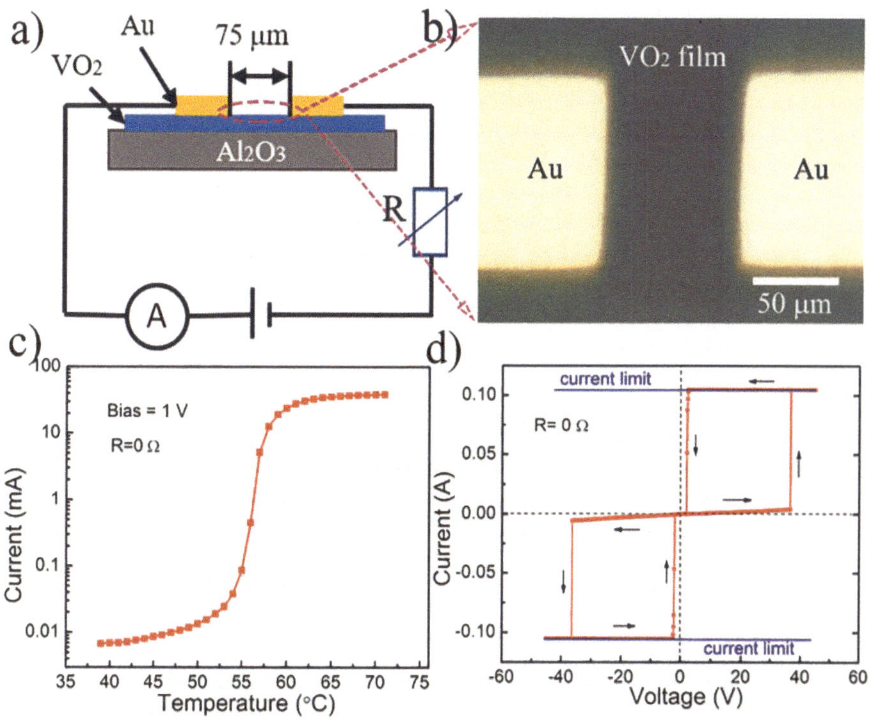

Fig. 4.11 Preparation of vanadium dioxide (VO_2) device and phase-transition characterization under the effect of temperature and voltage. (A) Schematic diagram of device for the measurement of phase transition. R is external resistor. (B) Optical microscope image of the prepared VO_2 sample. The Au nanofilm as electrodes is deposited on VO_2 nanofilm. (C) The current of VO_2 film versus temperature without external resistor at a voltage of 1 V. (D) The evolution of current versus voltage at room temperature without external resistor [66].

Table 4.7 Mode of operation, properties, and schematic of thermal detectors [30].

Mode of operation	Schematic of detector	Operation and properties
Thermopile	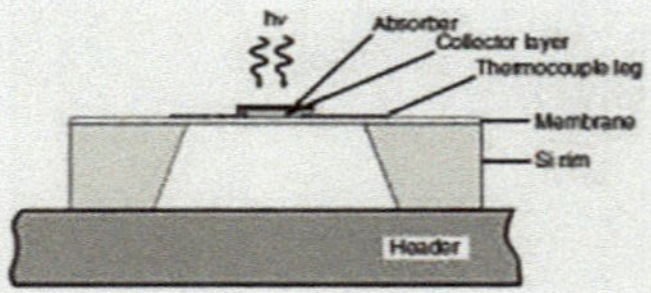	The thermocouple is usually a thin, blackened flake connected thermally to the junction of two dissimilar metals or semiconductors. Heat absorbed by the flake causes a temperature rise of the junction, and hence a thermoelectric electromotive force is developed that can be measured. Although thermopiles are not as sensitive as bolometers and pyroelectric detectors, they will replace these in many applications due to their reliable characteristics and good cost/performance ratio. Thermocouples are widely used in spectroscopy
Bolometer: Metal Semiconductor Superconductor hot electron	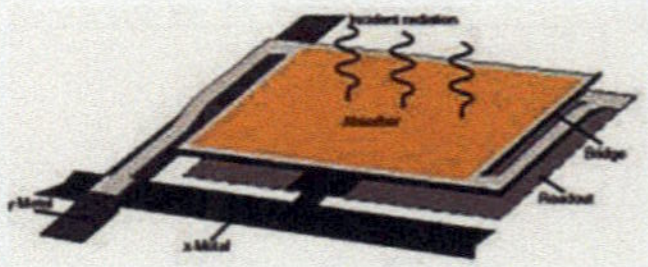	The bolometer is a resistive element constructed from a material with a very small thermal capacity and large temperature coefficient so that the absorbed radiation produces a large change in resistance. The change in resistance is like to the photoconductor: however, the basic detection mechanisms are different. In the case of a bolometer, radiant power produces heat within the material, which in turn produces the resistance change. There is no direct photon-electron interaction. Initially, most bolometers were the thermistor type made from oxides of manganese, cobalt, or nickel. At present, microbolometers are fabricated in large format arrays for thermal imaging applications. Some extremely sensitive low-temperature semiconductor and superconductor bolometers are used in the THz region

Pyroelectric detector	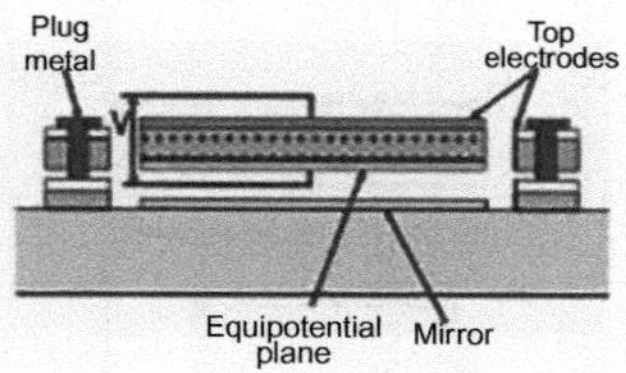	The pyroelectric detector can be considered as a small capacitor with two conducting electrodes mounted perpendicularly to the direction of spontaneous polarization. During incident radiation, the change in polarization appears as a charge on the capacitor and a current is generated, the magnitude of which depends on the temperature rise and the pyroelectrical coefficient of the material. The signal, however, must be chopped or modulated. The detector sensitivity is limited either by amplifier noise or by loss-tangent noise. Response speed can be engineered making pyroelectric detectors useful for fast laser pulse detection, however, with a proportional decrease in sensitivity
Golay cell	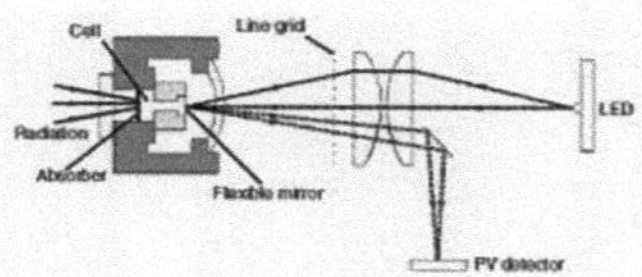	The Golay cell consists of a hermetically sealed container filled with gas (usually xenon for its low thermal conductivity) and arranged so that expansion of the gas under heating by a photon signal distorts a flexible membrane on which a mirror is mounted. The movement of the mirror is used to deflect a beam of light shining on a photocell and so producing a change in the photocell current as the output. In modern Golay cells the photocell is replaced by a solid-state photodiode and a light emitting diode is used for illumination. The performance of the Golay cell is only limited by the temperature noise associated with the thermal exchange between the absorbing film and the detector gas: consequently the detector can be extremely sensitive with $D^* \approx 3 \times 10^9$ cm $Hz^{1/2}$ W^{-1}, and responsivities of 10^5 to 10^6 V/W. The response time is quite long, typically 15 ms

Vanadium oxide (VO_x) has been the thermosensing material of choice in a variety of microbolometers [68] with temperature coefficient of resistance in the range of −2%/K to −3%/K [40,41,50,69] at room temperature [70]. However, it is not a standard material in the integrated circuit (IC) fabrication process technology. Vanadium oxide is not a single crystal material and its compositional variations have to be controlled very precisely; it has some problems such as unstable phases and poor reproducibility [71,72]. Thin films of vanadium oxide can be prepared by a variety of techniques, including reactive RF sputtering [73–75], pulsed laser deposition [35,52,76–85], ion beam sputtering, and magnetron controlled sputtering that are high cost methods [77,78] (Table 4.8).

Amorphous silicon (a-Si) is an alternative material that is compatible with the silicon manufacturing technology. It requires high values of electrical resistivity in order to achieve a high TCR [79–81]. In order to achieve a TCR of −3%/K to −4%/K [82] at room temperature, amorphous silicon should be exposed to high annealing temperatures resulting in a high electrical resistivity and a high pixel resistance in the order of Mega-Ohms (MΩ). The resulting high value of resistance causes large Johnson noise and consequently low detector sensitivity as well as a mismatch with the input impedance of the CMOS read-out circuits.

Hydrogenated amorphous silicon (a-Si:H), deposited by plasma enhanced chemical vapor deposition (PECVD), is very attractive to be used as IR sensing material in uncooled microbolometers, due to its very high activation energy ($E_a \approx 0.8$–1 eV), very high temperature coefficient of resistance (TCR ≈ 10–12%K^{-1}) and its compatibility with the standard silicon complementary metal-oxide-semiconductor (CMOS) technology. However, it also has very high resistivity (~109 Ω cm), which results in a mismatch with the input impedance of the CMOS integrated read-out circuits [56]. In commercial microbolometer arrays, boron doped a-Si:H (a-Si:H,B) has been adopted as thermosensing film, due to its improved σ_{RT} in several orders of magnitude in comparison with intrinsic a-Si:H. However, a-Si:H,B also has a reduced E_a (~0.22 eV) and TCR (~2.8%K^{-1}) [56], which is transduced in low film sensitivity to IR radiation [52].

Poly SiGe, amorphous SiGe alloys, and amorphous $Ge_xSi_{1-x}O_y$ are attractive materials of the SiGe family for bolometers mainly because of their high temperature

Table 4.8 Important properties of commonly used thermosensing materials in microbolometers [67].

Material	TCR (K^{-1})	E_a (eV)	σ_{RT} $(\Omega\ cm)^{-1}$
VO_x	0.021	0.16	2×10^{-1}
a-Si:H (PECVD)	0.1–0.13	0.8-1	$\sim 1 \times 10^{-9}$
a-Si:H,B (PECVD)	0.028	0.22	5×10^{-3}
a-Ge_xSi_y:H (PECVD)	0.043	0.34	1.6×10^{-6}
Poly-SiGe	0.024	0.18	9×10^{-2}
$Ge_xSi_{1-x}O_y$	0.042	0.32	2.6×10^{-2}
YBaCuO	0.033	0.26	1×10^{-3}

coefficient of resistance, although they exhibit 1/*f*-noise like all materials used in microbolometers [83].

New IR radiation absorber materials have emerged with extremely low reflectance exhibited across the IR spectra. Vantablack, a composite of thin film of carbon nanotubes and black silicon (BSi), a surface modification of silicon, exhibits absorptance close to unity [35,47,48,84,85]. Black gold with infrared absorptance close to unity has been developed [86]. However, the thin film fabrication process technology that is required for the deposition of these materials may not be compatible with all semiconductor processes and it does not allow the mass production of uncooled microbolometers [49]. Black gold, obtained by thermal evaporation methods, exhibits an absorptance close to unity [87–89]; but the fragile structure of this material and the degradation in absorption properties, when it is exposed to temperatures higher than 300°C, does not make it feasible for mass production [90].

From an industrial perspective, three materials, vanadium oxide, amorphous silicon and barium strontium titanate continue to be the dominant materials of choice for the manufacture of uncooled microbolometers. Vanadium oxide accounts for a major component of the market—this is illustrated in Fig. 4.12 [91,92].

From a device commercialization point of view, the global uncooled infrared camera market is expected to grow at the compound annual growth rate (CAGR) of 8% during the years 2016–2022 to US $4.4 billion at the end of 2022 [93]. This is summarized in Fig. 4.13 [93].

In Table 4.9, some of the physical properties of materials of interest to the manufacture of microbolometers are summarized [37,94].

The critical performance parameters that need to be considered in the design, fabrication, and manufacture of microbolometers are listed in Table 4.10 [95].

The simulations of thermal conductance of thermal infrared detectors, using Finite Element Method (FEM), have led to some interesting results [96]. A schematic diagram of a standard thermal infrared detector, considered in this study, is presented in Fig. 4.14a.

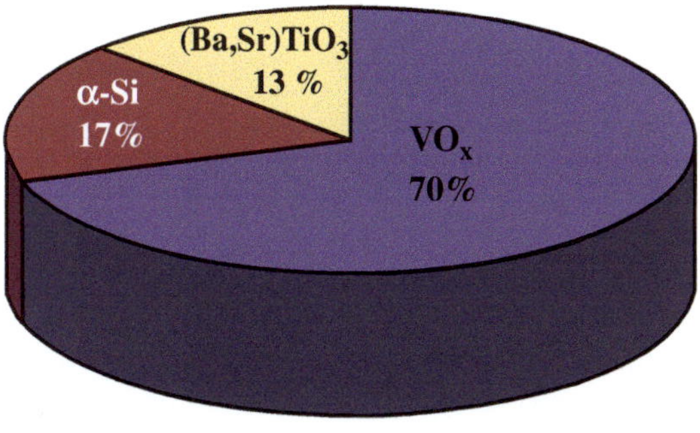

Fig. 4.12 Market share of materials for the manufacture of uncooled microbolometers [91,92].

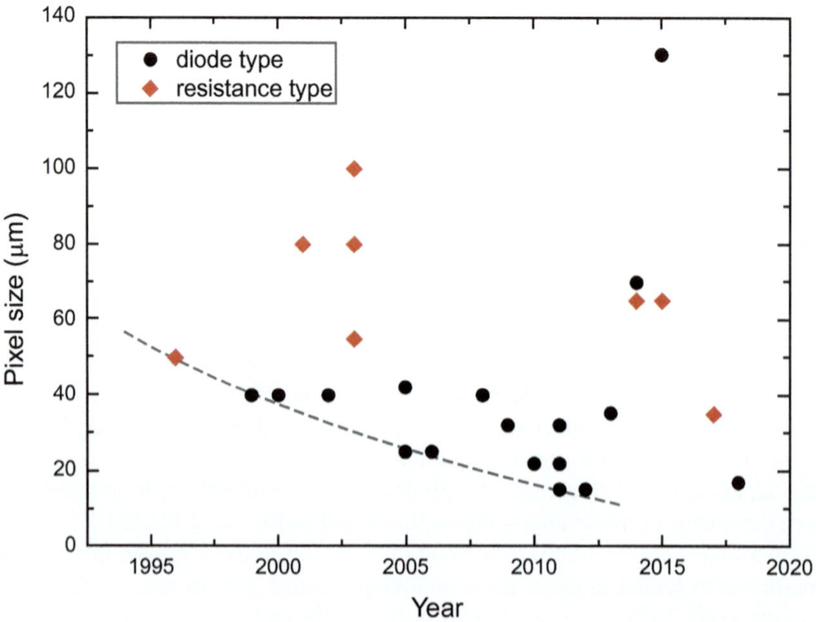

Fig. 4.13 Anticipated growth of uncooled IR camera market from 2016 to 2022 [93]. From Yu, L.; Guo, Y.; Zhu, H.; Luo, M.; Han, P.; Ji, X. Low-Cost Microbolometer Type Infrared Detectors. Micromachines 2020, 11, 800. https://doi.org/10.3390/mi11090800.

Table 4.9 Physical properties of microbolometer materials [37,94].

Material	Heat capacity (J/Kg K)	Density (Kg/m^3)	Thermal conductivity (W/m K)
Amorphous silicon	700	2330	1.8
Au	129	19,300	317
Epoxy SU-8		1200	0.2
NiCr	450	8300	13
Polycrystalline silicon	700	2330	13.8
Silicon	700	2330	149
Si_3N_4	700	3100	29, 0.3
SiO_2	703	2650	1.38
Ti	522	4509	21.9
VO_x	689.28	3360	30.7

In Fig. 4.14b, the thermal time constant vs thermal conductivity for bi-layer FEM model are plotted. Boundary conditions at the bottom and top of the substrate are two different conditions (air thermal conduction vs convection) and vacuum, respectively. The simulated value of the time constant of 68 ms corresponds to a substrate of thickness 300 μm and a frequency of 0.25 Hz, which is close to the static frequency [96].

Table 4.10 Performance considerations in the design, fabrication, and manufacture of microbolometers [95].

Critical performance parameters	Description
Joule effect	Self-heating effect (I^2R) due to influence of current flow on resistivity of microbolometer
Material properties	Physical properties of active absorbing layer: in particular, electrical properties (resistivity, TCR), mechanical properties (Young's modulus), thermal properties (thermal conductivity)
Thermal isolation	Thermal isolation of absorbing layer from substrate layer; it is a function of design, construction, and geometry of the microbolometer
Time constant	Time required by ROIC to display the correct value—in case of instantaneous changes in the incident infrared radiation
Noise	Noise generated during operation and signal processing in CMOS ICs

4.4 Amorphous silicon

Several approaches have been reported in the literature for producing thin conductive films for microbolometer and photoconductor applications [97]. In addition to being completely compatible with silicon processing and technology, amorphous Si (a-Si) films are attractive for producing low-cost focal plane arrays [98]. Hydrogenated amorphous silicon films, deposited by plasma-enhanced chemical vapor deposition on thin silicon nitride membranes, are being used to form microbolometers for IR imaging (resonant quarter-wave pixel structures) and for multichannel gas sensing [99,100]. Deposited amorphous $Ge_xSi_{1-x}O_y$ films are also being investigated for resistive microbolometer sensors, due to their compatibility of the processing steps with IC fabrication [101].

From a historical perspective, the research, invention, and initial development of the fundamental technology for the silicon microbolometer IR detectors was conducted at DSTO, the Defense Science and Technology Organization (now, Defense Science & Technology, Since July 1, 2015), Australia, in the late 1970s [38]. These studies led to the utilization and growth of bulk and surface micromachining in silicon to be able to create thermally isolated detector structures on silicon substrates [38]. The early studies by Spear and LeComber [39], at the University of Dundee, led to the selection of amorphous silicon as a potential material for bolometers in 1979.

A schematic of the amorphous silicon microbolometer is presented in Fig. 4.15 [42].

US Patent # 5,367,167, issued to Texas Instruments in 1993, describes an amorphous silicon-based uncooled infrared detector and method for forming the same [44]. The detector was proposed to operate in the wavelength range of 7 to 12 μm.

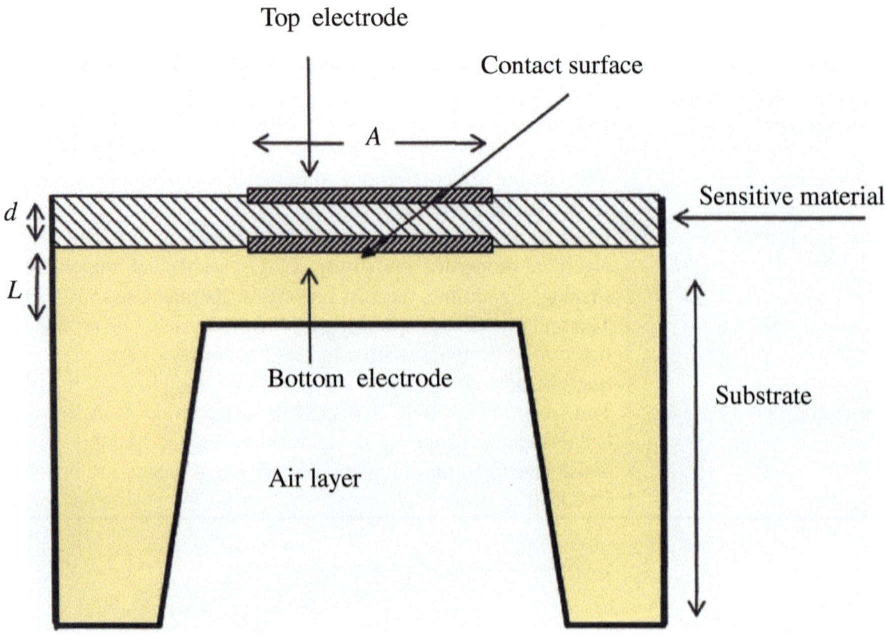

(A)

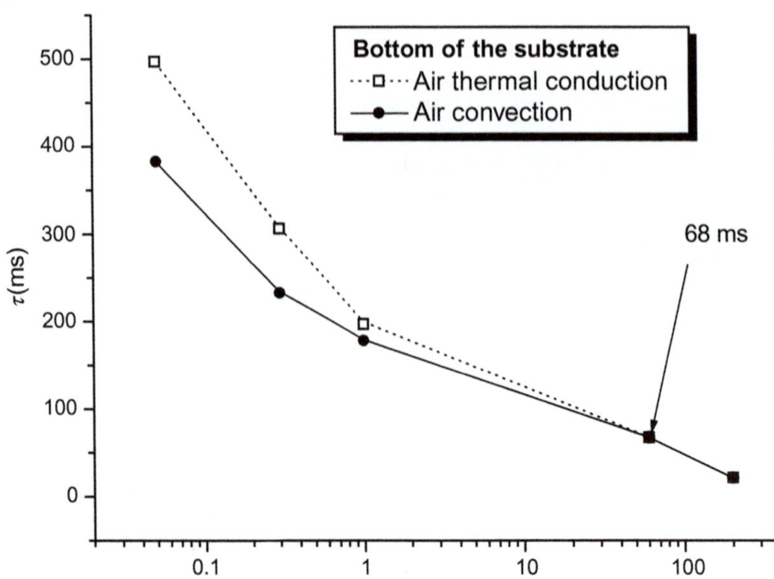

(B)

Fig. 4.14 (a). A schematic diagram of a standard thermal infrared detector [96]. (b). Thermal time constant τ vs thermal conductivity k for bilayer FEM model. Boundary conditions at the bottom and top of the substrate are two different conditions (air thermal conduction vs convection) and vacuum, respectively [96].

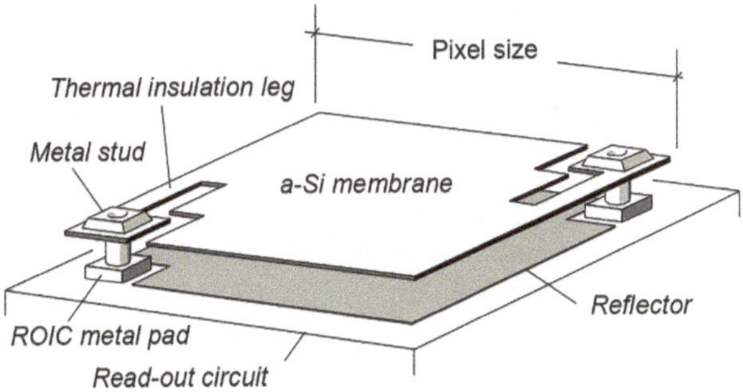

Fig. 4.15 Schematic of the amorphous silicon microbolometer pixel [42].

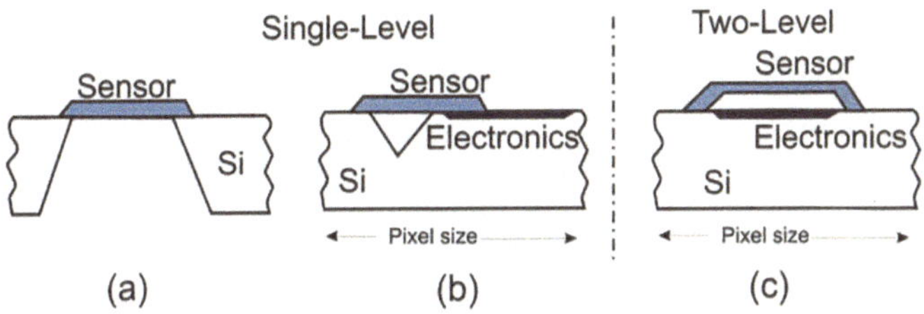

Fig. 4.16 Two categories of the microbolometer designs: (a) one-level, complete etch under the area of sensor, (b) Type-V structure, and (c) two level configuration by surface micromachining [45].

An illustration of the two categories of microbolometer designs that can be easily accomplished using MEMS technology is presented in Fig. 4.16 [45].

Fig. 4.17 [46] shows the cross-sectional images of designs of two bolometers, with resonant optical cavities. These resonant cavities facilitate in enhanced absorption of the incident radiation.

Hydrogenated amorphous silicon, a-Si:H, has low noise properties, high TCR, and can be prepared with a range of electrical resistivities to meet sensor resistance specifications. Hydrogenated amorphous silicon (a-Si:H), deposited by plasma enhanced chemical vapor deposition (PECVD), is very attractive as thermo-sensing material in uncooled micro-bolometers [51]. This is because of the fact that amorphous silicon

(a)
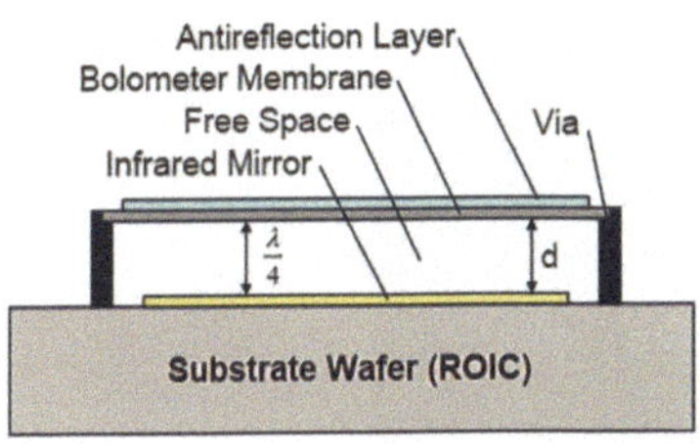

(b)
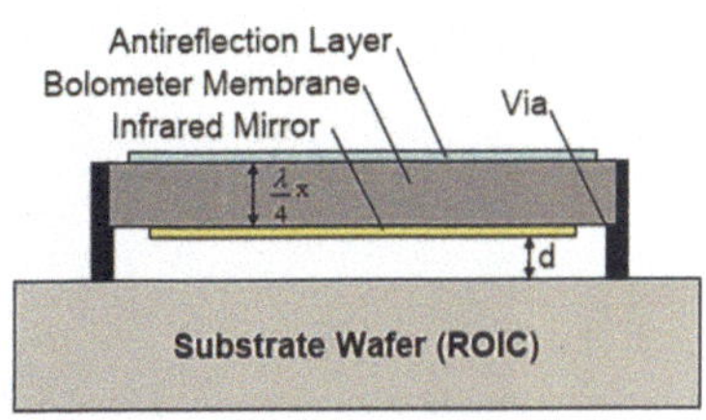

Fig. 4.17 Cross-sectional image of two bolometer designs with resonant optical cavities for high absorption of the incident radiation [46].

films have the additional advantage of low-cost and are completely compatible with the standard silicon CMOS technology. The high TCR and low sheet resistivity are critical parameters that need to be achieved simultaneously. The TCR of undoped amorphous silicon film can be high as −2%/K to −8%/K, but its high resistance also results in high noise.

Transition metals (Ti, Co, Ni, etc.), with their excellent thermosensing properties, can be used to dope in a-Si:H-based microbolometer. Co implanted in a-Si has a maximum TCR of −1.06%/K with dopant dose of 5×10^{15} ions cm^{-2} before annealing. The TCR can increase to −2.5%/K after 500°*C* annealing. Ti implanted in a-Si:H has an average TCR value and can achieve −0.8%/K with 19.8 Ω cm resistivity. The average TCR value of samples with Ti content of 3.5% can achieve −1.134%/K with a resistivity of 47 Ω cm. The TCR values continue incrementing with decreasing Ti content and increasing resistivity. Doping has effectively decreased the resistivity; however, the TCR also decreases compared to that of intrinsic a-Si:H.

NiCr may be used as the contact and electrode material as it makes the best contact to a-Si:H. This is due to its relatively low thermal conductivity. Therefore, it provides a better thermal isolation. Amorphous silicon carbide is an IC compatible material with high TCR of −2%/K to −4%/K. SiC requires high annealing temperature of ~1000°*C* to achieve stability of microstructures [102]. In a study of a new structure of an uncooled a-Si microbolometer for infrared detection, Liu et al. [102] propose polyimide (PI) as a thermal isolation layer. A cross-section of the proposed structure is shown in Fig. 4.18 [102].

The main process steps of the microbolometer, as considered in this study, are shown in Fig. 4.19. The fabrication process starts with a 3-inch single-side polished silicon wafer. The PI is spin coated, solidified and cured to achieve thick PI film with high stability as a thermal isolation layer. Aluminum is sputtered and patterned to form bottom IR reflective layer. Plasma-enhanced chemical vapor deposition system is used to deposit SiN_x, SiO_x and a-Si active layer. Deposition is carried out at a temperature of 300°C and the thin film is doped by boron to obtain the desired resistivity and TCR.

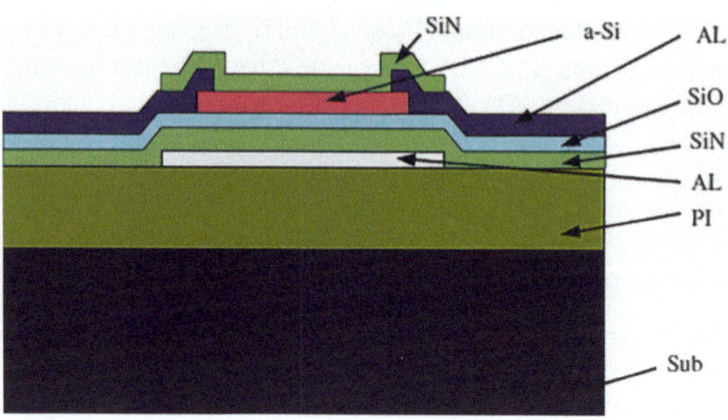

Fig. 4.18 The structure schematic of a-Si microbolometer use PI as thermal isolation layer [102].

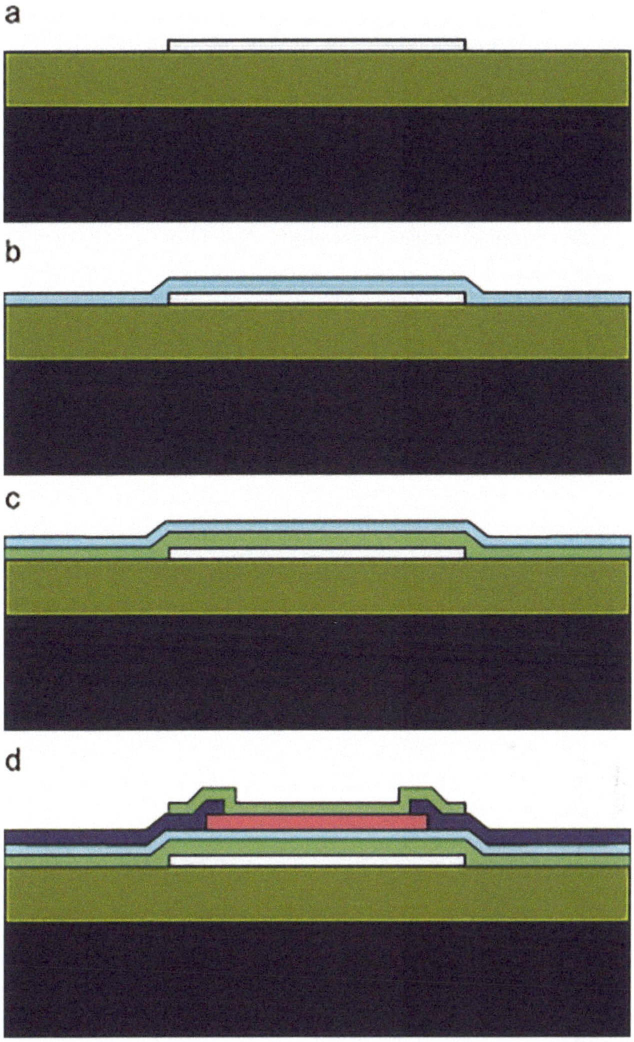

Fig. 4.19 Fabrication process of the a-Si microbolometer based on PI thermal isolation structure [102].

The a-Si resistor is patterned by reactive ion etching (RIE) using SF_6 as reactive gas, and SiO_x is used as etch stopping layer. Aluminum of thickness 1.0 μm is sputtered, and the contacts of the electrodes are patterned and metallized. SiN_x of 100 nm is deposited by PECVD and patterned as passivation layer to form sandwich IR absorption structure and complete the detector fabrication. The incident IR radiation is absorbed by SiN_x passivation layer first and passes through the active a-Si layer, then is absorbed by the complex SiO_x-SiN_x layer. The radiation that reaches the bottom metal layer is reflected and is again absorbed by the above sandwich structure.

This structure fabricated with amorphous silicon has simpler fabrication process when compared to conventional structure. It has the advantages of lower cost and higher yield as well [102]. The amorphous Si microbolometer, investigated in this case study, has a detectivity that depends strongly on frequency, increasing from over $1.0 \times 10^8 cmHz^{1/2}W^{-1}$ at 10 Hz to a peak of about $1.7 \times 10^8 cm\,Hz^{1/2}\,W^{-1}$ at 170 Hz, then remaining unchanged on the whole and decreasing at 460 Hz.

The results of temperature coefficient of resistance of boron-doped mixed-phase hydrogenated a-Si, from another case study by Shin et al., are presented in Fig. 4.20 [103]. As can be seen in this figure, the TCR is increased from 0.41%/K to 7.15%/K as B_2H_6/SiH_4 increases—this corresponds to $H_2/SiH_4=80$. A similar trend is observed for different values of H_2/SiH_4. These results demonstrate the ability to obtain high TCR by choosing the appropriate ratios of H_2/SiH_4 and B_2H_6/SiH_4.

In Fig. 4.21 [104], TCR is plotted as a function of the resistivity of amorphous silicon. As can be seen in this figure, TCR values of 2.5%/K can be obtained for heavily doped layers, while for less doped layers, 5%/K are feasible [104].

A simulation of the thermal response of conventional microbolometer design, as function of pixel size, is presented in Fig. 4.22 [105]. In utilizing conventional pixel design, at small pixel dimensions, the short thermal conduction paths lead to poor responsivity.

The design, fabrication, and characterization of amorphous silicon based microbolometers, with shared-anchor structure, have been demonstrated to show enhancements in responsivity from 1.08×10^5 to 1.23×10^5 V/W and an increased thermal time constant of 10 ms at a 30-Hz chopping frequency [105]. The process flow of the proposed microbolometer with 25 μm pixel size is shown in Fig. 4.23 [105].

An SEM image of the released microbolometers is shown in Fig. 4.24 [105].

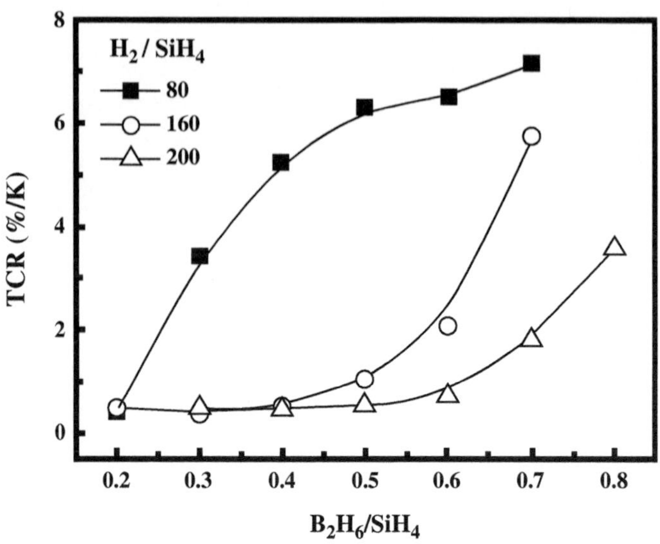

Fig. 4.20 Temperature coefficient of resistance (TCR) of boron-doped mixed-phase hydrogenated silicon (BMP-Si: H) with respect to changes in B_2H_6/SiH_4 and H_2/SiH_4. The power density is 0.04 W/cm^2 [103].

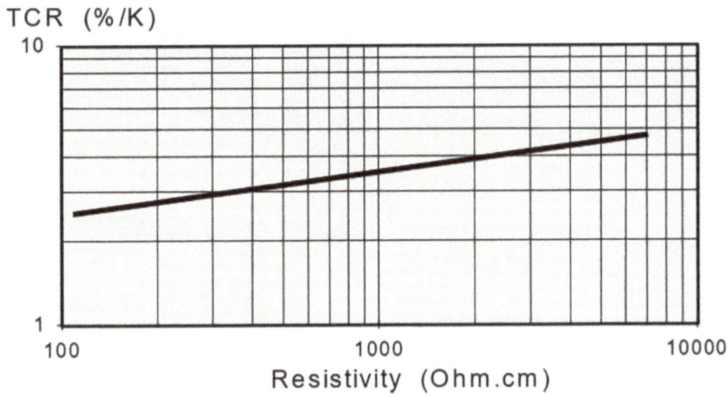

Fig. 4.21 Evolution of TCR as function of electrical resistivity of a-Si [104].

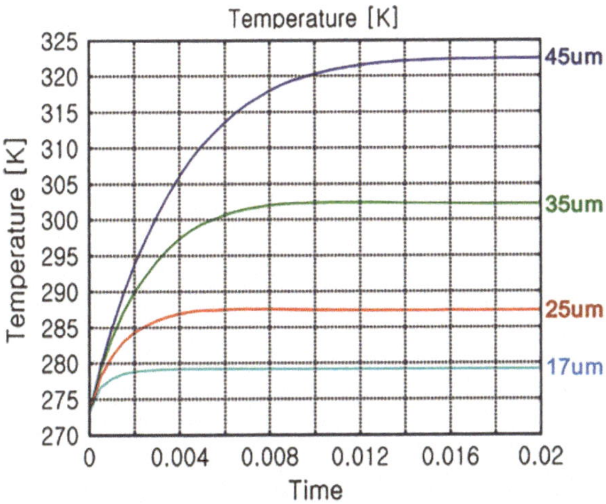

Fig. 4.22 Simulation of thermal response of conventional microbolometer design, as function of pixel size [105].

The improvements in the mechanical and thermal properties of the microbolometers, by implementing the proposed anchor design, are summarized in Table 4.11 [105].

In recent years, there has been significant interest in infrared imagers that are fabricated on nonplanar surfaces [106]. Examples of these include applications such as automotive, smart textiles, robots, space, and aeronautics. A schematic of the cross-section of an infrared detector on flexible substrate is presented in Fig. 4.25 [106]. In this study, amorphous silicon has been used as the sensing material and the detector is packaged, at the device level, in a vacuum cavity with an aluminum oxide packaging layer.

In a study of the reliability of the microbolometer thermal imagers using chip-scale packaging, Elßner and Vogt report a significant reduction in the performance of the detector [107]. The authors attribute this reduction to outgassing processes inside the package of the detector. The principle construction of the imager is shown in Fig. 4.25.

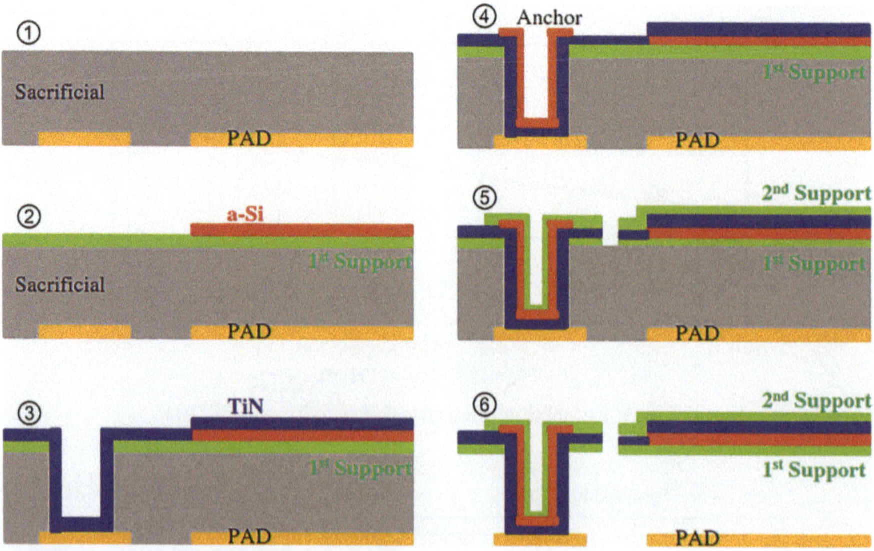

Fig. 4.23 Process flow of the proposed microbolometer with 25 μm pixel size [105].

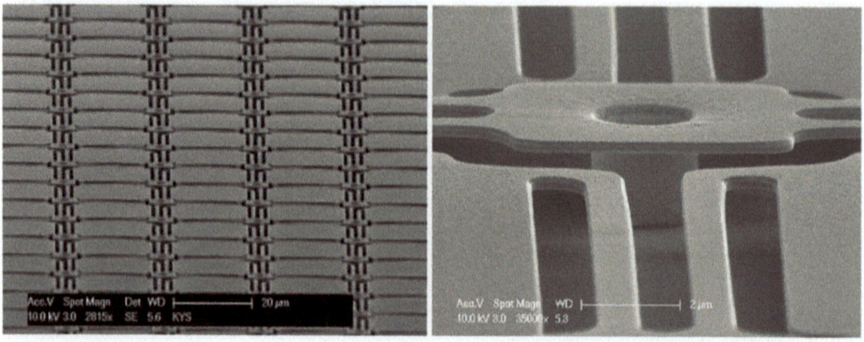

Fig. 4.24 SEM image of the released microbolometers [105].

Table 4.11 Summary of performance characteristics in the proposed anchor design [105].

Characteristics	Unit	2 Anchor	Shared design 1	Shared design 2
ε (fill factor)	%	53	73	62
R_B (resistance)	ohm	5.17E+05	5.36E+05	4.85E+05
τ (thermal time constant)	ms	4.03	4.01	7.16
G (thermal conductance)	W/K	5.02E−08	3.8E−08	2.07E−08
C (heat capacitance)	J/K	2.02E−10	1.43E−10	1.48E−10
Rv (responsivity)	V/W	8.09E+04	1.23E+05	1.18E+05

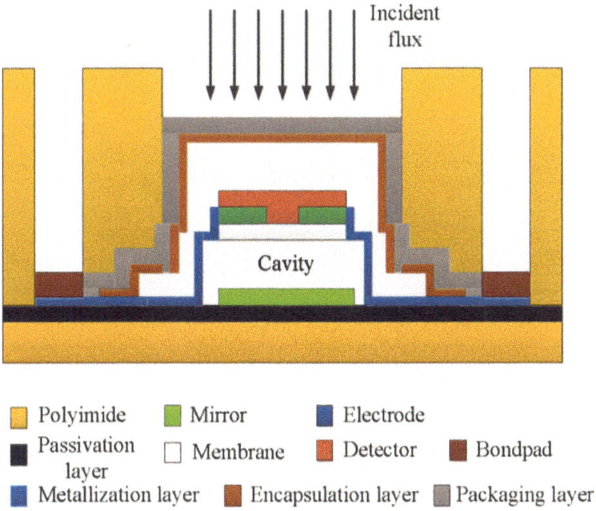

Fig. 4.25 A schematic of the cross-section of an infrared detector on flexible substrate [106].

4.5 Vanadium oxide

Vanadium oxide (VO_x) is the most widely known and used material in the design and manufacture of uncooled microbolometers (Fig. 4.26). VO_x has a high TCR of about −2%/K to −3%/K. There are different stoichiometric phases in VO_x, such as V_2O_5, VO_2, and V_2O_3; among them, V_2O_5 exhibits high TCR, but its resistance is very high for practical use. Therefore, VO_2 or V_2O_3 phases have been mixed with the V_2O_5 phase to yield a desired low resistance. The first results of the metal-to-insulator transition in vanadium oxides, in the literature, was reported over 60 years ago [108]. It is presented in Fig. 4.27 [108]. As can be seen in this figure, the electrical conductivity in VO, VO_2, V_2O_3, and Ti_2O_3 decreases by several orders of magnitude with decrease in temperature—across the transition temperature.

Due to their unusual electronic, optical, magnetic, and physical properties, vanadium oxides as well as other binary transition metal oxides represent excellent examples of strongly correlated materials. Their metal-insulator transition behavior has been utilized in numerous applications. These applications are illustrated in Fig. 4.28 [109].

The phase diagram of vanadium oxide is presented in Fig. 4.29a [110]. The Ellingham diagrams representing the various oxides of vanadium in oxygen, as function of temperature and pressure, are shown in Fig. 4.29b [110]. Resistivity of various phases of the vanadium—oxygen system, as function of temperature, are summarized in Fig. 4.29c [110].

The mixed valence compounds of vanadium-oxygen, described by the Magnéli phases, are defined by the following general stoichiometric formula [111]:

$$V_nO_{2n-1} = V_2O_3 + (n-2)\,VO_2, \text{where } 3 \le n \le 9. \tag{4.8}$$

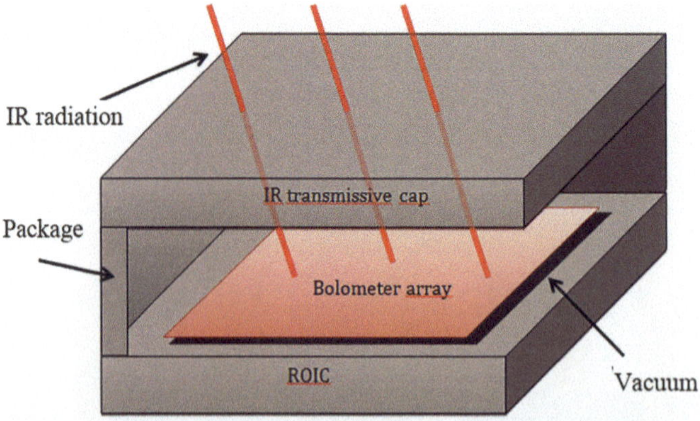

Fig. 4.26 Principle construction of an infrared imager for reliability studies [107].

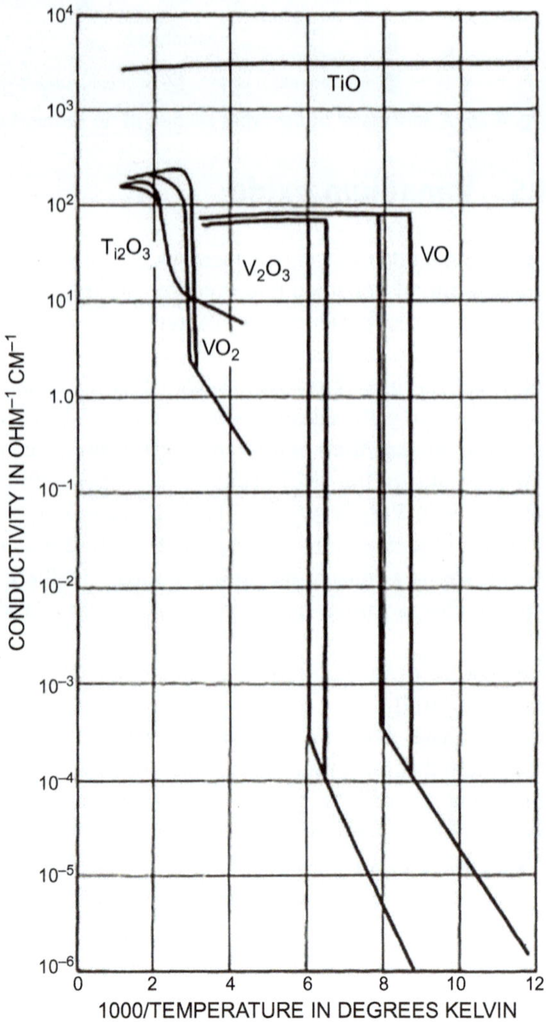

Fig. 4.27 Electrical conductivity as a function of reciprocal temperature for the lower oxides of titanium and vanadium. Measurements were made along the [46] direction in VO, and along the *c* axis in V_2O_3 and VO_2 [108].

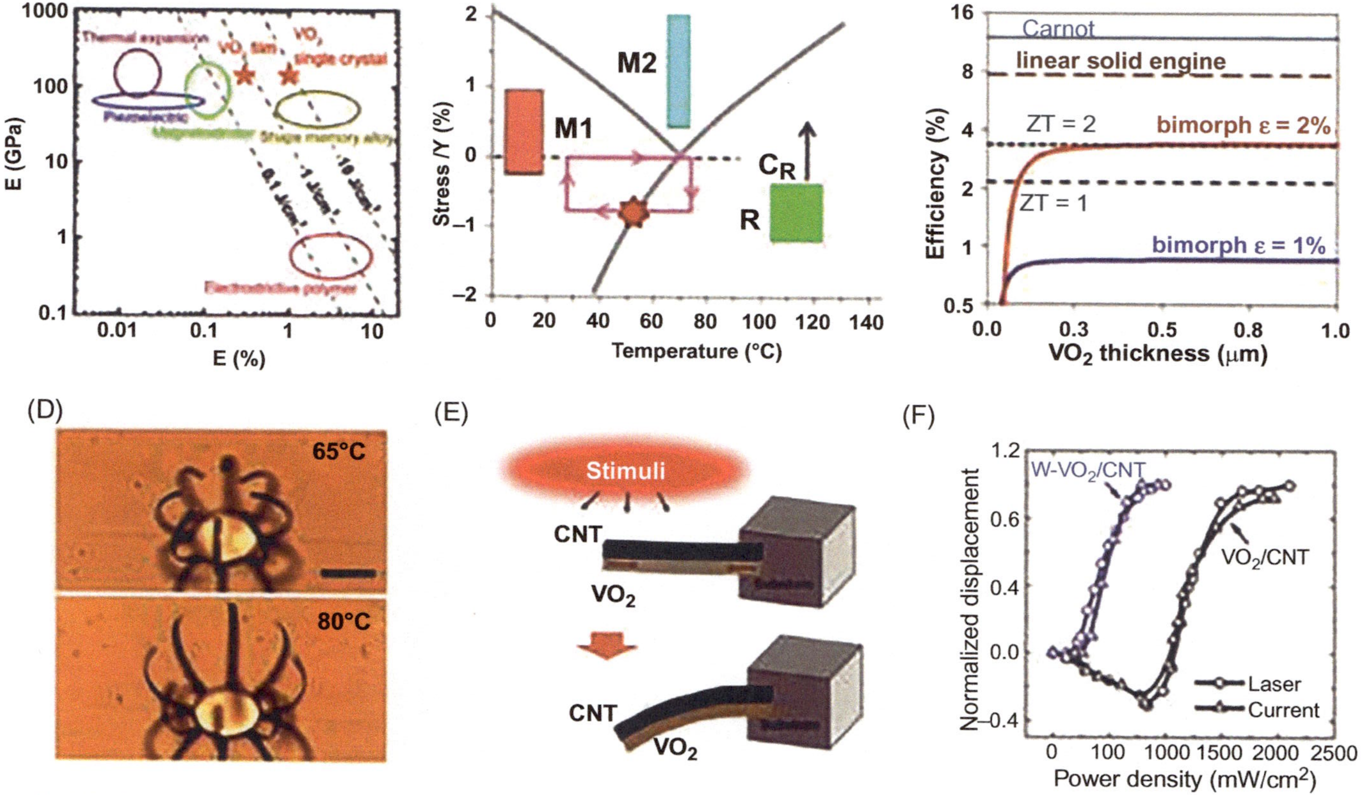

Fig. 4.28 MIT in oxides: a compelling case for materials physics and solid-state electronics research [109].
Panels (A and D) reproduced with permission from K. Liu, C. Cheng Z. Cheng, K. Wang, R. Ramesh, J. Wu, Giant-amplitude, high-work density microactuators with phase transition activated nanolayer bimorphs, Nano Lett. 12 (12) (2012) 6302–6308. https://doi.org/10.1021/nl303405g. Copyright © 2012 American Chemical Society. Panels (B and C) from K. Wang, C. Cheng, E. Cardona, J. Guan, K. Liu, J. Wu, Performance limits of microactuation with vanadium dioxide as a solid engine, ACS Nano 7 (3) (2013) 2266–2272. https://doi.org/10.1021/nn305419e. Copyright © 2013 American Chemical Society. Panels (E and F) from H. Ma, J. Hou, X. Wang, J. Zhang, Z. Yuan, L. Xiao, Y. Wei, S. Fan, K. Jiang, K. Liu, Flexible, all-inorganic actuators based on vanadium dioxide and carbon nanotube bimorphs, Nano Lett. 17 (1) (2017) 421–428. https://doi.org/10.1021/acs.nanolett.6b04393. Copyright © 2016 American Chemical Society.

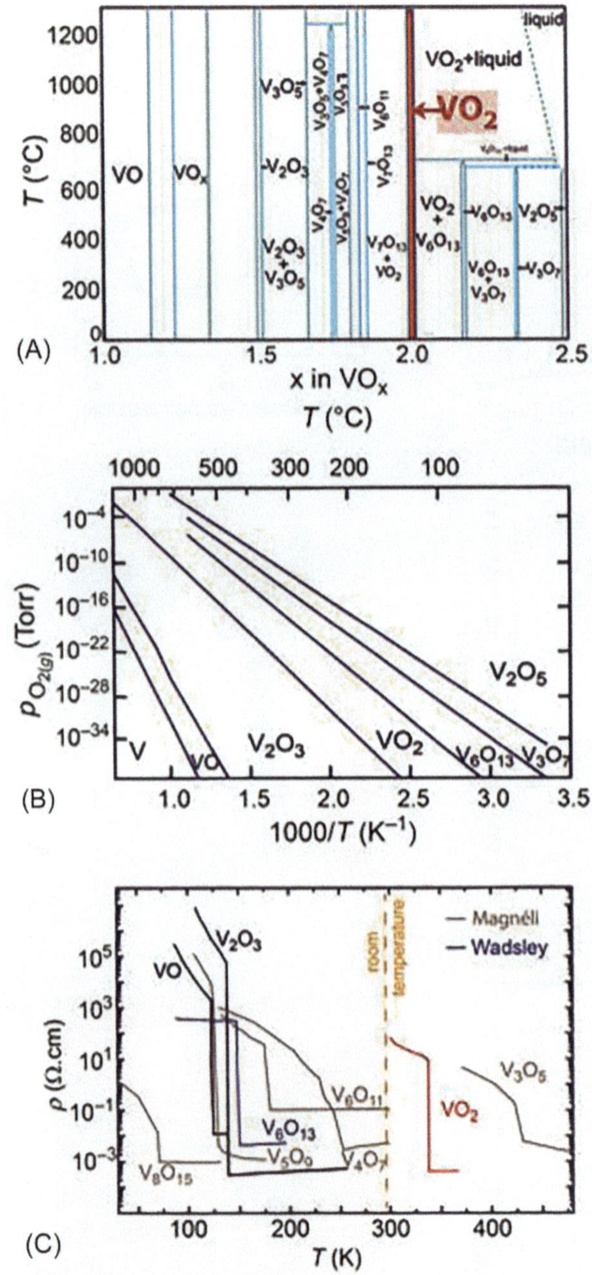

Fig. 4.29 (a) Phase diagram of vanadium oxide; (b) The Ellingham diagram for vanadium oxide phases; (c) Resistivity versus temperature for various V-O phases, with a MIT at different temperatures. While V_2O_3 exhibits the largest change in resistivity over eight orders of magnitude, but below room temperature, VO_2 has the largest resistivity ratio above room temperature of about five orders of magnitude (Ref. [110] and references therein).

This is in addition to the oxides of vanadium that consist of various formal valency stages of vanadium such as two in VO, three in V_2O_3, and four in VO_2 five in V_2O_5 [111–114]. VO_2 and V_2O_5 represent the end members in the V_nO_{2n+1} phase [115–117]. A theory of morphotropic transformations in vanadium oxides has been discussed in detail by Katzke, Tolédano, and Depmeier [117]. Some of the earlier studies on phase analysis [118] and crystal structure [119] of vanadium oxides has been reported in the 1950s.

The metal-insulator-transition temperatures of several vanadium oxides, in the series V_nO_{2n-1}, are summarized in Table 4.12.

An illustration of the major oxidation states as well as the other intermediate states of vanadium oxide is presented in Fig. 4.30 [120].

Table 4.12 Metal insulator transition temperatures of vanadium compounds in the series V_nO_{2n-1}.

Compound V_nO_{2n-1}	MIT temperature (K)
V_2O_3	168
V_3O_5	430
V_4O_7	250
V_5O_9	135
V_6O_{11}	170
V_7O_{13}	Metallic
V_8O_{15}	70
V_9O_{17}	–
VO_2	340

Modified from SchwingenschlÖgl U., Eyert V., The vanadium magnéli phases V_nO_{2n-1}, Ann. Phys., (2004), 13, 9 p. 475-510 https://doi.org/10.1002/andp.200410099.

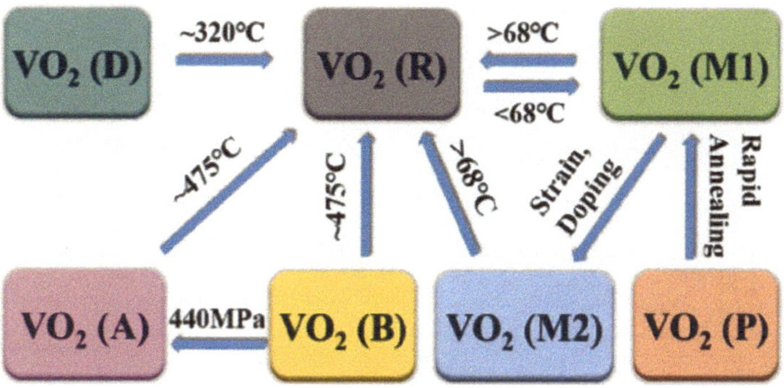

Fig. 4.30 Major oxidation states and other intermediate states of vanadium oxide [120]. From Zhang, Y.; Xiong, W.; Chen, W.; Zheng, Y. Recent progress on vanadium dioxide nanostructures and devices: fabrication, properties, applications and perspectives. Nanomaterials 2021, 11, 338. https://doi.org/10.3390/nano11020338.

V_2O_5 is known to be a stable layered anisotropic semiconductor with van der Waals structure [121,122]. Its energy gap has been reported to be 2.17 eV (E||c) and 2.54 eV (E||c) [122]. Self-assembled nanoflakes of VO_2 and V_2O_5 have been demonstrated for their gas sensing properties [123].

The reduction of V_2O_5 to VO_2 is illustrated in Fig. 4.31 [120].

In general, transition metal oxides exhibit diverse properties [110–114]. Due to the several oxidation states of vanadium that are associated with a variety of chemical, electronic, magnetic, mechanical, optical, physical, structural, and thermal properties, the oxidation of vanadium is one of the most studied topic in condensed matter physics, and materials science and engineering. Some of the important applications of the oxides of vanadium include the production of sulfuric acid and several other chemicals as a catalyst, in electrochromic devices, optical switching, reversible cathode materials for lithium rechargeable batteries, sensors, and smart windows.

Examples of the interesting properties of vanadium oxides are illustrated in Fig. 4.32 and 4.33. In Fig. 4.32, a comparison of the metal-insulator-transition (MIT) temperatures and the magnetic ordering temperatures of the Magnéli phases in vanadium oxides is presented. As can be seen in the figure, MIT and magnetic

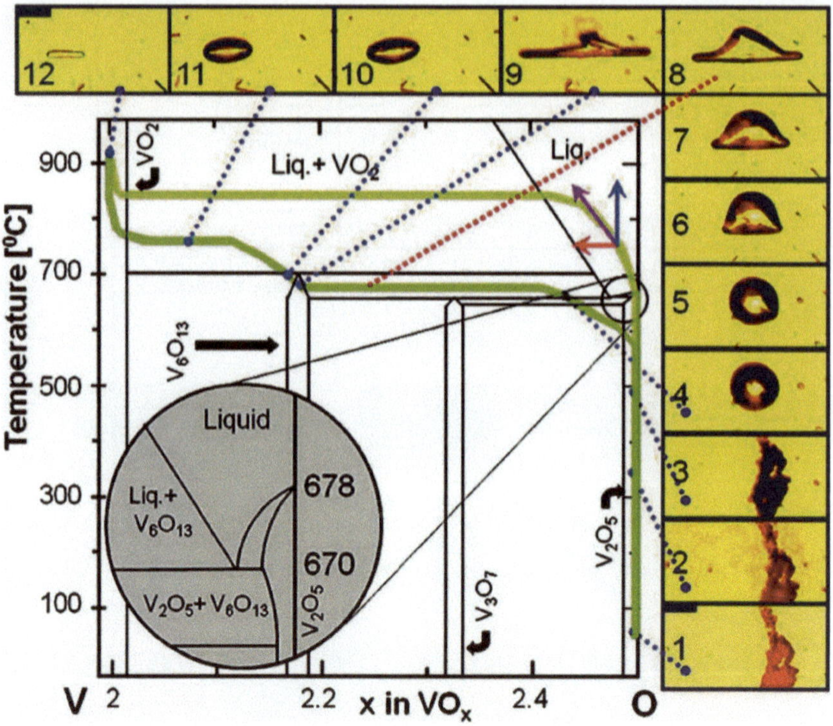

Fig. 4.31 Schematic of reduction steps from V_2O_5 to VO_2 [120].
From Zhang, Y.; Xiong, W.; Chen, W.; Zheng, Y. Recent progress on vanadium dioxide nanostructures and devices: fabrication, properties, applications and perspectives. Nanomaterials 2021, 11, 338. https://doi.org/10.3390/nano11020338.

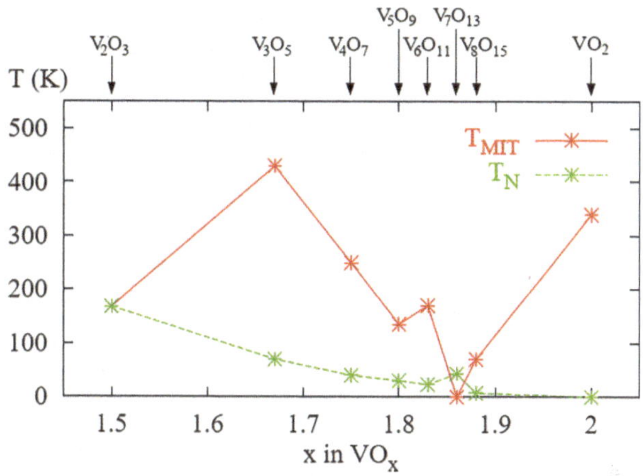

Fig. 4.32 Comparison of the MIT temperatures and the magnetic ordering temperatures of the Magnéli phases. Besides V_7O_{13}, all members of the series undergo an MIT. Except for VO_2, each compound develops an antiferromagnetic order at low temperatures. In the case of V_2O_3, the MIT and the magnetic ordering coincide at a temperature of 168 K [111].

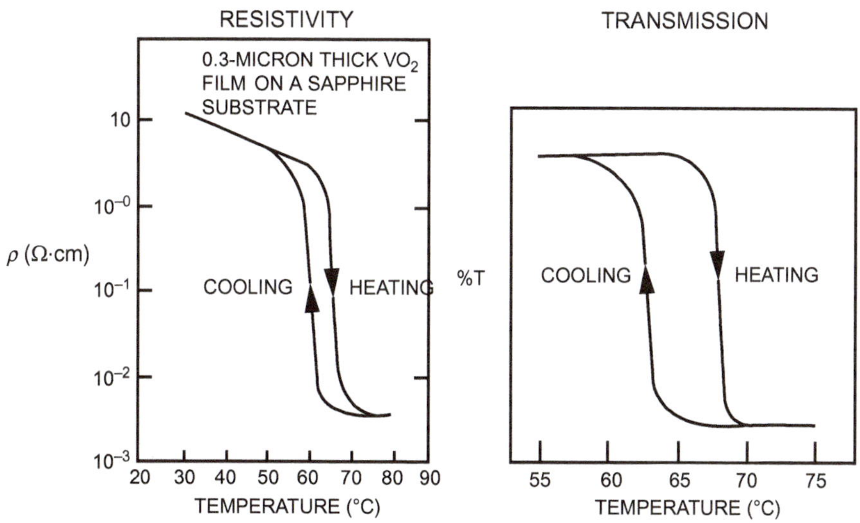

Fig. 4.33 Optical and electronic switching properties of thin film VO_2 [124].

ordering coincide at 168 K in V_2O_3 [111]. In Fig. 4.33, the optical and electronic switching properties of thin film VO_2 are illustrated [124]. As can be seen in the figure, the transition in the optical properties follow the resistivity versus temperature behavior in VO_2.

Similar studies of the MIT on cantilevers of VO_2/TiO_2 are presented in Fig. 4.34 [125]. The results are consistent with those in Fig. 4.33.

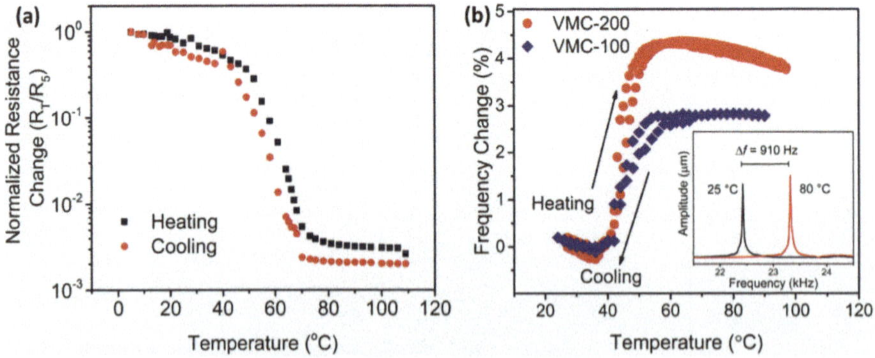

Fig. 4.34 Metal-insulator transition for VO_2 cantilevers measured (A) electrically and (B) mechanically. The transition was sharper in mechanical detection, and more pronounced in VMC-200, likely due to lower stiffness of the cantilever (VMC-200 was 28 μm wide, 200 μm long) [125].

An example of the results of X-ray diffraction of V_2O_3, VO_2, and V_2O_5, synthesized from vanadium ethylene glycolate, is presented in Fig. 4.35 [126].

Nanostructured vanadium oxides obtained by hydrothermal synthesis lead to very interesting structures. The results of this study are illustrated in Fig. 4.36 [127].

The uncooled microbolometer design takes advantage of the smart properties of VO: it can switch reversibly from an IR-transparent to an opaque thin film when properly triggered. This optical behavior is exploited in a case study as a smart mirror that can modify the Fabry-Perot resonant cavity depth between the suspended thermistor material VO and a mirror patterned on the substrate, thereby altering the wavelength sensitivity and creating a dual-band microbolometer.

A schematic of a dual-band microbolometer, along with the structural view and operational view, is shown in Fig. 4.37 [128]. The fixed reflective mirror is switched between the smart material (VO) and Au layer. In the first case, the VO is used in the metallic phase by heating the detector to $68°C$. Thus, the IR radiation is reflected back to the suspended VO enhancing the IR absorption in the first spectral band (9.4–10.8 μm). In the second case, VO is switched to the semiconducting phase by cooling the detector to room temperature. Thus, most of the IR radiation is transmitted through the transparent VO layer and through a SiO layer that is used as a spacer material, and then reflected by the gold (Au) thin film mirror layer, which is patterned underneath the SiO_2 spacer layer on the substrate. Thus, the gap underneath the microbolometer pixel, which initially contained only vacuum, is increased by the thicknesses of the SiO_2 and VO transparent materials. Therefore, the IR absorption is shifted to a second spectral band (8–9.4 μm). The resonant cavity depths are optimized to maximize IR absorption, thereby increasing the responsivity and detectivity of the microbolometer in the 8–9.4 and 9.4–10.8 μm spectral bands. The two-band design enhances the probability of detecting and identifying objects in a scene [128].

Uncooled vanadium oxide microbolometer arrays are quite common, commercially available and used as digital infrared imagers; they are typically fabricated using

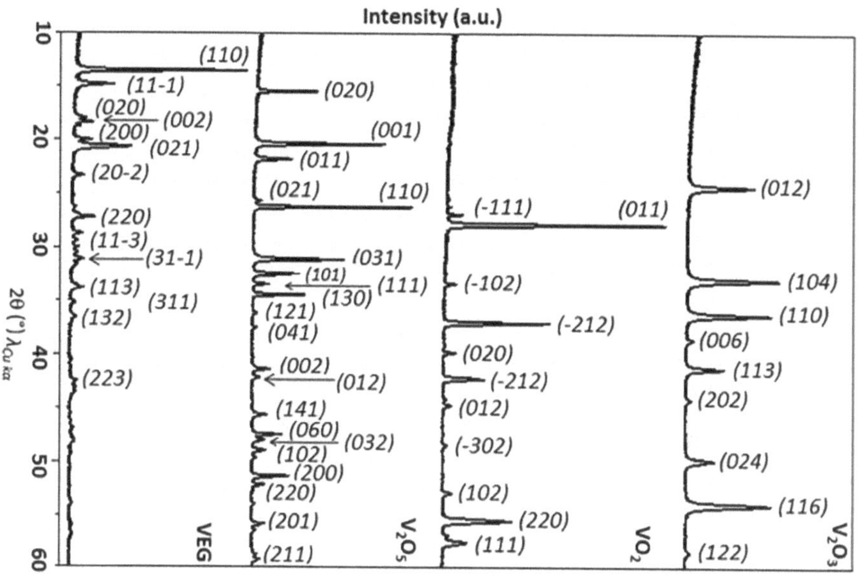

Fig. 4.35 X-ray diffraction patterns of the as-synthesized VEG (vanadium ethyleneglycolate) powders and the vanadium pentoxide, V_2O_5, vanadium dioxide VO_2, and vanadium sesquioxide V_2O_3, obtained at 500°C post-treatment under air, post-treatment under vacuum and Ar-H_2 atmosphere, respectively [126].

Fig. 4.36 Nanostructured vanadium oxides obtained via the self-assembling of nanoparticles: (A) and (B) Nano-urchin from VO_x nanotubes, (C) and (D) nanospheres from platelets hexavanadate nanoparticles $Cs_2V_6O_{16}$ [127].

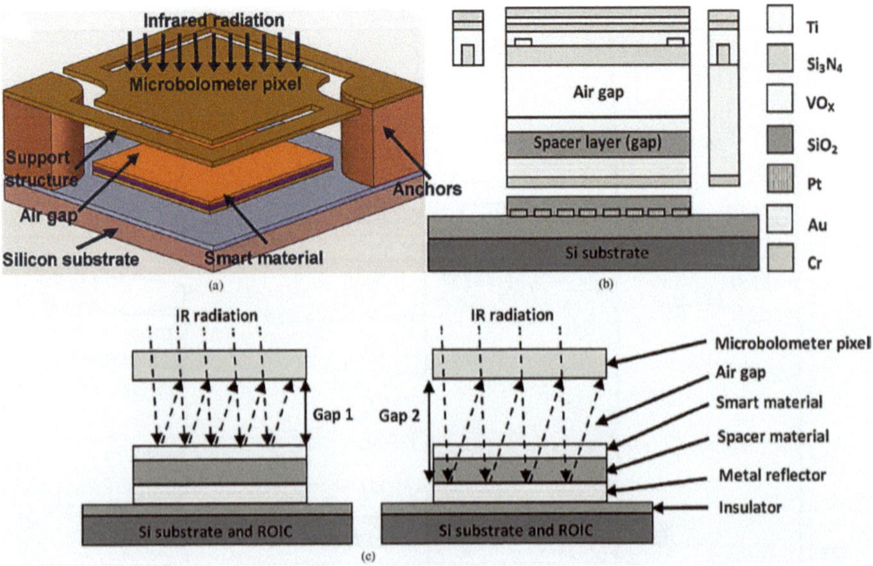

Fig. 4.37 Schematics showing a dual-band microbolometer in (A) 3D view, (B) side view, and (C) operational view. IR radiation is reflected by the smart material when it is used in the metallic phase [(C), *left*], and is reflected by the gold layer when the smart material is used in the semiconducting phase [128].

silicon micromachining techniques and are integrated with CMOS technology. The most widely used heat sensitive layer material in microbolometers is vanadium-oxide (VO_x). These films are fabricated using low-cost techniques. VO_x thin-films show high temperature coefficients of resistance (TCR) and exhibit suitable pixel resistances for CMOS readout circuits.

In most microbolometers, the detector elements are thermally insulated from their environment by micromachining thin membranes or cantilevers on which the detectors are located, thereby greatly reducing thermal losses through conduction and enabling high responsivity. Instead of relying on thin micromachined membranes to provide thermal insulation from the substrate, the approach, in a case study [129], consists of fabricating detectors on thin (12.5 μm), thermally insulating polyimide substrates. Besides simplicity and low-cost, this approach results in more robust and reliable devices of utmost importance in space missions. The main challenge is to produce high quality and high sensitivity detectors on these thin polyimide films, as they are nonplanar, imperfect, and porous substrates.

In a study of "High temperature coefficients of resistance of VO_2 films grown by excimer-laser-assisted metal organic deposition process for bolometer application," Nishikawa et al. [130] report the electrical performance properties of laser irradiated VO_2 films. Fig. 4.38 shows the temperature dependences of electrical resistivity of the films laser irradiated after preheating at 300°C. The films irradiated at 60 mJ/cm^2 for 10 min, which exhibited no XRD peaks, are semiconducting. In contrast, the films irradiated at 100 mJ/cm^2 for 5 or 10 min, which also show no XRD peaks, exhibit broad metal insulator (MI) transitions at around 300 K with negligible hysteresis,

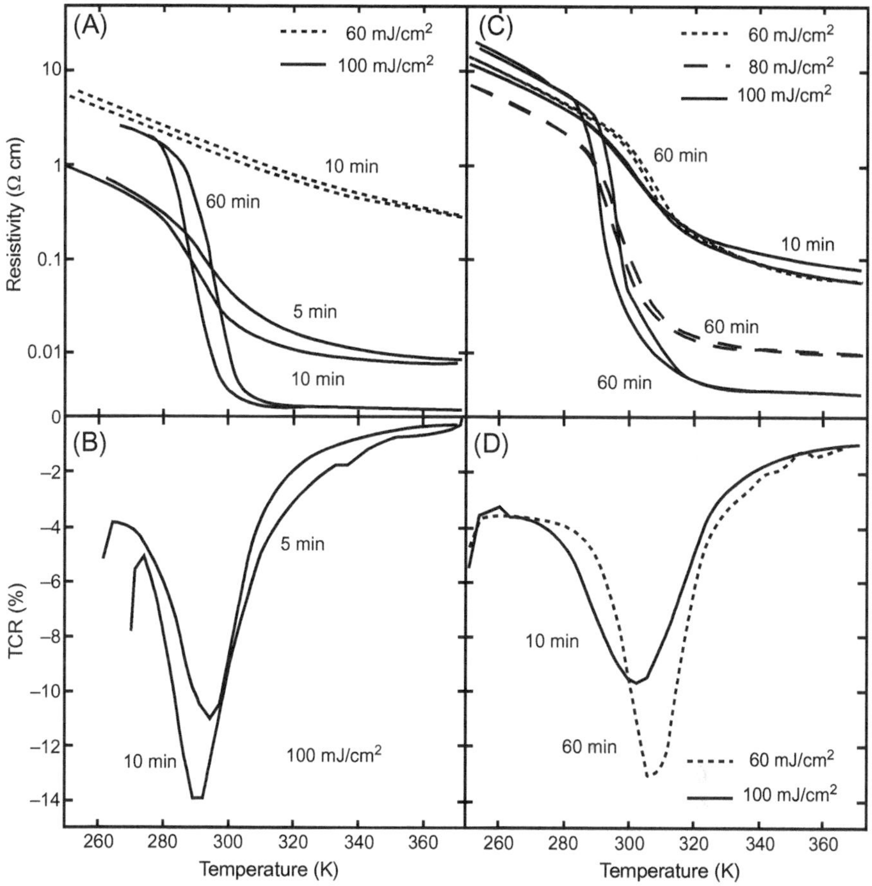

Fig. 4.38 Temperature dependences of electrical resistivity and TCR values of films irradiated with the laser (a) and (b) after preheating at 300°C and (c) and (d) after UV-lamp irradiation [130].

which are essential characteristics for the IR sensors. The film irradiated at 100 mJ/cm^2 for 60 min, in which the VO_2 phase is confirmed by XRD, exhibits a sharp MI transition accompanied by a hysteresis width of 10°C.

In another case study [131], as illustrated in Fig. 4.39, (V_2O_3/V)/V_2O_3 multilayer thin film structure was sputter deposited on a silicon substrate on which a 300 nm layer of silicon dioxide (SiO_2) was deposited for thermal insulation. This insulation layer is required for performing sheet resistance versus temperature measurements. The thin film structure is composed of 18 interchanged layers of V_2O_3 and V—each having a thickness of 5 nm above which a final V_2O_3 layer was deposited. The V_2O_3 layers were RF sputter deposited from a 99.5% pure V_2O_3 target at 150 W of power and a deposition rate of 0.68 nm/min. The V layers were DC sputter deposited from a 99.9% pure V target at 150 W of power and a deposition rate of 4 nm/min. Both V_2O_3 and V layers were sputter deposited at 3 mTorr of argon pressure and at a sputter-chamber base pressure of 4.48×10^{-6} Torr. Two different samples of the

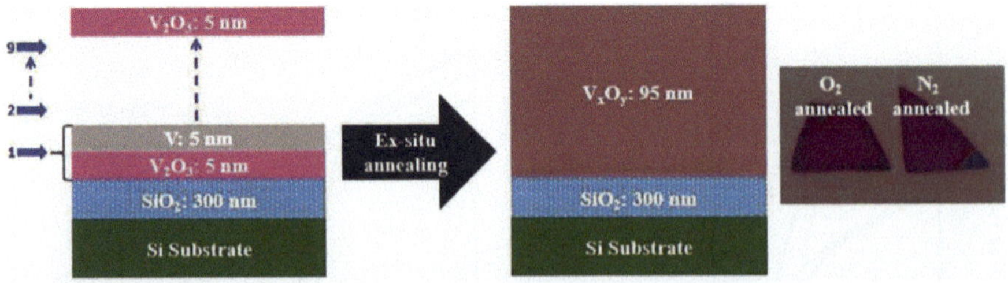

Fig. 4.39 Schematic diagram of the synthesis process: ex situ annealing the $(V_2O_3/V)^9/V_2O_3$ multilayer thin film resulting in a V_xO_y thin film [131]. And a photograph of the thin film samples (far right).

deposited multilayer thin film structure were then separately ex-situ annealed in a horizontal tube furnace at 300°C for 30 min in nitrogen and oxygen atmospheres at a gas flow rate of 120 mL/min. The annealing process is meant to cause oxygen atoms to diffuse from oxygen rich V_2O_3 layers to the thin metallic V layers; and as a result, a V_xO_y thin film is formed containing multiple phases of vanadium oxide compounds. In addition, annealing in O_2 atmosphere is proposed to introduce more O_2 atoms within the thin film structure, thus producing V_xO_y thin film with different thermoelectric and thermo-optic properties as compared to the N_2 annealed V_xO_y thin film.

Fig. 4.40 shows the resistance thermal loop of a nanostructured VO_2 thin film [132].

A bolometer suspended structure, from another example in the literature [133], is shown in Fig. 4.41. Vanadium oxide is utilized as the thermal sensitive material. The

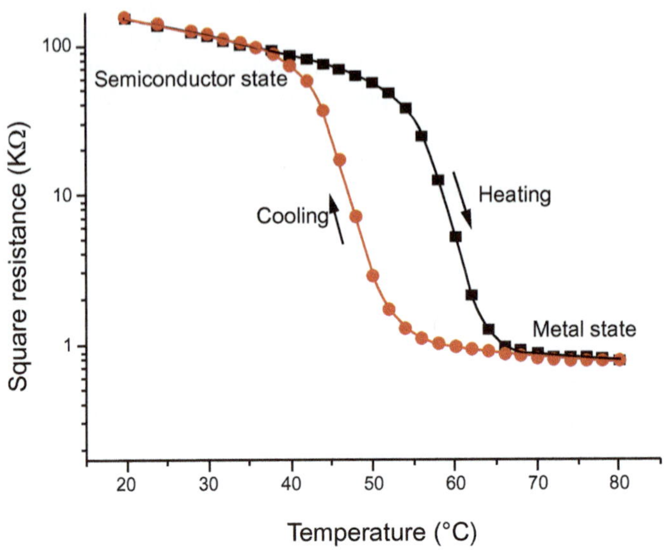

Fig. 4.40 Resistance thermal hysteresis loop of nanostructured VO_2 thin film [132].

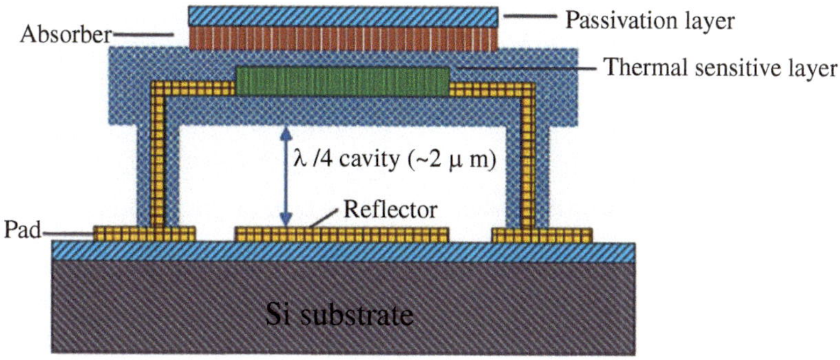

Fig. 4.41 The structure of the designed microbolometer. $\lambda/4$ cavity structure is adopted to improve infrared absorption [133].

suspended $\lambda/4$ cavity structure is designed to enhance infrared absorption. The height of the gap between silicon nitride membrane and the bottom reflector is 2 μm, corresponding to the wavelength range of 8–14 μm. TiN thin film is applied as the absorber to obtain a perfect absorption combining the effect of $\lambda/4$. The top silicon nitride acts as passivation layer. The metal pad is connected through outer gold wire welding.

A schematic diagram describing the suspended vanadium oxide microbolometer fabrication is shown in Fig. 4.42 [133]. Au thin film with thickness of 250 nm, which plays the role of a reflector, is sputtered and patterned by photolithography on 300 nm thick silicon nitride layer. Then, polyimide which acts as a sacrificial layer is spin coated. It helps to pattern and make the anchor areas. The subsequent step is the deposition of silicon nitride and silicon dioxide with thicknesses of 350 and 200 nm, respectively, by PECVD. The contact holes are etched to be open by Reactive Ion Etching (RIE). After formation of the contact holes, Cr/Au electrodes with thickness of 10 nm/80 nm are deposited for electrical connection. The electrodes are of very low resistivity and good adhesion. By utilizing lift-off process, electrode areas are patterned to connect the bottom pad. Vanadium oxide thin film of thickness 100 nm is deposited and patterned by ion beam sputtering to form a thermal sensitive material. Then, silicon dioxide (100 nm) and silicon nitride (100 nm) are deposited by PECVD; they function as insulation layers between TiN and VO_x. TiN layer with a sheet resistance of 376 Ω and thickness of 60 nm is deposited by sputtering. TiN acts as the absorbing layer. Then, another layer of silicon nitride with thickness of 200 nm is deposited to act as passivation layer. Finally, removal of polyimide sacrificial layer is completed by plasma asher in an oxygen atmosphere.

The process steps of "Fabrication of vanadium oxide microbolometers on thin polyimide films" are detailed in Fig. 4.43 [129].

Fig. 4.44 shows the experimental sheet resistance-temperature characteristics of annealed magnetron controlled sputter deposited VO_2 thin films [134]. In this study, Chen et al. [134] report a relatively low post annealing temperature (400°C) for the phase transition to occur in the deposited films. The resistance ratio of low temperature to high temperature is very low, which means the transition process has happened

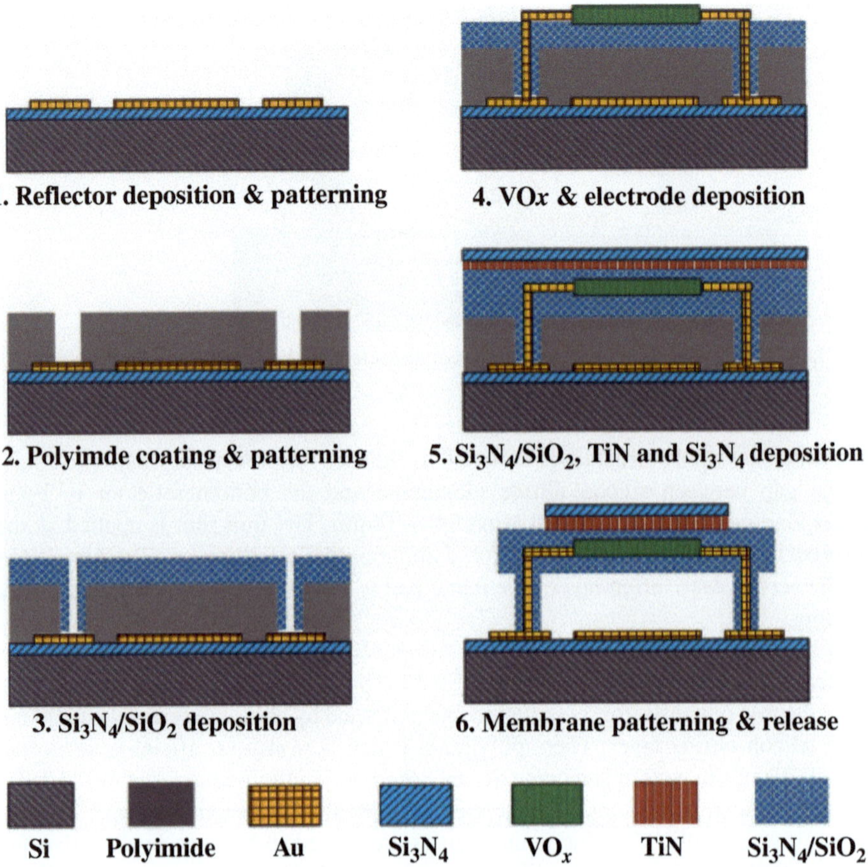

Fig. 4.42 Schematic diagram showing the process of fabrication of a microbolometer suspended structure [133].

partially. With increasing annealing temperature, this resistance ratio becomes larger. The resistance ratio reaches a maximum value at the temperature of 475°C. Although the resistance ratios are different, all the phase transitions occur at $T_c \approx 68°C$.

In general, during use, VO_x microbolometers are maintained under vacuum conditions in order to be able to achieve high sensitivity [135]. However, when they are utilized to monitor high temperatures in applications such as imaging of explosions, forest fire, fuselage, sun, etc. [136,137], the imagers saturate and require a long time to recover. In Ref. [135], Schoeman and du Plessis report a model to determine the gaseous thermal conductance of uncooled VO_x microbolometers [135]. A generic VO_x device design, considered by the authors [135], is shown in Fig. 4.45.

In Figs. 4.46a and b, the simulated results of temperature, at different power levels, for devices under vacuum and at atmospheric pressure, respectively, are presented. The authors report prototype devices with 0.5 μW/K under vacuum conditions and 15 μW/K at atmospheric pressure with a TCR of −1%/K.

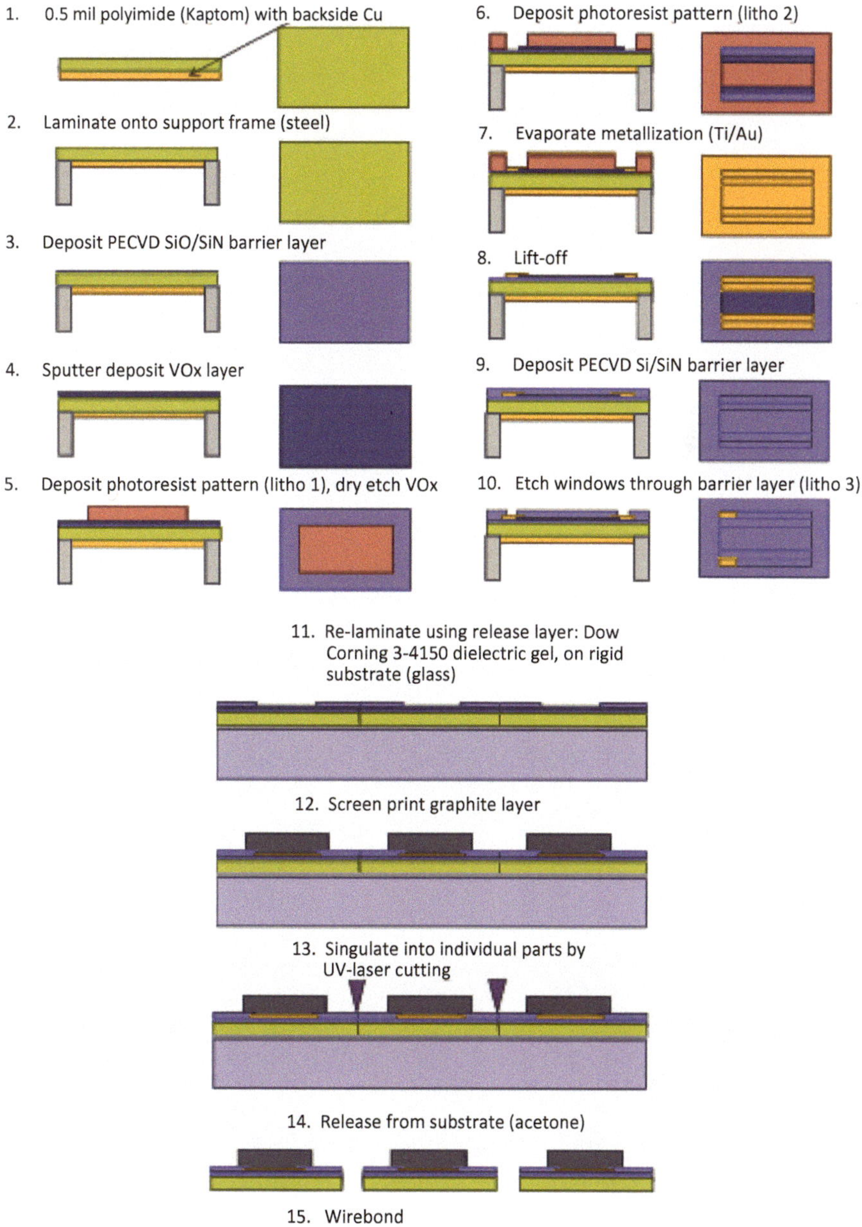

Fig. 4.43 Schematic fabrication process sequence of vanadium oxide microbolometers on thin polyimide films [129].

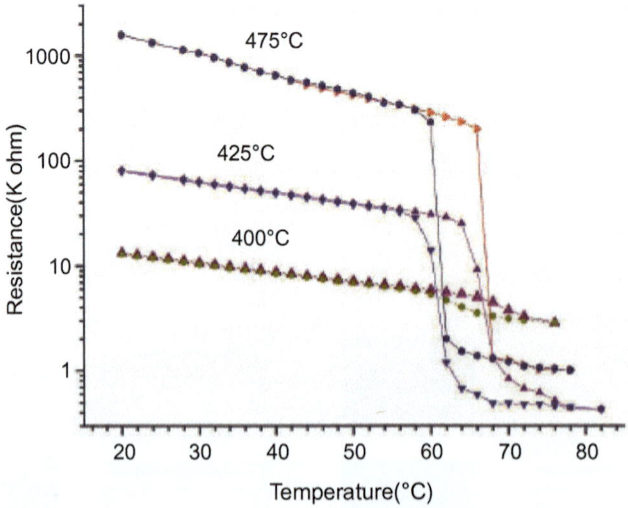

Fig. 4.44 Sheet resistance vs. temperature for VO_2 thin films after annealing at different temperatures [134].

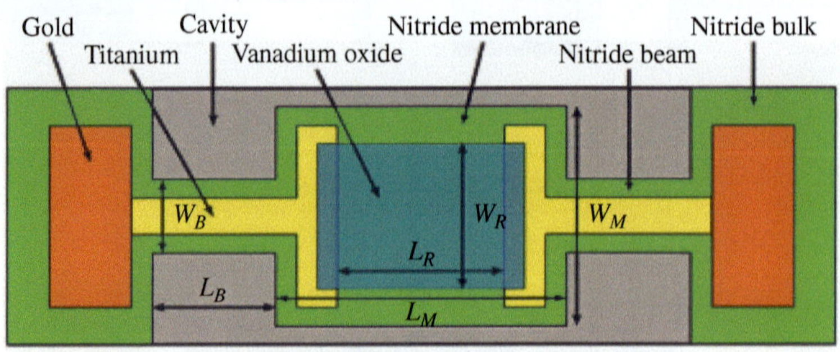

Fig. 4.45 Generic VO_x device design [135].

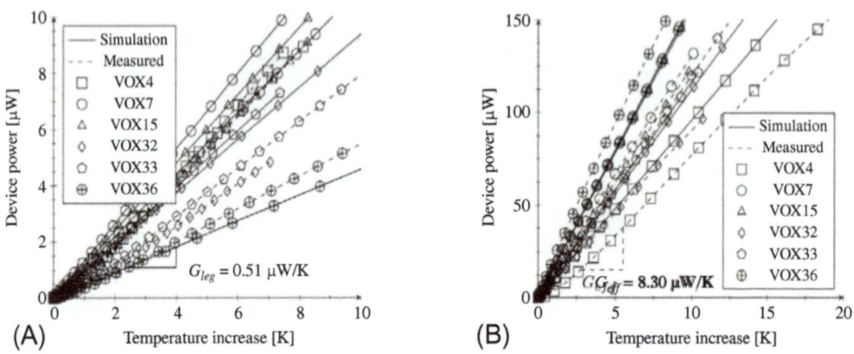

Simulated temperature curves at different power levels for devices under vacuum conditions.

Simulated temperature curves at different power levels for devices at atmospheric pressure.

Fig. 4.46 Simulated temperature response of VO_x devices at different power levels [135].

Table 4.13 Summary of the most important material properties that are used in the manufacture of microbolometers [135].

Material	Density (Kg/m^3)	Specific heat (J/g K)	Electrical conductivity (S/m)	Thermal conductivity (W/mK)
SiO	2150–2200	1.0	10^{-16}–10^{-14}	0.76–1.42
Si_3N_4	2400–3200	0.69–1.73	10^{-10}	1.65–5.1
Ti	4510	0.5	2.56×10^6	16.3
VO_x	4570	1.0	203.6	15
Air	1.16	1.007	0	0.0284

In considering simulations, it is important to have access to reliable data of the material properties. While there is a general agreement in the material properties in the literature, sometimes, access to reliable data is not always easy. From that perspective, some of the material properties, considered in Ref. [135], are summarized in Table 4.13.

4.6 Black metals as infrared absorbers: Integration with microbolometers, polycrystalline silicon, black silicon, pyroelectrics

The ability to tailor the optical properties of materials by (a) modifying their surface texture [138], (b) doping [139], (c) alloying [140], (d) annealing [141], (e) surface defects and microstructure [142], especially in the desired range of wavelengths, has significant implications and applications. This is particularly the case in optics, photonics, and energy related applications.

In general, highly absorbing materials such as black coatings [143] or blackening of metals [144] or black oxide coatings [145] have significant applications including the following:

1. Prevention of hydrogen uptake in printed circuitry in order to enhance bonding in multilayers (for example, black oxide coatings on copper);
2. reduction in glare and reflectivity in optical systems, cameras, and defense related hardware systems;
3. solar collectors to facilitate to maximize the absorption of the incident solar thermal radiation and minimize the reflection or emission.

The performance of infrared detectors is improved when a major fraction of the incident radiation is absorbed by the detector. A minimal component of the incident radiation is reflected when the detector materials are coated with uniformly high absorbing metals [86,143–152]. Metal absorbers such as gold (Au) black [140,143–145], silver black (Ag) [143], and platinum (Pt) black [140,145–147] are most commonly used for this purpose. In recent years, materials such as graphene

[149] and black arsenic phosphorous [150] have been shown to exhibit absorption in the infrared range of wavelengths.

Thermal infrared detectors, such as bolometers, measure the intensity of radiation via the increase in temperature of an active element. For the transformation of infrared radiation into heat, the active element is covered with an absorbing layer, but this layer must have high absorptance and small mass or small heat capacity, so that the responsivity of the sensor will not decrease, i.e., so that the time constant will not increase [153]. An efficient infrared absorber layer must therefore combine high absorption with low film mass. The most effective mechanism for the promotion of infrared absorption is high porosity, since this corresponds to low heat capacity. Gold black films offer a possible option, since they are extremely porous and have a low thermal mass. However, highly porous films are difficult to handle, since they are extremely vulnerable and brittle and can be easily destroyed, even by a simple touch. One solution to the problem is the deposition of porous films as the final step in the technological process, but this is not always feasible. On-going research is in the process of developing new material/device possibilities that are compatible with the silicon microelectronics process.

A potential candidate for the active material in a microbolometer is polycrystalline silicon [154,155]. Its TCR and electrical conductivity can be controlled by doping and it is completely compatible with silicon integrated circuit technology. A large TCR leads to a significant change in bolometer resistance with only a small temperature increase. The structure of a bolometer is shown in Fig. 4.47. Thermal insulation is defined by the thermal conductivity and the geometry of the arms support. The TCR values of polysilicon bolometer have been reported to be slightly above 1%/K [155]. In a case study, Gorgulu et al. have reported "all-silicon ultra-broadband infrared light absorbers" [156]. The authors propose the implementation of periodically arranged silicon gratings as shown in Fig. 4.48. They are fabricated using optical lithography and deep reactive ion etching techniques.

The measured and simulated optical properties of the proposed absorber structure is summarized in Fig. 4.49 [156]. The absorbance of the structure is >90% in the wavelength range of 5–20 μm.

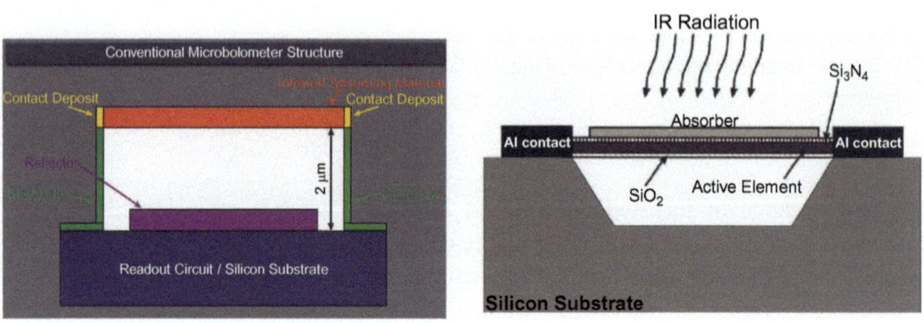

Fig. 4.47 Structure of microbolometer

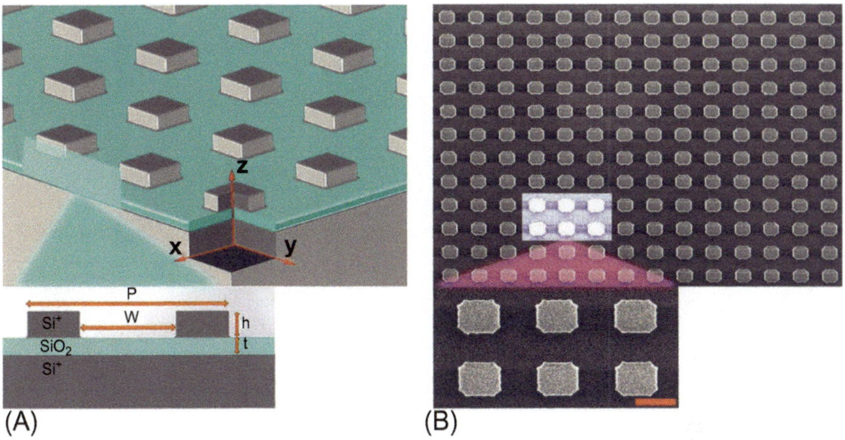

Fig. 4.48 (A) Schematic representation of a three-layer silicon-silicon dioxide-silicon structure with the top silicon layer patterned as 2D gratings of periodicity P, (B) SEM image of one of the fabricated samples with periodicity, $P=8$ μm. Scale bar is 4 μm [156].

The production of a silicon bolometer, integrated with gold black as the absorber layer, is illustrated in Fig. 4.50 [157]. It involves various steps, including the formation of the following layers: (1) reflecting layer, (2) deposition of polyimide on the reflecting layer, (3) deposition of silicon nitride and silicon dioxide, (4) deposition of metal contact and VO_x layer, (5) deposition of SiO_2, gold-black layer and silicon nitride, and (6) sacrificial layer patterning and release.

In recent years, there has been significant interest in black silicon [60], a surface modification of crystalline silicon, for the manufacture of CMOS imagers with improved performance in the near-infrared spectrum [158,159]. An illustration of

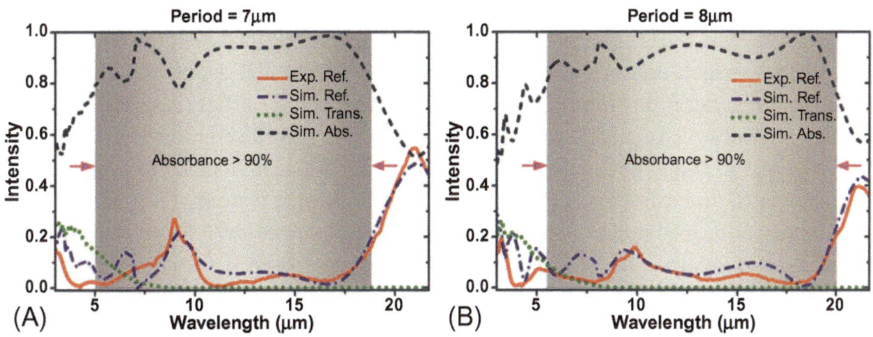

Fig. 4.49 (A,B) Simulated reflection (Sim. Ref.), transmission (Sim. Trans.), and absorption (Sim. Abs.) spectra and measured reflection (Exp. Ref.) spectrum for the samples with periodicities of $P=7$ μm and $P=8$ μm [156].

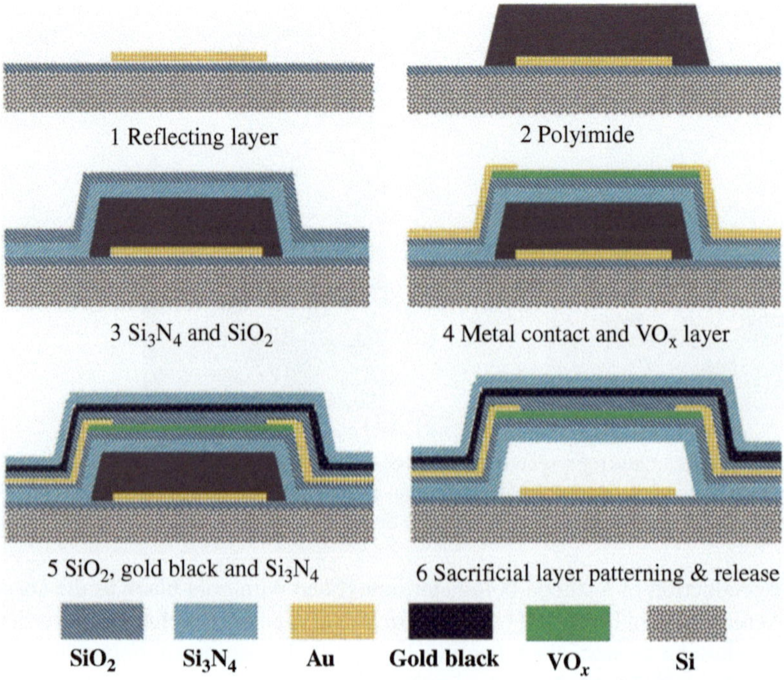

Fig. 4.50 Schematic flow chart of VO_x microbolometers fabrication, showing the main process steps [157].

the responsivity of black silicon, in comparison to standard silicon, indium gallium arsenide, and germanium, is presented in Fig. 4.51 [160].

Vacuum deposition techniques such as thermal evaporation and sputtering are commonly used to process porous metals. Femtosecond lasers can also be utilized to blacken metals [151]. For surface structuring, amplified Ti:sapphire femtosecond laser system, which consists of a mode-locked oscillator and a two-stage amplifier including a regenerative amplifier and a two-pass power amplifier, is used. The laser system generates 65-fs pulses with energy around 1.1 mJ/pulse at a maximum repetition rate of 1 kHz and with a central wavelength $\lambda = 800$ nm. In order to produce surface structures, direct femtosecond laser ablation technique is used. The laser beam is horizontally polarized and is focused normally onto the samples. To produce a large blackened surface area, faster scanning of the sample is applied. Before laser surface treatment, the samples are mechanically polished and further cleaned with methanol. Surface structuring is performed in air at a pressure of 1 atm [161].

Reflectance of blackened metals, Ti90/Al6/V4 alloy, gold, and tungsten are shown in Fig. 4.52 [151,162], Fig. 4.53 [151,162], and Fig. 4.54 [151], respectively. As can be seen in these figures, the reflectance of the metal, in each case is reduced significantly after femtosecond laser treatment of the metal.

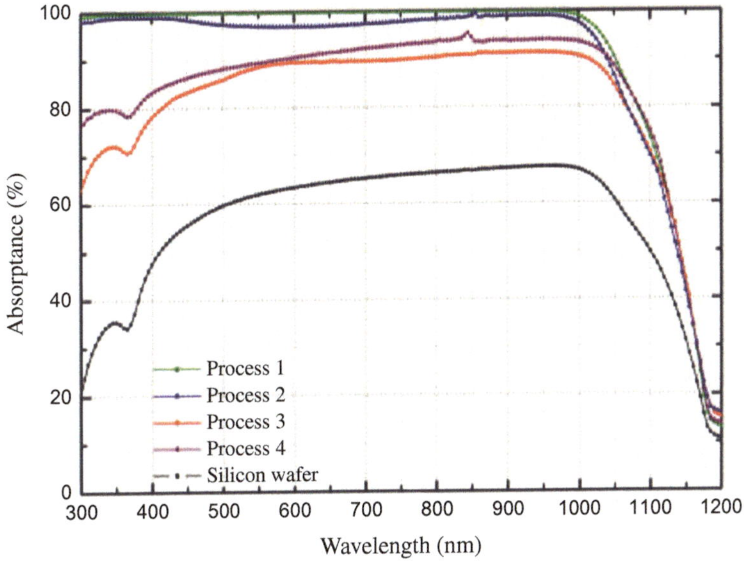

Fig. 4.51 The responsivity of a black-silicon detector exceeds that for standard silicon in the visible and NIR and is competitive with InGaAs and germanium in the SWIR [160].
From Morphologies and optical properties of black silicon by room temperature reactive ion etching; Mater. Res. Bull. 131, 2020, 110973.

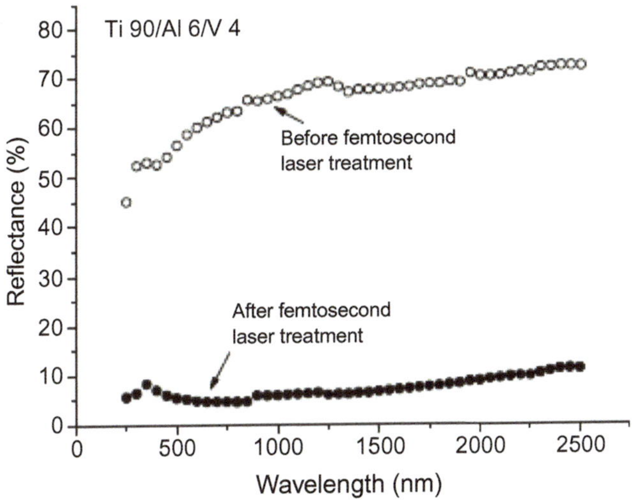

Fig. 4.52 Spectral reflectance of the black Ti90/Al6/V4 alloy as a function of wavelength before and after laser treatment [151,162].

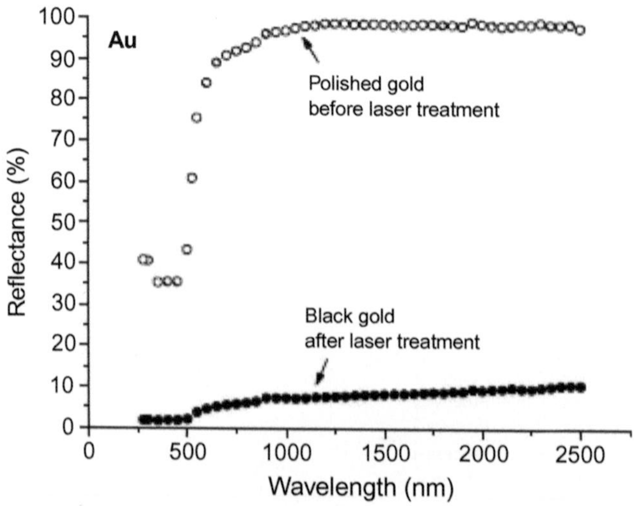

Fig. 4.53 Spectral reflectance of the black gold as a function of wavelength before and after laser treatment [151,162].

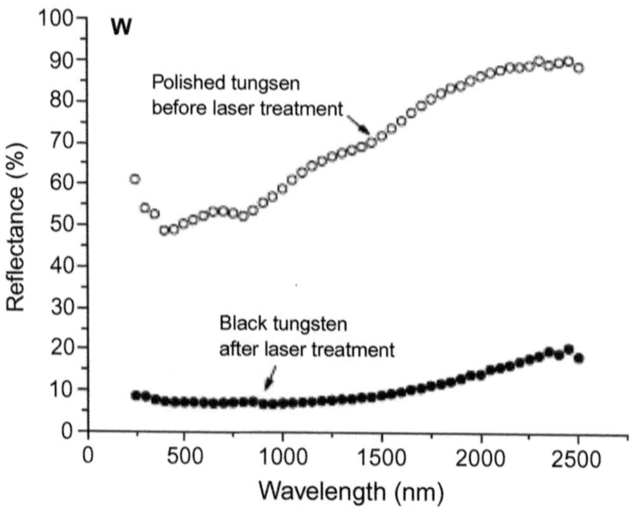

Fig. 4.54 Spectral reflectance of black tungsten as a function of wavelength before and after laser treatment [151].

The spectral reflectance of black platinum is presented in Fig. 4.54 [161]. As can be seen in the figure, subsequent to laser processing, the reflectance decreases significantly (Fig. 4.55).

A "Linear bolometer array using a high TCR VO_x-gold film" has been demonstrated by Smith et al. [162]. Their results of the TCR of the VO_x-gold film are shown in Fig. 4.56 [163].

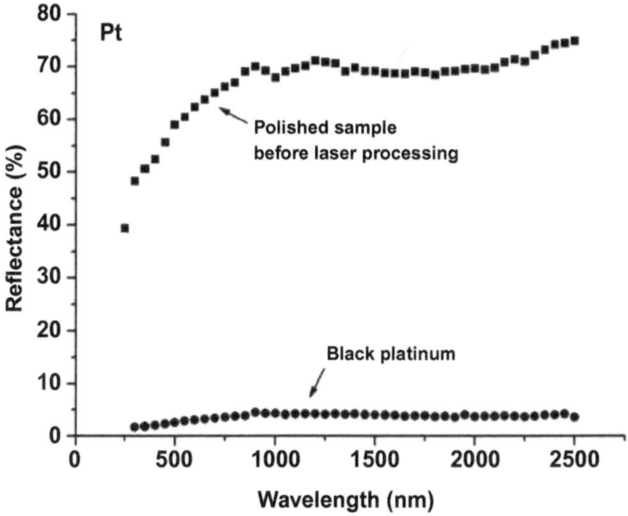

Fig. 4.55 Spectral reflectance of black platinum as a function of wavelength. For comparison, the reflectance of the polished sample before laser processing is also shown [161].

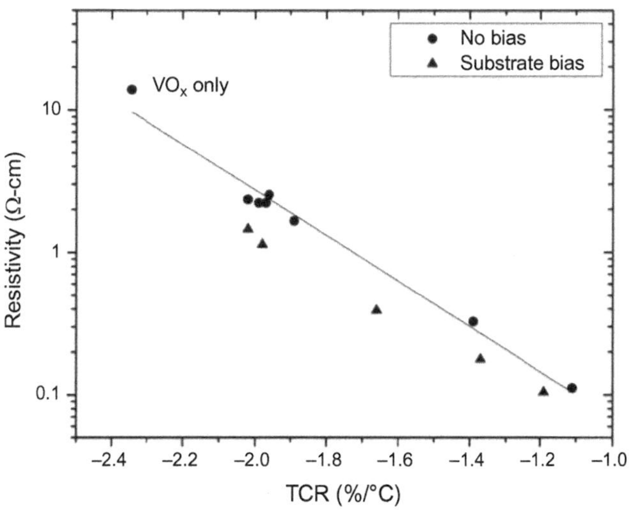

Fig. 4.56 TCR versus room temperature resistivity of VO_x-gold film [163].

Metals such as titanium [164,165] and aluminum [166] have been investigated for potential applications in the manufacture of uncooled microbolometers. A schematic of a Wheatstone bridge based measurement circuit for the determination of TCR of microbolometer elements is presented in Fig. 4.57 [165].

The process flow chart of an aluminum microbolometer is illustrated in Fig. 4.58 [166].

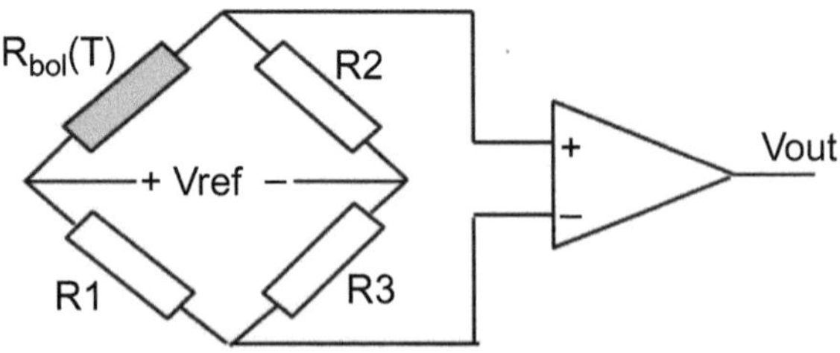

Fig. 4.57 Measurement circuit based on Wheatstone bridge [165].

The resistivity and TCR of polysilicon, Al, and Ti are summarized in Table 4.14 [166].

Amorphous germanium based materials have also drawn the attention for the fabrication of uncooled microbolometers [167–169]. An example of the process flow of Si_xGe_{1-x}:H-based microbolometers is shown in Fig. 4.59 [168].

A comparison of the device performance of amorphous germanium based microbolometers is presented in Table 4.15.

Various materials are being tested for potential applications in the fabrication of microbolometers. These include the following:

(a) LPCVD Si_3N_4 as absorber membrane [170],
(b) GeSn alloy as a possible sensing layer [171],
(c) TiO_2 [172], ZnO [173], and MEMS-based cantilever thermal imagers [174,175].

Pyroelectric detectors operate on the principle of temperature induced change in polarization in the material. This is illustrated in Fig. 4.60 [176]. The temperature dependence of the polarization is the pyroelectric coefficient—it determines the current responsivity of the detector material [176,177]. While pyroelectric detectors account for a small share of the global market of thermal detectors, there is continued interest in the manufacture of these devices for applications in security, traffic management/transportation, health care, and related areas [178].

In Table 4.16, properties of three pyroelectric detectors are summarized [179].

Properties of some pyroelectric materials are summarized in Table 4.17 [177].

An example of the design principle of a pyroelectric detector, with a tunable filter, is presented in Fig. 4.61 [180]. In this study, the authors report the design, fabrication, and test results of a tunable pyroelectric detector with an integrated micromachined Fabry-Perot filter for analyzing gases in the midwave infrared range of wavelengths [180].

In summary, in recent years, several materials have been investigated for their applications as thermosensing films in microbolometers. Some of the recent results of the TCR of these films is summarized in Table 4.18. Commercial uncooled microbolometers will continue to benefit from the reliability and high TCR of vanadium oxide and doped amorphous silicon, although amorphous silicon has the inherent advantages of its complete compatibility with silicon microelectronics manufacturing.

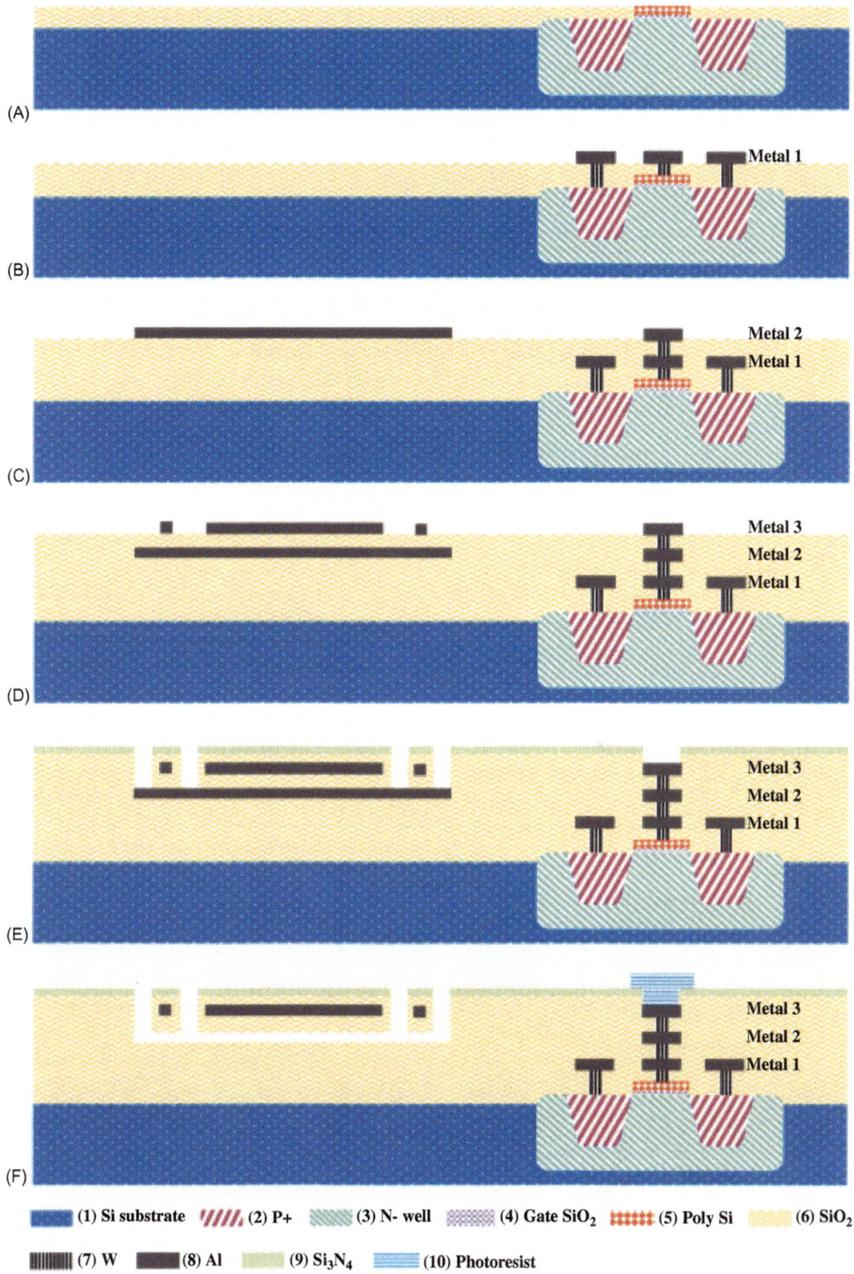

Fig. 4.58 The process flow for an Al microbolometer: (A–E) are in a standard CMOS process and (F) is in a subtractive microelectromechanical system (MEMS) process [166].

Table 4.14 Bolometer related properties of polysilicon, Al, and Ti [166].

Material	Resistivity (10^{-4} Ω cm, at 300 K)	TCR (%/K)
Undoped polysilicon	199	−0.085
n-Polysilicon	62.2	−0.016
Al	0.03	0.38
Ti	1.2	0.25

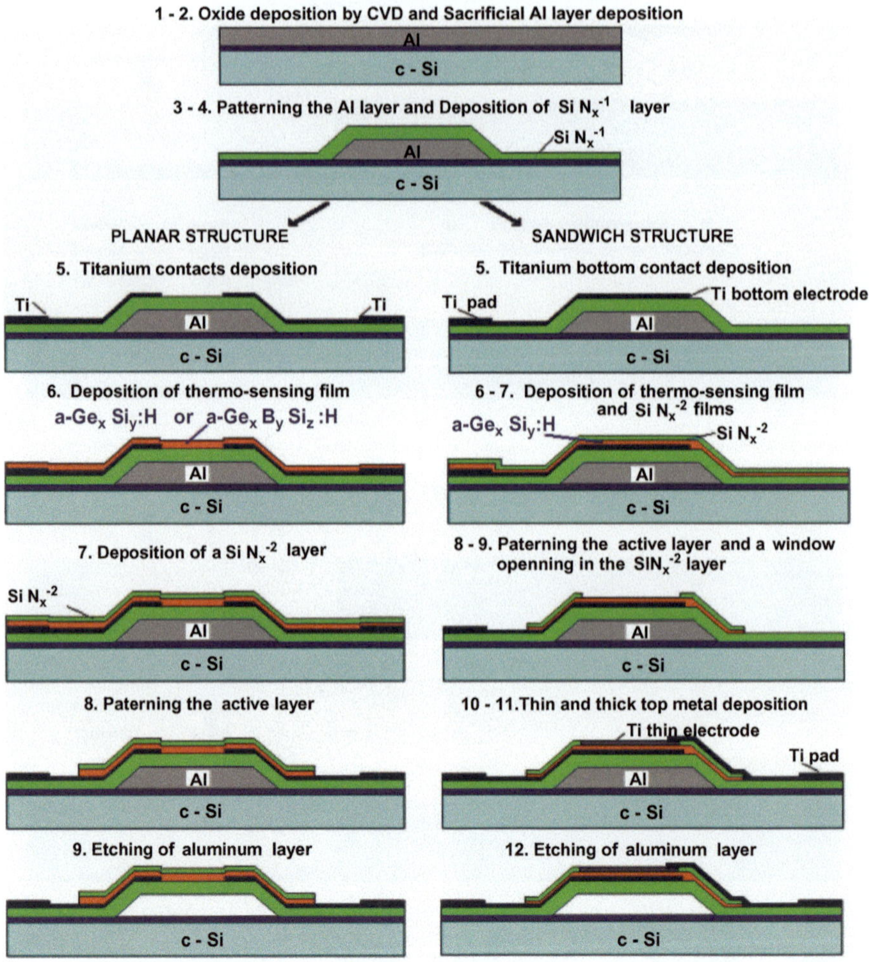

Fig. 4.59 Planar and sandwich microbolometers fabrication process flow [168].

Table 4.15 Comparison of hydrogenated polymorphous silicon-germanium alloy (pm-Si_xGe_{1-x}:H) microbolometer performance with other devices at 30 Hz (Ref. [169] and references therein).

Material	Pm-$SiGe_{1x}$:H	VO_x/ZnO	a-Si:H	VO_x	Poly-SiGe
Size (μm^2)	50×50	50×50	35×35	25×25	50×50
$\Re_v$ (V/W)	2×10^4	2×10^4	3.7×10^3	2.44×10^5	1.64×10^4
NEP ($W/Hz^{1/2}$)	2×10^{-10}	1×10^{-10}	7.7×10^{-10}	1.4×10^{-10}	6×10^{-10}
D^* (cm $Hz^{1/2}$/W)	2×10^7	4×10^7	4.5×10^6	1.55×10^7	2.26×10^9
TCR (%/K)	4.08	3.12	2.27	2.6	0.68
R (Ω)	15×10^3	–	200×10^9	200×10^3	10×10^3

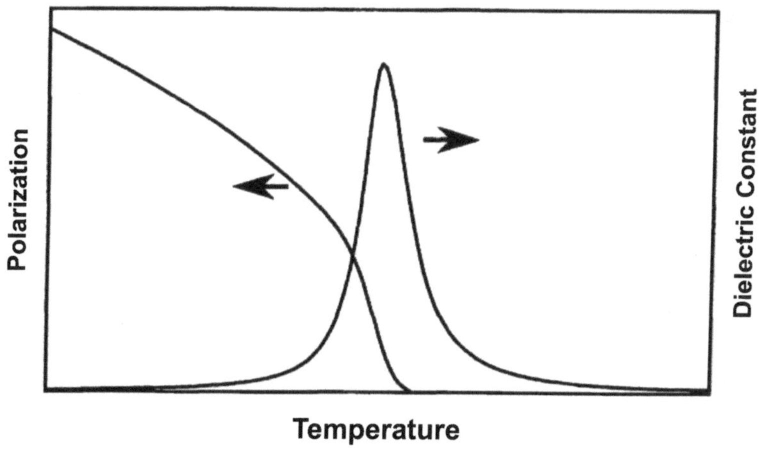

Fig. 4.60 Temperature dependence of important pyroelectric properties near the phase transition [176].

Table 4.16 Properties of three commercial pyroelectric detectors [179].

Material	DLaTGS	$LiTaO_3$	PZT
Pyroelectric coefficient p (Coulomb/cm^2 K)	2.5×10^{-8}	1.7×10^{-8}	4.4×10^{-8}
Dielectric constant ε	18	51	180
Specific heat capacity C (J/cm^3 K)	2.7	3.2	2.6
Resistivity ρ (Ω cm)	2.4×10^{10}	4.2×10^{10}	4.2×10^{10}
Figure of merit ($p\rho/\varepsilon C$)	12.3	4.3	0.4

Source: GIRMET Company.

Table 4.17 Properties of some pyroelectric materials [177].

Material	Volume specific heat C (J/cm^3 K)	Pyroelectric coefficient p (C/cm^2 K)	Dielectric constant ε
TGS	1.70	3.5×10^{-8}	35
$LiTaO_3$	3.19	1.9×10^{-8}	58
$LiNbO_3$	2.32	4.0×10^{-9}	75
$Li_2SO_4 \times H_2O$	0.82	1.0×10^{-8}	10
$BaTiO_3$	3.01	2.0×10^{-8}	4100
$NaNO_2$	2.016	1.2×10^{-8}	8
PVF_2	2.40	3.0×10^{-9}	11
SBN	2.34	6.0×10^{-8}	400
SbSi	2.378	2.6×10^{-7}	10^4

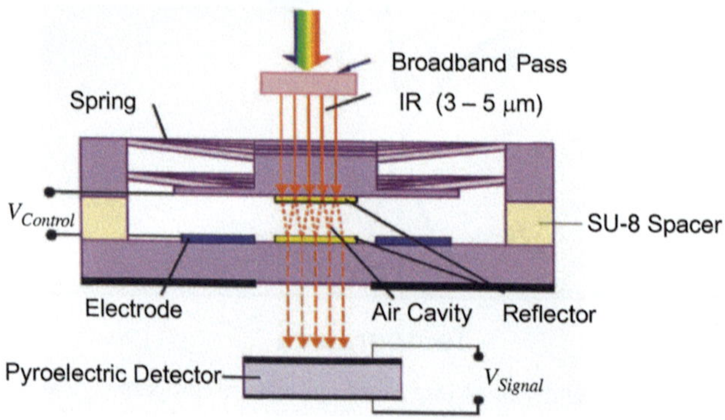

Fig. 4.61 Design principle of a pyroelectric detector with tunable filter [180].

Table 4.18 TCR of important thermosensing materials.

Material	TCR (%/K)	Ref.
a-Si-doped	2–5	[51,181–185]
CNTs-carbon nanotubes	2.1–3.1	[186–188]
Poly Si-Ge	0.68–2	[189]
Pt	0.14	[190]
Ti	0.35	[191,192]
VO_x	~2	[133,193–198]
Y-Ba-Cu-O	3.3	[199]
ZnO	10.4	[200]

References

[1] K.A. Zimmermann, Temperature: Facts, History & Definition, 2013. https://www.livescience.com/39841-temperature.html#:~:text=History%20of%20the%20thermometer,side%20of%20this%20neutral%20temperature. (Accessed 9 March 2021).

[2] Thermoscope, From Wikipedia, the Free Encyclopedia, https://en.wikipedia.org/wiki/Thermoscope, Accessed 9 March 2021.

[3] E. Grodzinsky, M.S. Levander, History of the thermometer, in: Understanding Fever and Body Temperature, 2019, pp. 23–35, https://doi.org/10.1007/978-3-030-21886-7_3.

[4] Christiaan Huygens, in Famous Scientists The Art of Genius, https://www.famousscientists.org/christiaan-huygens/#:~:text=In%201656%20he%20patented%20his,position%20of%20ships%20at%20sea, Accessed 9 March 2021.

[5] Eighteenth-Century Development of Temperature Scales, https://www.encyclopedia.com/science/encyclopedias-almanacs-transcripts-and-maps/eighteenth-century-development-temperature-scales, Accessed 9 March 2021.

[6] A. Rubin, History of Thermometers, 2018. https://jamaicahospital.org/newsletter/history-of-thermometers/#:~:text=The%20more%20modern%20thermometer%20wasthermometer%20using%20the%20same%20scale. (Accessed 9 March 2021).

[7] Andres Celsius and the History of the Celsius Scale The Life of the Swedish Astronomer Who Invented the Centigrade Scale, https://www.thoughtco.com/history-of-the-thermometer-p3-1991492, Accessed 9 March 2021.

[8] J.M.S. Pearce, A brief history of the clinical thermometer, historical note, Q.J. Med. 95 (2002) 251–252 (references therein).

[9] History of the Thermometer Lord Kelvin Invented the Kelvin Scale in 1848, https://www.thoughtco.com/history-of-the-thermometer-p2-1992034#:~:text=Lord%20Kelvin%20invented%20the%20Kelvin%20Scale%20in%201848%20used%20on,the%20dynamical%20theory%20of%20heat, Accessed 9 March 2021.

[10] Thomas Johann Seebeck German Physicist, https://www.britannica.com/biography/Thomas-Johann-Seebeck, Accessed 9 March 2021.

[11] Thermoelectrics Generators Market, https://www.marketsandmarkets.com/Market-Reports/thermoelectric-generators-market-91553904.html?gclid=CjwKCAiAhbeCBhBcEiwAkv2cY2cyv0vFZJaUvaHI94MKMOUa60XqO_lr3b3Pl1HnMYLykqjCmgTNJhoCnn0QAvD_BwE, Accessed 14 March 2021.

[12] C. Corsi, Infrared: a key technology for security systems, Adv. Opt. Technol. 838752 (2012) 15.

[13] A. Rogalski, Recent progress in infrared detector technologies, Infrared Physics & Technology 54 (3) (2011) 136–154.

[14] J. Beletic, R. Blank, D. Gulbransen, D. Lee, M. Loose et.al., Teledyne imaging sensors: Infrared imaging technologies for astronomy and civil space, in: Proc. SPIE 7021, High Energy, Optical, and Infrared Detectors for Astronomy III, 70210H, 22 July (2008), https://doi.org/10.1117/12.790382.

[15] W.L. Wolfe, G.J. Zissis (Eds.), The Infrared Handbook, Environmental Research Institute of Michigan, 1978.

[16] M. Henini, M. Razeghi, Handbook of Infrared Detection Technologies, Elsevier Science, 2002.

[17] A. Rogalski, Infrared Detectors, Second ed., CRC Press, Taylor & Francis Group, 2020.

[18] E.L. Dereniak, Infrared detectors & arrays, in: Proceedings of SPIE, vol. 0930, 1988.

[19] J.D. Vincent, S. Hodges, J. Vampola, M. Stegall, G. Pierce, Fundamentals of Infrared and Visible Detector Operation and Testing, second ed., Wiley Series in Pure and Applied Optics, 2015.
[20] M.A. Kinch, Fundamentals of Infrared Detector Materials, SPIE Tutorial Text Volume TT76, SPIE Press Book, 2007.
[21] E.L. Dereniak, G.D. Boreman, Infrared Detectors & Systems, first ed., Wiley Interscience, 1996.
[22] R.J. Keyes, Optical and Infrared Detectors, Series: Topics in Applied Physics, second ed., vol. 19, Springer, 1980.
[23] P.W. Kruse, L.D. McGlauchlin, R.B. McQuistan, Elements of Infrared Technology Generation, Transmission and Detection, Wiley, 1962.
[24] W. Smith, Effect of light on selenium during the passage of an electric current, Nature 7 (1873) 303.
[25] History of Infrared Detectors, Xenics Infrared Solutions, https://www.xenics.com/history-of-infrared-detectors/#:~:text=Lead%20sulfide%20(PbS)%20was%20the,were%20developed%20for%20IR%20sensing, Accessed 15 March 2021.
[26] W.D. Lawson, S. Nielson, E.H. Putley, A.S. Young, Preparation and properties of HgTe and mixed crystals of HgTe − CdTe, J. Phys. Chem. Solids 9 (1959) 325–329.
[27] A. Rogalski, History of infrared detectors, Opto-Electron. Rev. 20 (3) (2012) 279–308.
[28] W.-C. Wang, Optical Detectors, 2021. https://depts.washington.edu/mictech/optics/me557/detector.pdf. (Accessed 17 March 2021).
[29] E. Mounier, Technical and market trends for microbolometers for thermography and night vision, in: Proc. SPIE 8012, Infrared Technology and Applications XXXVII, 80121U (20 May 2011), 2011, pp. 80121U-1–80121U-6, https://doi.org/10.1117/12.877050.
[30] A. Rogalski, Graphene-based materials in the infrared and terahertz detector families: a tutorial, Adv. Opt. Photon. 11 (2) (2019) 316–372.
[31] A. Rogalski, P. Martyniuk, InAs/GaInSb superlattices as a promising material system for third generation infrared detectors, Infrared Phys. Technol. 48 (2006) 39–52.
[32] D. Morozov, M. Tarkhov, P. Mauskopf, N. Kaurova, O. Minaeva, V. Seleznev, B. Vornov, G. Gol'tsman, Single photon counting detector for THz radioastronomy, in: 17th International Symposium on Space Terahertz Technology, WE2-3, 10-12 May, 2006, Paris, France, 2006.
[33] A. Rogalski, M. Kopytko, P. Martyniuk, Antimonide-Based Infrared Detectors: A New Perspective, SPIE—The International Society for Optical Engineering, 2018.
[34] F. Zhuge, Z. Zheng, P. Luo, L. Lv, Y. Huang, H. Li, T. Zhai, Nanostructured materials and architectures for advanced infrared photodetection, Adv. Mater. Technol. 1700005 (2017) 26.
[35] S.R. Marthi, et al., Optical properties of black silicon: an analysis, J. Miner. Met. Miner. Soc. 67 (2015) 2154–2159.
[36] J. Goldsmith, et al., Black aluminum: a novel anti-reflective absorbing coating, in: NAECON—IEEE National Aerospace and Electronics Conference, 2014, pp. 69–70.
[37] J.-L. Lai, C.-J. Liao, G.-D.J. Su, Using an SU-8 photoresist structure and cytochrome c thin film sensing material for a microbolometer, Sensors 12 (2012) 16390–16403, https://doi.org/10.3390/s121216390.
[38] L. Kevin, The active microbolometer: a new concept in infrared detection, in: Proc. SPIE 5274, Microelectronics: Design, Technology, and Packaging, Event: Microelectronics, MEMS, and Nanotechnology, 2003, Perth, Australia, 2004, https://doi.org/10.1117/12.530832.

[39] W.E. Spear, P.G. Le Comber, Substitutional doping of amorphous silicon, Solid State Commun. 17 (1975) 1193–1196.
[40] P. Wang, et al., High sensitivity 17 μm pixel pitch 640 × 512 uncooled infrared focal plane arrays based on amorphous vanadium oxide thin films, IEEE Electron Device Lett. 36 (2015) 923–925.
[41] C. Chang-Hong, et al., Infrared responsivity of uncooled VO_2 based thin films bolometer, Acta Phys. Polon. 50 (2001) 450–452.
[42] J.-J. Yon, E. Mottin, J.-L. Tissot, Latest amorphous silicon microbolometer developments at leti-lir, infrared technology and applications XXXIV, in: B.F. Andresen, G.F. Fulop, P.R. Norton (Eds.), Proc. of SPIE Vol. 6940, 69401W, 2008, https://doi.org/10.1117/12.780538 (0277-786X/08).
[43] A.H. Pfund, The optical properties of metallic and crystalline powders, J. Opt. Soc. Am. 23 (1933) 375–378.
[44] W.F. Keenan, Uncooled Infrared Detector and Method for Forming the Same, US Patent # 5,367,167, issued on November 22, 1994.
[45] R.A. Wood, Monolithic silicon microbolometer arrays, in: P.W. Kruse, D.D. Skatrud (Eds.), Semiconductors and Semimetals, vol. 47, Academic Press, San Diego, 1997. Chapter 3.
[46] F. Niklaus, C. Vieider, H. Jakobsen, MEMS-based uncooled infrared bolometer arrays—a review, MEMS/MOEMS technologies and applications III, in: J.-C. Chiao, X. Chen, Z. Zhou, X. Li (Eds.), Proc. of SPIE, vol. 6836, 2007, p. 68360D, https://doi.org/10.1117/12.755128. 0277-786X/07.
[47] N.M. Ravindra, et al., Modeling of optical properties of black silicon/crystalline silicon, J. Sci. Ind. Metrol. 1 (2015) 1–7.
[48] X. Liu, et al., Black silicon: fabrication methods, properties and solar energy applications, Energy Environ. Sci. 10 (2014) 1–95.
[49] R.R. Neli, et al., Characterization of fast-response and low noise poly Si uncooled far infrared sensor, in: 2014 IEEE 5th Latin American Symposium on Circuits and Systems, 2014, pp. 1–4.
[50] C. Chen, et al., Micromachined uncooled IR bolometer linear array using VO_2 thin films, Int. J. Infrared Millimeter Waves 22 (2001) 53–58.
[51] T. Zhou, Y. Zuo, K. Qiu, J. Zheng, Q. Wang, Potential of Ti doped hydrogenated amorphous Si film with suitable resistivity and high TCR for microbolometer applications, Vacuum 119 (2015) 30–33.
[52] M. Moreno, N. Delgadillo, A. Torres, R. Ambrosio, P. Rosales, A. Kosarev, C. Reyes-Betanzo, J. de la Hidalga-Wade, C. Zuniga, W. Calleja, Boron doping compensation of hydrogenated amorphous and polymorphous germanium thin films for infrared detection applications, Thin Solid Films 548 (2013) 533–538.
[53] https://www.horiba.com/en_en/technology/measurement-and-control-techniques/spectroscopy/detectors/detectors/, Accessed 17 March 2021.
[54] L. Harris, et al., The preparation and optical properties of gold blacks, J. Opt. Soc. Am. 38 (1938) 582–589.
[55] W. Lang, et al., Absorbing layers for thermal infrared detectors, Sensors Actuators A Phys. 34 (1992) 243–248.
[56] F.J.J. Clarke, J.A. Larkin, Measurement of total reflectance, transmittance and emissivity over the thermal IR spectrum, Infrared Phys. 25 (1985) 359–367.
[57] F.J.J. Clarke, J.A. Larkin, Low reflectance coatings: platinum black for infrared technology, Platin. Met. Rev. 30 (1986) 21–22.

[58] A.Y. Vorobyev, C. Guo, Colorizing metals with femtosecond laser pulses, Appl. Phys. Lett. 92 (2008) 1–3.
[59] A.Y. Vorobyev, C. Guo, Femtosecond laser blackening of metals, in: 52nd IEEE International Midwest Symposium on Circuits and Systems, 2009, pp. 905–908.
[60] N.M. Ravindra, S.R. Marthi, S. Sekhri, Black Silicon Processing, Properties, and Applications, Emerging Materials Collection, Momentum Press, 2016.
[61] F. Forsberg, et al., Very large scale heterogeneous integration (VLSHI) and wafer-level vacuum packaging for infrared bolometer focal plane arrays, Infrared Phys. Technol. 60 (2013) 251–259.
[62] B.E. Cole, et al., Monolithic two-dimensional arrays of micromachined microstructures for infrared applications, Proc. IEEE 86 (1998) 1679–1686.
[63] A. Rogalski, Infrared detectors for the future, Acta Phys. Pol. A 116 (2009) 389–406.
[64] A. Rogalski, Recent progress in infrared detector technologies, Infrared Phys. Technol. 54 (2011) 136–154.
[65] B. Wang, J. Lai, H. Li, H. Hu, S. Chen, Nanostructured vanadium oxide thin film with high TCR at room temperature for microbolometer, Infrared Phys. Technol. 57 (2013) 8–13.
[66] P.-F. Wang, Q. Hu, T. Zeng, Y. Liu, X. Xu, J.-L. Sun, Optically monitored electric-field-induced phase transition in vanadium dioxide crystal film, Crystals 10 (2020) 764, https://doi.org/10.3390/cryst10090764.
[67] M. Moreno, A. Torres, R. Ambrosio, A. Kosarev, Un-cooled microbolometers with amorphous germanium-silicon (a-Ge_xSi_y:H) thermo-sensing films, in: U. Perera (Ed.), Bolometers, InTech, 2012, pp. 23–50, ISBN: 978-953-51-0235-9. http://www.intechopen.com/books/bolometers/un-cooledmicrobolometers-with-amorphous-germanium-silicon-a-gexsiy-h-thermo-sensing-films.
[68] P.W. Kruse, D.D. Skatrud, Uncooled Infrared Imaging Arrays and Systems, Semiconductors and Semimetals, Academic Press, 1997, ISBN: 978-0127521558.
[69] C. Posch, et al., A microbolometer asynchronous dynamic vision sensor for LWIR, IEEE Sensors J. 9 (2009) 654–664.
[70] S. Chen, et al., Characterization of nanostructured VO2 thin films grown by magnetron controlled sputtering deposition and post annealing method, Opt. Express 17 (2009) 24153–24161.
[71] S. Moon, Novel infrared absorbing material coupled uncooled microbolometer, in: Proceedings of IEEE, vol. 2, 2004, pp. 658–660.
[72] N.H. Azhan, K. Okimura, Role of strain and oxygen vacancies in modifying the transition temperature of VO2 films deposited via Rf biased sputtering, in: International Conference on Engineering Technology and Technopreneurship (ICE2T), 2017, pp. 1–5.
[73] S. Chen, et al., Fabrication and performance of microbolometer arrays based on nanostructured vanadium oxide thin films, Smart Mater. Struct. 16 (2007) 696–700.
[74] F. Guinnetona, et al., Comparative study between nanocrystalline powder and thin film of vanadium dioxide VO2: electrical and infrared properties, J. Phys. Chem. Solids 62 (2001) 1229–1238.
[75] R.T.R. Kumar, et al., Study of a pulsed laser deposited vanadium oxide based microbolometer array, Smart Mater. Struct. 12 (2003) 188–192.
[76] G. Li, et al., Low temperature deposited nano-structured vanadium oxide thin films for uncooled infrared detectors, in: 2nd IEEE International Nanoelectronics Conference (INEC), 2008, pp. 921–923.
[77] S.V. Grayli, et al., Room temperature deposition of highly sensitive vanadium oxide films for infrared light sensing applications, IEEE Sens. (2015) 1–4.

[78] M. Almasri, et al., Amorphous silicon two-color microbolometer for uncooled IR detection, IEEE Sensors J. 6 (2006) 239–300.
[79] X. Liu, et al., A novel uncooled a-Si microbolometer for infrared detection, in: 8th International Conference on Solid-State and Integrated Circuit Technology Proceedings, 2006, pp. 626–628.
[80] F. Simoens, et al., Active imaging with THz fully-customized uncooled amorphous-silicon microbolometer focal plane arrays, in: International Conference on Infrared, Millimeter, and Terahertz Waves, 2011, pp. 1–2.
[81] A. Heredia, et al., Low temperature electrical properties of a boron-doped amorphous silicon bolometer, Proc. Electrochem. Soc. (2003), Abs.881.
[82] J.-H. Su, et al., Temperature coefficient of resistance related to amorphous silicon/metal contact design of microbolometers, in: 3rd International Conference on Devices, Circuits and Systems, 2016, pp. 63–67.
[83] M.M. Rana, D.P. Butler, Noise reduction of a-$Si_{1-x}Ge_xO_y$ microbolometers by forming gas passivation, Thin Solid Films 516 (2008) 6499–6503.
[84] P.G. Maloney, et al., Emissivity of microstructure silicon, Appl. Opt. 49 (2010) 1065–1068.
[85] T.P. Pasanin, et al., Impact of standard cleaning on electrical and optical properties of phosphorous-doped black silicon, IEEE J. Photovoltaics 8 (2018) 697–702.
[86] R.R. Neli, et al., Development of process for far infrared sensor fabrication, Sensors Actuators A 132 (2006) 400–406.
[87] C.N. Chen, W.C. Huang, A CMOS-MEMS thermopile with low thermal conductance and a near-perfect emissivity in the 8–14-μm wavelength range, IEEE Electron Device Lett. 32 (2011) 96–98.
[88] N. Nelms, J. Dowson, Goldblack coating for thermal infrared detectors, Sensors Actuators A 120 (2005) 403–407.
[89] S. Ilias, et al., Deposition and characterization of gold black coatings for thermal infrared detectors, in: Proceedings of SPIE, 7750, 2010 (77501J-10).
[90] Y. Fang, et al., Preparation and characterization of ultra-low density porous gold foams, Mater. Lett. 67 (2012) 46–48.
[91] FLIR Commercial Vision Systems B.V, Uncooled detectors for thermal imaging cameras, in: FLIR Commercial Vision Systems B.V. FLIR Technical Notes, Breda, Netherlands. Web, 2011.
[92] http://www.flir.com/uploadedFiles/Eurasia/Cores and Components/Technical Notes/uncooled%20detectors.
[93] L. Yu, Y. Guo, H. Zhu, M. Luo, P. Han, X. Ji, Low-Cost Microbolometer Type Infrared Detectors, Micromachines. 11 (2020) 800, https://doi.org/10.3390/mi11090800.
[94] M. Abdel-Rahman, N. Al-Khalli, M.F. Zia, M. Alduraibi, B. Ilahi, E. Awad, N. Debbar, Fabrication and design of vanadium oxide microbolometer, AIP Conf. Proc. 1809 (2017), https://doi.org/10.1063/1.4975416, 020001.
[95] J. Nazdrowicz, M. Szermer, C. Maj, W. Zabierowski, A. Napieralski, A study on microbolometer electro-thermal circuit modelling, in: Proceedings of the 22nd International Conference "Mixed Design of Integrated Circuits and Systems", June 25-27, 2015, Toru, Poland, 2015.
[96] P. Ramos, A. Andrés, A. López, A. Manzanares, Study and Calculation of Thermal Conductance of Thermal Infrared Detectors Using Finite Element Method, arXiv:1712.02691v1, 2017.
[97] V.R. Mehta, S. Shet, N.M. Ravindra, A.T. Fiory, M.P. Lepselter, Silicon-integrated uncooled infrared detectors: perspectives on thin films and microstructures, J. Electron. Mater. 34 (5) (2005) 484–490.

[98] A. Bain, J.L. Martin, E. Mottin, J.-L.O. Buffet, J.-L. Tissot, R. Tronel, M. Vilain, J.-J. Yon, Amorphous silicon-based uncooled microbolometer technology for low-cost IRFPA, Sens. Mater. 12 (6) (2000) 365–373.
[99] A.J. Syllaios, T.R. Schimert, R.W. Gooch, W.L. McCardel, B.A. Ritchey, J.H. Tregilgas, Amorphous silicon microbolometer technology, Mater. Res. Soc. Conf. Proc. 609 (2000). A14.4.1-6.
[100] T.R. Schimert, N. Cunningham, G.L. Francisco, R.W. Gooch, J. Gooden, P. McCardel, B.E. Neal, B. Ritchey, J. Rife, A.J. Syllaios, J.H. Tregilgas, J.F. Brady III, J. Gilstrap, S.J. Ropson, Low-cost low-power uncooled 120x160 a-si-based microinfrared camera for law enforcement applications, in: Proc. SPIE 4232, Enabling Technologies for Law Enforcement and Security, 2001, pp. 187–194.
[101] E. Iborra, M. Clement, L.V. Herrero, J. Sangrador, IR uncooled bolometers based on amorphous Ge/sub x/Si/sub1-x/O/sub y/ on silicon micromachined structures, J. Microelectromech. Syst. 11 (2002) 322–329.
[102] X.M. Liu, H.J. Fang, L.T. Tiu, Study on new structure uncooled a-Si microbolometer for infrared detection, Microelectron. J. 38 (2007) 735–739.
[103] C. Shin, D.P. Pham, J. Park, S. Kim, Y.-J. Lee, J. Yi, Structure and electrical properties of boron doped hydrogenated mixed-phase silicon films for uncooled microbolometer, Infrared Phys. Technol. 96 (2019) 84–88.
[104] E. Mottin, A. Bain, J.-L. Martin, J.-L. Ouvrier-Buffet, S. Bisotto, J.-J. Yon, J.-L. Tissot, Uncooled amorphous silicon technology enhancement for 25-μm-pixel pitch achievement, in: Proc. SPIE 4820, Infrared Technology and Applications XXVIII, 2003, https://doi.org/10.1117/12.464848.
[105] H. Jung, T.H. Kim, G. Kim, C.M. Yang, J.H. Park, H.Y. Kim, Design and characteristics of a-Si-based micro-bolometers with shared-anchor structure in vacuum packaged systems, Microsyst. Technol. 20 (2014) 899–905, https://doi.org/10.1007/s00542-014-2073-3.
[106] M. Ahmed, D.P. Butler, Z. Celik-Butler, Low-profile, self-packaged uncooled microbolometer on a flexible substrate towards an infrared radiation sensitive skin, in: Sensors, IEEE, 2014, pp. 1093–1095, https://doi.org/10.1109/ICSENS.2014.6985465.
[107] M. Elßner, H. Vogt, Reliability of microbolometer thermal imager sensors using chip-scale packaging, Procedia Eng. 120 (2015) 1191–1196.
[108] F.J. Morin, Oxides which show a metal-to-insulator transition at the Neel temperature, Phys. Rev. Lett. 3 (1) (1959) 34–36.
[109] Z. Yang, C. Ko, S. Ramanathan, Oxide electronics utilizing ultrafast metal-insulator transitions, Annu. Rev. Mater. Res. 41 (2011) 337–367, https://doi.org/10.1146/annurev-matsci-062910-100347.
[110] M. Brahlek, L. Zhang, J. Lapano, H.-T. Zhang, R. Engel-Herbert, N. Shukla, S. Datta, H. Paik, D.G. Schlom, Opportunities in vanadium-based strongly correlated electron systems, MRS Commun. 7 (2017) 27–52, https://doi.org/10.1557/mrc.2017.2. references therein.
[111] U. SchwingenschlÖgl, V. Eyert, The vanadium magnéli phases V_nO_{2n-1}, Ann. Phys. 13 (9) (2004) 475–510, https://doi.org/10.1002/andp.200410099.
[112] J.B. Goodenough, Metallic oxides, Prog. Solid State Chem. 5 (1971) 145.
[113] W. Bruckner, H. Oppermann, W. Reichelt, J.I. Terukov, F.A. Tschudnovski, E. Wolf, Vanadiumoxide, Akademie-Verlag, Berlin, 1983.
[114] M. Imada, A. Fujimori, Y. Tokura, Metal-insulator transitions, Rev. Mod. Phys. 70 (1998) 1039.
[115] T. Reeswinkel, D. Music, J.M. Schneider, Ab initio calculations of the structure and mechanical properties of Vanadium oxides, J. Phys. Condens. Matter 21 (2009) 145404.

[116] W. Tillman, L. Hagen, D. Kokalj, M. Paulus, M. Tolan, Temperature-induced formation of lubricous oxides in vanadium containing iron-based arc sprayed coatings, Coatings 9 (2019) 18, https://doi.org/10.3390/coatings9010018.

[117] H. Katzke, P. Tolédano, W. Depmeier, Theory of morphotropic transformations in vanadium oxides, Phys. Rev. B 68 (2003) 024109-1–024109-7.

[118] G. Andersson, Studies on vanadium oxides I. Phase analysis, Acta Chem. Scand. 8 (1954) 1599–1606.

[119] G. Andersson, Studies on vanadium oxides II. The crystal structure of vanadium dioxide, Acta Chem. Scand. 10 (1956) 623–628.

[120] Y. Zhang, W. Xiong, W. Chen, Y. Zheng, Recent Progress on Vanadium Dioxide Nanostructures and Devices: Fabrication, Properties, Applications and Perspectives, Nanomaterials. 11 (2021) 338, https://doi.org/10.3390/nano11020338.

[121] S. Sucharitakul, G. Ye, W.R.L. Lambrecht, C. Bhandari, A. Gross, R. He, H. Poelman, X. P.A. Gao, V2O5: a 2D van der Waals oxide with strong in-plane electrical and optical anisotropy, ACS Appl. Mater. Interfaces 9 (2017) 23949–23956, https://doi.org/10.1021/acsami.7b05377.

[122] K. Schneider, Optical properties and electronic structure of V_2O_5, V_2O_3 and VO_2, J. Mater. Sci. Mater. Electron. 31 (2020) 10478–10488, https://doi.org/10.1007/s10854-020-03596-0.

[123] H. Yin, C. Song, Z. Wang, H. Shao, Y. Li, H. Deng, Q. Ma, K. Yu, Self-assembled vanadium oxide Nanoflakes for p-type ammonia sensors at room temperature, Nanomaterials 9 (2019) 317, https://doi.org/10.3390/nano9030317 (12 pages).

[124] E.E. Chain, Optical properties of vanadium dioxide and vanadium pentoxide thin films, Appl. Opt. 30 (19) (1991) 2782–2787.

[125] R. McGee, A. Goswami, R. Abraham, S. Bukhari, T. Thundat, Phase transformation induced modulation of the resonance frequency of VO_2/TiO_2 cantilevers, MRS Adv. (2018), https://doi.org/10.1557/adv.2018.140.

[126] I. Mjejri, A. Rougier, M. Gaudon, Low cost and facile synthesis of various vanadium oxides: V_2O_3, VO_2 and V_2O_5 and their magnetic, thermochromic and electrochromic properties, inorganic chemistry, Am. Chem. Soc. 56 (3) (2017) 1734–1741.

[127] J. Livage, Hydrothermal synthesis of nanostructured vanadium oxides, Materials 3 (2010) 4175–4195, https://doi.org/10.3390/ma3084175.

[128] Q. Cheng, S. Paradis, T. Bui, M. Almasri, Design of dual-band uncooled infrared microbolometer, IEEE Sensors J. 11 (1) (2011) 167–175.

[129] D. De Bruyker, B. Xu, Fabrication of vanadium oxide microbolometers on thin polyimide films, Transducers (2013) 62–65.

[130] M. Nishikawa, T. Nakajima, T. Manabe, T. Okutani, T. Tsuchiya, High temperature coefficients of resistance of VO_2 films grown by excimer-laser-assisted metal organic deposition process for bolometer application, Mater. Lett. 64 (2010) 1921–1924.

[131] M.R. Abdel-Rahman, Vanadium oxide thin films with high mid-wave & long wave infrared thermo-optic coefficients and high temperature coefficients of resistance, Optik 159 (2018) 79–86.

[132] X. Chen, Q. Lv, Resistance hysteresis loop characteristic analysis of VO_2 thin film for high sensitive microbolometer, Optik 126 (2015) 2718–2722.

[133] B. Wang, J. Lai, H. Li, H. Hu, S. Chen, Nanostructured vanadium oxide thin film with high TCR at room temperature for Microbolometer, Infrared Phys. Technol. 57 (2013) 8–13.

[134] S. Chen, J. Lai, J. Dai, H. Ma, H. Wang, X. Yi, Characterization of nanostructured VO_2 thin films grown by magnetron controlled sputtering deposition and post annealing method, Opt. Express 17 (26) (2009) 24153–24161.

[135] J. Schoeman, M. du Plessis, An analytic model employing an elliptical surface area to determine the gaseous thermal conductance of uncooled VO_x microbolometers, Sensors Actuators A Phys. 250 (2016) 229–236.

[136] W.-B. Song, J.J. Talghader, Design and characterization of adaptive microbolometers, J. Micromech. Microeng. 16 (5) (2006) 1073.

[137] N. Topaloglu, P.M. Nieva, M. Yavuz, J.P. Huissoon, A pixel-by-pixel thermal conductance tuning mechanism for uncooled microbolometers, in: 1st IEEE Microsystems and Nanoelectronics Research Conference, 2008, MNRC 2008, 2008, pp. 201–204.

[138] G.B. Smith, M.P. RC, G.H. Derrick, Surface structure and the optical properties of black chrome, Appl. Phys. A Mater. Sci. Process. 36 (1985) 193–204.

[139] S. Burdyukh, O. Berezina, A. Pergament, Modification of the properties of vanadium oxide thin films by plasma-immersion ion implantation, Adv. Condens. Matter Phys. 2018 (2018) 14, https://doi.org/10.1155/2018/9789370. Article ID 9789370.

[140] M.-H. Chiu, J.-H. Li, T. Nagao, Optical properties of Au-based and Pt-based alloys for infrared device applications: a combined first principle and electromagnetic simulation study, Micromachines 10 (2019) 73, https://doi.org/10.3390/mi10010073.

[141] H. Reddy, et al., Temperature-dependent optical properties of gold thin films, Opt. Mater. Express 6 (9) (2016) 2776–2802.

[142] F. Guinneton, et al., Role of surface defects and microstructure in infrared optical properties of thermochromic VO_2 materials, J. Phys. Chem. Solids 66 (1) (2005) 63–73, https://doi.org/10.1016/j.jpcs.2004.08.032.

[143] J. Takadoum, Black coatings: a review, Eur. Phys. J.: Appl. Phy., EDP Sci. 52 (3) (2010), https://doi.org/10.1051/epjap/2010155.hal-00646463.

[144] R.W. Farrell Jr., Surface Treatments—Blackening of Ferrous Metals, 2000. *www.hubbardhall.com*. (Accessed 11 June 2020).

[145] Z. Su, J. Yin, X. Zhao, Soft and broadband infrared metamaterial absorber based on gold nanorod/liquid crystal hybrid with tunable total absorption, Sci. Rep. 5 (2015) 16698, https://doi.org/10.1038/srep16698.

[146] J.W. Cleary, et al., Gold-Black as an IR Absorber, MRS Online Proceeding Library Archive, 2009, https://doi.org/10.1557/PROC-1208-O21-04.

[147] D. Panjwani, Infrared Detector With Metal-Black Coating Having Dielectric Overlayer and Related Methods, US Patent 9,518,868 B2, 2016.

[148] S.E. Stanca, et al., Optical assets of *in situ* electro-assembled platinum black nanolayers, Sci. Rep. 7 (2017) 14955, https://doi.org/10.1038/s41598-017-14630.

[149] S. Lee, H. Heo, S. Kim, Graphene perfect absorber of ultra-wide bandwidth based on wavelength-insensitive phase matching in prism coupling, Sci. Rep. 9 (2019) 11967, https://doi.org/10.1038/s41598-019-48501-w.

[150] B. Liu, et al., Black arsenic–phosphorus: layered anisotropic infrared semiconductors with highly tunable compositions and properties, Adv. Mater. 27 (2015) 4423–4429.

[151] A.Y. Vorobyev, C. Guo, Femtosecond Laser Blackening of Metals, IEEE, 2009, pp. 905–908.

[152] W. Lang, K. Kuhl, H. Sandmaier, Absorbing layers for thermal infrared detectors, IEEE, 1991, pp. 635–638.

[153] E.L. Dereniak, G.D. Boreman, Infrared Detectors and Systems, John Wiley & Sons, 1996, pp. 395–430.

[154] W.-T. Chang, Y. Liang, Geometric design of microbolometers made from CMOS polycrystalline silicon, IEEE Sensors J. 15 (1) (2015) 264–268.

[155] E. Socher, Y. Sinai, Y. Nemirovsky, A low-cost CMOS compatible serpentine-structured polysilicon-based microbolometer array, in: Proceedings Transducers '03 The 12th

International Conference on Solid State Sensors, Actuators and Microsystems, Boston, June 8-12, IEEE, 2003, pp. 320–323.

[156] K. Gorgulu, A. Gok, M. Yilmaz, K. Topalli, N. Biuikli, A.K. Okyay, All-silicon ultra-broadband infrared light absorbers, Sci. Rep. 6 (2016) 38589, https://doi.org/10.1038/srep38589.

[157] B. Wang, J. Lai, E. Zhao, H. Hu, Q. Liu, S. Chen, Vanadium oxide microbolometer with gold black absorbing layer, Opt. Eng. 51 (7) (2012) 074003-1–074003-6.

[158] vision-systems, Image Sensors: Black Silicon Sensors Target Low-Light Level, IR imaging, Vision Systems Design, 2013. https://www.vision-systems.com/cameras-accessories/article/16736090/image-sensors-black-silicon-sensors-target-lowlight-level-ir-imaging. (Accessed 7 April 2021).

[159] Experience the World's Most Advanced Color Night Vision, https://www.sionyx.com/?gclid=Cj0KCQjwsLWDBhCmARIsAPSL3_022UfCkQ2M1VcXLJ8MKDSVbHChJrCCPPOHfuCTFUHGKMtdoLBNHxcaAsflEALw_wcB, Accessed 7 April 2021.

[160] F. Atteia et al., Morphologies and optical properties of black silicon by room temperature reactive ion etching, Materials Research Bulletin 131(10):110973, June (2020).

[161] A.Y. Vorobyev, C. Guo, Femtosecond laser blackening of platinum, J. Appl. Phys. 104 (2008) 053516-1–053516-5.

[162] C. Guo, A.Y. Vorobyev, Black metals produced by femtosecond laser pulses, AIP Conf. Proc. 1278 (2010) 838–842, https://doi.org/10.1063/1.3507179.

[163] E.M. Smith, J.C. Ginn, A.P. Warren, C.J. Long, D. Panjwani, R.E. Peale, D.J. Shelton, Linear bolometer arrays using a high TCR VO_x-Au film, in: Proceedings Volume 9070, Infrared Technology and Applications XL, 90701Z, 2014, https://doi.org/10.1117/12.2050434.

[164] A. Gaitas, W. Zhu, N. Gulari, E. Covington, C. Kurdak, Characterization of room temperature metal microbolometers near the metal-insulator transition regime for scanning thermal microscopy, Appl. Phys. Lett. 95 (2009) 153108-1–153108-3, https://doi.org/10.1063/1.3250434.

[165] P. Zajac, C. Maj, M. Szermer, W. Zabierowski, A. Napieralski, M. Lobour, Electrothermal FEM simulation of uncooled titanium-based microbolometer, in: MEMSTECH 2012, Lviv, Ukraine, 2012.

[166] L. Yu, Y. Guo, H. Zhu, M. Luo, P. Han, X. Ji, Low-cost microbolometer type infrared detectors, Micromachines 11 (2020) 800, https://doi.org/10.3390/mi11090800.

[167] E. Iborra, M. Clement, L.V. Herrero, J. Sangrador, IR uncooled bolometers based on amorphous $Ge_xSi_{1-x}O_y$ on silicon micromachined structures, IEEE J. Microelectromech. Syst. 11 (4) (2002) 322–329.

[168] M. Moreno, A. Torres, R. Ambrosio, A. Kosarev, in: U. Perera (Ed.), Un-Cooled Microbolometers With Amorphous Germanium-Silicon (a-GexSiy:H) Thermo-Sensing Films, Bolometers, InTech Open, 2012, ISBN: 978-953-51-0235-9.

[169] R. Jimenez, M. Moreno, A. Torres, A. Morales, A. Ponce, D. Ferrusca, J. Rangel-Magdaleno, J. Castro-Ramos, J. Hernandez-Perez, E. Cano, Fabrication of microbolometer arrays based on polymorphous silicon–Germanium, Sensors 20 (9) (2020) 2716, https://doi.org/10.3390/s20092716.

[170] F. Jutzi, D.H.B. Wicaksono, G. Pandraud, N. de Rooij, P.J. French, Far infrared sensor with LPCVD-deposited low-stress Si-rich nitride absorber membrane. Part 2: thermal property and sensitivity, Sensors Actuators A Phys. 152 (2009) 126–138.

[171] M. Abdel-Rahman, M. Alduraibi, M. Hezam, B. Ilahi, Sputter deposited GeSn alloy: a candidate material for temperature sensing layers in uncooled microbolometers, Infrared Phys. Technol. 97 (2019) 376–380.

[172] Y.J. Jeong, S.G. Kang, M.H. Kwon, H. Jung, Development of titanium oxide based 12μm pixel pitch uncooled infrared detector, in: Proc. SPIE 10624, Infrared Technology and Applications XLIV, 106241F, 2018, https://doi.org/10.1117/12.2309764.

[173] Y.E. Kesim, E. Battal, M.Y. Tanrikulu, A.K. Okyay, An all-ZnO microbolometer for infrared imaging, Infrared Phys. Technol. 67 (2014) 245–249.

[174] F. Dong, Q. Zhang, D. Chen, L. Pan, Z. Guo, W. Wang, Z. Duan, X. Wu, An uncooled optically readable infrared imaging detector, Sensors Actuators A Phys. 133 (1) (2007) 236–242, https://doi.org/10.1016/j.sna.2006.04.031.

[175] C.D.W. Jones, C.A. Bolle, R. Ryf, M.E. Simon, F. Pardo, V.A. Aksyuk, W.Y.-C. Lai, J.E. Bower, J.F. Miner, F.P. Klemens, R.A. Cirelli, T.W. Sorsch, E.J. Ferry, L.A. Fetter, C.-S. Pai, J.A. Taylor, B. Vyas, G.P. Watson, B. Stekas, M.R. Baker, A.R. Papazian, N.R. Basavanhally, W.M. Mansfield, A. Kornblit, R.C. Keller, J.V. Gates, A.P. Ramirez, MEMS thermal imager with optical readout, Sensors Actuators A Phys. 155 (2009) 45–57.

[176] J.F. Belcher, C.M. Hanson, H.R. Beratan, K.R. Udayakumar, K.L. Soch, Uncooled monolithic ferroelectric IRFPA technology, in: Proc. SPIE 3436, Infrared Technology and Applications XXIV, 1998, pp. 611–622, https://doi.org/10.1117/12.328061.

[177] G. Hyseni, N. Caka, K. Hyseni, Infrared thermal detectors parameters: semiconductor bolometers versus pyroelectrics, WSEAS Trans. Circuits Syst. 9 (4) (2010) 238–247.

[178] A.J. Holden, Applications of pyroelectric materials in array-based detectors, IEEE Trans. Ultrason. Ferroelectr. Freq. Control 58 (9) (2011) 1981–1987.

[179] Pyroelectric Detectors: Materials, Applications and Working Principle, Photonics News No. 71, https://www.lasercomponents.com/us/news/pyroelectric-detectors-materials-applications-and-working-principle/, Accessed 8 April 2021.

[180] N. Neumann, M. Ebermann, S. Kurth, K. Hiller, Tunable infrared detector with integrated micromachined Fabry-Perot filter, J. Micro/Nanolith. MEMS MOEMS 7 (2) (2008) 021004-1–021004-9.

[181] A.H.Z. Ahmed, R.N. Tait, T.B. Oogarah, H.C. Liu, M.W. Denhoff, G. Sproule, M. Graham, A surface micromachined amorphous $Ge_xSi_{1-x}O_y$ bolometer for thermal imaging applications, in: Proceedings of the Photonic Applications in Astronomy, Biomedicine, Imaging, Materials Processing, and Education, Ottawa, ON, Canada, 26–29 September 2004, vol. 5578, 2004, pp. 298–309.

[182] T. Schimert, J. Brady, T. Fagan, M. Taylor, W. McCardel, R. Gooch, S. Ajmera, C. Hanson, A.J. Syllaios, Amorphous silicon based large format uncooled FPA microbolometer technology, Infrared Technol. Appl. 6940 (2008) 694023.

[183] S. Ajmera, J. Brady, C. Hanson, T. Schimert, A.J. Syllaios, M. Taylor, Performance improvement in amorphous silicon based uncooled microbolometers through pixel design and materials development, Infrared Technol. Appl. 8012 (2011) 80121.

[184] J.L. Tissot, P. Robert, A. Crastes, S. Tinnes, E. Bercier, A. Durand, Status of uncooled infrared detector technology at ULIS, France, Def. Sci. J. 63 (2013) 545–549.

[185] J.-H. Su, W.-K. Yeh, W.-T. Chang, B.-L. Wu, J.-M. Lu, J.-H. Wang, Y.-K. Fang, P.-Y. Chen, Temperature coefficient of resistance related to amorphous silicon/metal contact design of microbolometers, in: Proceedings of the 3rd International Conference on Devices, Circuits and Systems (ICDCS), Coimbatore, India, 3–5 March 2016, 2016, pp. 63–67.

[186] K. Narita, R. Kuribayashi, E. Altintas, H. Someya, K. Tsuda, K. Ohashi, T. Tabuchi, S. Okubo, M. Imazato, S. Komatsubara, A plastic-based bolometer array sensor using carbon nanotubes for low-cost infrared imaging devices, Sens. Actuators A Phys. 195 (2013) 142–147.

[187] M. Liger, Y.-C. Tai, A 32*32 Parylene-pyrolyzed carbon bolometer imager, in: Proceedings of the 19th IEEE International Conference on Micro Electro Mechanical Systems, Istanbul, Turkey, 22–26 January 2006, 2006, pp. 106–109.

[188] A. Håkansson, M. Shahi, J.W. Brill, S. Fabiano, X. Crispin, Conducting-polymer bolometers for low-cost IR-detection systems, Adv. Electron. Mater. 5 (2019) 1–7.

[189] S. Sedky, P. Fiorini, K. Baert, L. Hermans, R. Mertens, Characterization and optimization of infrared polySiGe bolometers, IEEE Trans. Electron Devices 46 (1999) 675–682.

[190] S. Yoneoka, M. Liger, G. Yama, R. Schuster, F. Purkl, J. Provine, F.B. Prinz, R.T. Howe, T.W. Kenny, ALD-metal uncooled bolometer, in: Proceedings of the IEEE 24th International Conference on Micro Electro Mechanical Systems, Cancun, Mexico, 23–27 January 2011, 2011, pp. 676–679.

[191] R. Saxena, R. Bhan, P.S. Rana, A. Vishwakarma, A. Aggarwal, K. Khurana, S. Gupta, Study of performance degradation in Titanium microbolometer IR detectors due to elevated heating, Infrared Phys. Technol. 54 (2011) 343–352.

[192] A. Tanaka, S. Matsumoto, N. Tsukamoto, S. Itoh, K. Chiba, T. Endoh, A. Nakazato, K. Okuyama, Y. Kumazawa, M. Hijikawa, et al., Influence of bias heating on a titanium bolometer infrared sensor, AeroSense 3061 (1997) 198–209.

[193] T.D. Pope, H. Jerominek, C. Alain, F. Cayer, B. Tremblay, C. Grenier, P.A. Topart, S. LeClair, F. Picard, C. Larouche, et al., Commercial and custom 160x120, 256x1, and 512x3 pixel bolometric FPAs, AeroSense 4721 (2002) 64–75.

[194] C. Li, G.D. Skidmore, C.J. Han, Uncooled VOx infrared sensor development and application, in: SPIE Defense, Security, and Sensing, vol. 8012, 2011, p. 80121.

[195] O. Riou, P.-O. Logerais, J.F. Durastanti, Quantitative study of the temperature dependence of normal LWIR apparent emissivity, Infrared Phys. Technol. 60 (2013) 244–250.

[196] M.-S. Han, D.H. Kim, H.-J. Ko, J.C. Shin, A fabrication and characterictics of microbolometer detectors using VOx/ZnO/VOx multilayer thin film processing, in: SPIE Defense+Security, vol. 9070, 2014, p. 90701.

[197] S.V. Grayli, I. El-Chami, B. Bahreyni, G. Leach, Room temperature deposition of highly sensitive vanadium oxide films for infrared light sensing applications, in: Proceedings of the IEEE Sensors, Busan, South Korea, 1–4 November 2015, 2005, pp. 1–4.

[198] D. Gu, X. Zhou, R. Guo, Z. Wang, Y. Jiang, The microstructures and electrical properties of Y-doped amorphous vanadium oxide thin films, Infrared Phys. Technol. 81 (2017) 64–68.

[199] F.M. Tong, N.M. Ravindra, L. Ganapathi, S. Giles, R. Rao, High T_c $YBa_2Cu_3O_{7-\delta}$ transition-edge microbolometers, Infrared Phys. Technol. (1995) 1053–1058.

[200] M.Y. Tanrikulu, A.K. Okyay, C. Yildizak, T. Akin, O. Akar, A. Sarac, Single layer microbolometer detector pixel using ZnO material, Infrared Technol. Appl. 10624 (2018) 1062417.

Other materials

5

5.1 Introduction

From a global perspective, silicon microelectronics continues to be the basis for materials, processes, devices, and circuit integration in VLSI technology. The abundance, life expectancy, maturity, reliability, scaling, stability, and the cost of manufacturing of silicon-based devices, circuits, and systems are its rich attributes. Technologies for silicon process integration of infrared (IR) detector systems are driven by the pursuits of low cost and high performance that are required for commercial viability [1]. The high device yields that are commonplace in standard silicon integrated circuit (IC) manufacture continue to be very attractive when applied to fabricating optoelectronic components with comparably superior uniformity and yield. Infrared photodetector technologies, based on quantum detection in semiconductors or bolometric detection {thermoresistors, including transition-edge superconductors [2]}, have utilized cryogenic cooling by either Peltier effect or Joule-Thompson refrigeration for optimum low signal sensitivity at wavelengths beyond the near IR ($\lambda > 1.5$ μm). For example, Schottky-barrier focal plane array detectors [3, 4] have been developed for multi wavelength imaging pyrometry systems for military [5] and civilian applications [6]. Uncooled technologies, with reduced photon sensitivity, are also of value in systems requiring small physical dimensions, lightweight, reduced complexity, and low component costs. In telecommunications, IR detectors operating at optical fiber band passes at 1.3 μm and 1.55 μm are needed for photoreceivers and transreceivers [1, 7]. For massively parallel duplex data links, optical interconnects require small area and low power IR detector arrays operating at terabit/second system rates [8].

Uncooled focal plane arrays are used to produce low-cost and lightweight IR cameras for applications in astronomy, autonomous vehicles, drones, firefighting, forward-looking IR radar, healthcare including body temperature scanning and detection of tumors, inspection of roads, bridges and structures, law enforcement, machine vision, metrology, motion sensing, night vision, process monitoring and control, rescue operations, security, thermal imaging, transportation, unmanned airborne vehicles, and numerous other utilities. Infrared detector technologies employ several principles, including photovoltaics, photoconductors, pyroelectrics, and micromechanics [1]. For high-speed mid-IR devices (e.g., 8–12 μm, $>10^{10}$ Hz), inter-sub-band transitions in quantum well IR photodetectors (QWIPs) [9] are operated at cryogenic temperatures to suppress dark current or uncooled with heterodyne or photovoltaic detection [10]. At longer wavelengths, IR camera technologies are being adapted to extend their sensitivities beyond the IR into the terahertz region for see-through object detection, security, and defense applications [11, 12].

The ability to tailor the electrical, electronic, mechanical, and optical properties of materials has been made possible through bandgap engineering. Bandgap engineering has been an important enabler for the development of infrared imagers [13–20].

Microbolometers. https://doi.org/10.1016/B978-0-08-102812-4.00010-3

Examples include the use of group IV elements such as silicon and germanium [14], InAs/GaSb type-II superlattices [17], Cd_xPb_{1-x} S [15], and II–VI narrow bandgap semiconductors [16, 20]. Being noninvasive, infrared camera is being used in a variety of applications in medicine [21–24]. Since fever is a signature symptom of the Covid-19 virus, infrared cameras are being routinely used to measure body temperature to screen employees at workplaces [25].

Since the demonstration of the first infrared line scanner by the US military and Texas Instruments in 1947, the thermal imaging camera technology has evolved over four generations. The direct detection—photon counting—technology is based on the conversion of incident photons directly into electrons. Historically, these cameras utilized detectors made from materials such as HgCdTe, InSb, PbSe, and PtSi. Further, they needed cooling to cryogenic temperatures. Thermal imagers that operate in the short wavelength infrared range do not need cooling. Several papers and presentations have focused on the evolution of thermal imaging detectors [26–30]. The interest in the ability to detect objects, remotely, has grown exponentially over the years. Military intelligence, disaster rescue, and urban traffic management are just a few of the examples of applications of remote sensing target detection technology [31].

In general, the choice of the material for the infrared detector is a function of the range of wavelength of interest in the infrared. These range of wavelengths are broadly classified as summarized in Table 5.1.

In principle, for semiconductor-based infrared detectors, the energy gap of the semiconductor material or the built in potential of a semiconductor device, such as a p-n junction or Schottky barrier (i.e., a metal-semiconductor contact), is the primary determining factor for the choice of the semiconductor material or the semiconductor device, respectively.

A schematic of the historical development of infrared detectors and systems is presented in Fig. 5.1. From a materials perspective, it may be noted that the narrow band semiconductors, lead chalcogenides, PbS and PbSe, played a critical role in the development of the infrared detectors in the 1940s. It may also be noted that unlike most semiconductors, the lead chalcogenides, PbS, PbSe, and PbTe, exhibit an increased bandgap with increasing temperature $\{$i.e., $a + (dE_g/dT) \approx 4 \times 10^{-4}$ eV/K$\}$ [32].

Table 5.1 Classification of IR range of wavelengths.

IR spectrum	Range of wavelength (in μm)
Near infrared (NIR)	0.75–1.4
Short wave infrared (SWIR)	1.4–3.0
Mid wave infrared (MWIR)	3.0–5.0
Long wave infrared (LWIR)	8.0–12.0
Very long wave infrared (VLWIR)	12.0–25.0
Far wave infrared (FWIR)	25.0–1000

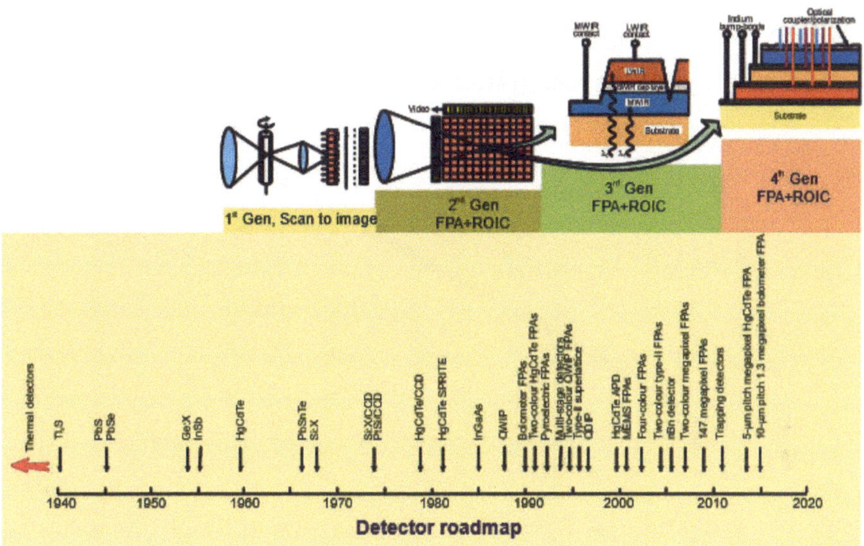

Fig. 5.1 History of the development of infrared detectors and systems. Four generation systems can be considered for principal military and civilian applications [27].

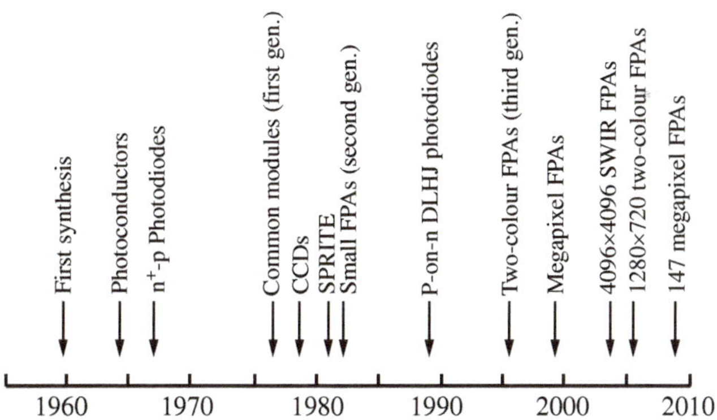

Fig. 5.2 History of the development of HgCdTe detectors [34].

5.2 HgCdTe

The discovery of variable band gap $Hg_xCd_{1-x}Te$ alloy in 1959 by Lawson and his coworkers [33] has provided an unprecedented degree of freedom to design cameras in a variety of range of wavelengths in the infrared. The bandgap energy tunability results in IR detector applications that span the short wavelength IR (SWIR: 1–3 μm), mid wavelength (MWIR: 3–5 μm), long wavelength (LWIR: 8–14 μm), and very long wavelength (VLWIR: 14–30 μm) ranges. HgCdTe technology development was and continues to be primarily for military applications. The history of the

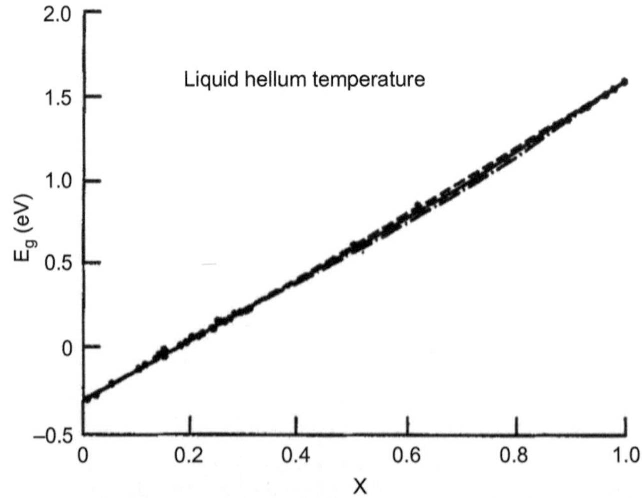

Fig. 5.3 Low temperature (~4 K) dependence of the energy gap of $Hg_xCd_{1-x}Te$ with composition, and comparison with calculated results by Hansen et al., Laurenti et al., and this work {See [35] and references therein}.

development of HgCdTe detectors is described in Fig. 5.2. $Hg_xCd_{1-x}Te$ alloy is the most widely used variable gap semiconductor for IR photodetectors [34–37]. Fig. 5.3 shows the evolution in the energy gap of $Hg_xCd_{1-x}Te$ with composition at ~4 K [35].

A detailed comparison of the detectivity of various commercial available infrared detectors, when operated at the indicated temperatures, is shown in Fig. 5.4. The fundamental figure of merit is another parameter of interest that is commonly utilized to evaluate the performance of materials for applications in infrared photodetection. Mathematically, the fundamental figure of merit is defined as the ratio of the absorption coefficient (α) to the thermal generation rate (G) of the material.

5.3 Silicon-germanium

Recent research has reported a promising possibility of using amorphous SiGe alloys and amorphous $Ge_xSi_{1-x}O_y$ compound for bolometer fabrication due to their high TCR, compatibility with current MEMS (microelectromechanical systems) process and relatively low $1/f$ noise compared to amorphous Si. However, they still exhibit high $1/f$ noise when compared to nonsemiconductor materials [39–41]. Since the melting point of silicon-germanium (SiGe) alloy is lower than that of Si, one advantage of utilizing polycrystalline silicon-germanium (poly-SiGe) film is that it requires lower thermal budget for crystallization, grain growth, and dopant activation; thus, these transitions occur at a temperature lower than that of the poly-Si film. Additionally, the thermal conductivity of the poly-SiGe alloy is lower than that of poly-Si. A schematic diagram of the poly-SiGe microbolometer is shown in Fig. 5.5.

In the example considered in this study, the poly-SiGe thin film resistor is located at a thermally well-isolated microbridge supported by two or four legs, which maximizes

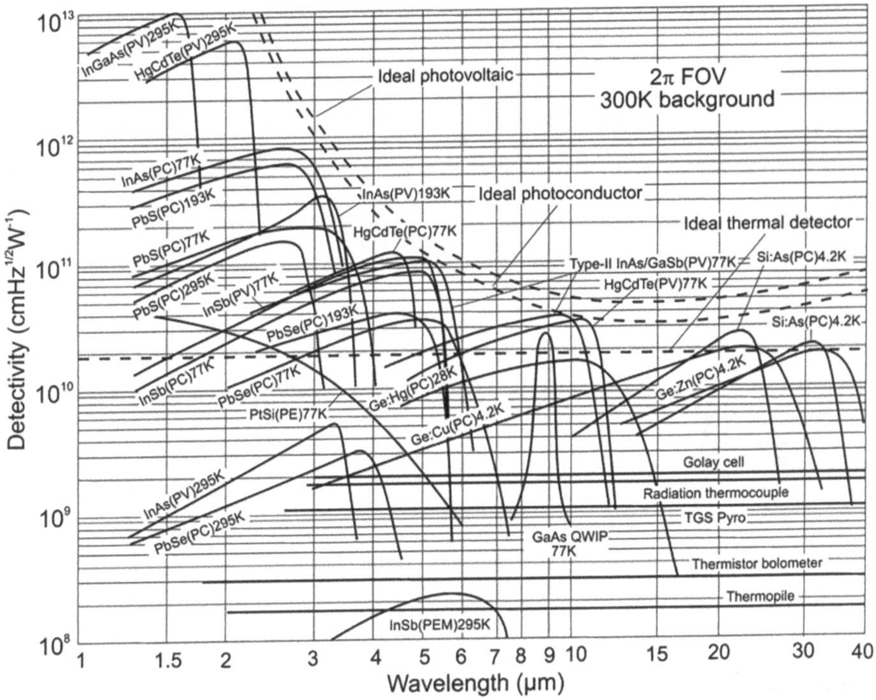

Fig. 5.4 Comparison of the detectivity D of various commercially available infrared detectors when operated at the indicated temperature. The chopping frequency is 1000 Hz for all detectors except the thermopile (10 Hz), thermocouple (10 Hz), thermistor bolometer (10 Hz), Golay cell (10 Hz), and pyroelectric detector (10 Hz). Each detector is assumed to view a hemispherical surround at a temperature of 300 K. Theoretical curves for the background-limited D for ideal photovoltaic and photoconductive detectors and thermal detectors are also shown [38].

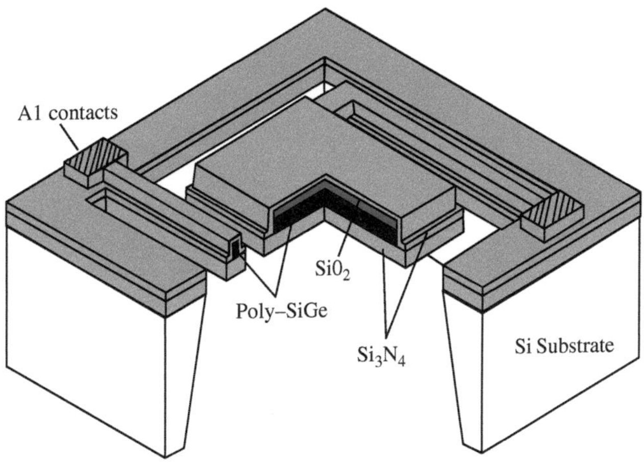

Fig. 5.5 Schematic diagram of a poly-SiGe microbolometer [40].

the temperature increase due to the absorption of IR radiation. The microbridge is formed by bulk micromachining. The IR absorber on the surface is a composite membrane made of SiO_2 and Si_3N_4, which can typically achieve an average absorptivity of 20%–30% in the wavelength range of 8–14 μm. The main fabrication process of the poly-SiGe microbolometer is shown in Fig. 5.6.

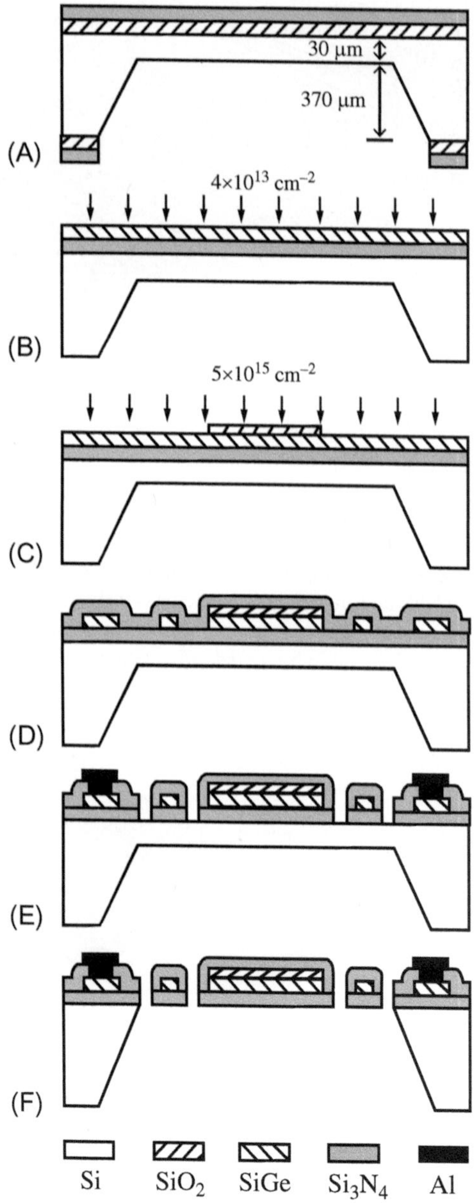

Fig. 5.6 Main fabrication process of the poly-SiGe microbolometer [40].

In a case study, beginning with the 75 mm diameter double-side polished (100) oriented silicon wafers, 500 nm thermally grown SiO_2 and 200 nm LPCVD Si_3N_4 are used as masking layer for bulk silicon anisotropic etching. Back windows in the masking layer are patterned and opened. Wafers then are rinsed in a TMAH (tetramethylammonium hydroxide) solution to etch a depth of 370 μm bulk silicon, leaving a silicon membrane of 30 μm. In short, the etching is performed at 80°C in a 10% TMAH solution, with an etch rate of ~45 μm/h in the (1 0 0) direction (Fig. 5.6a). After the masking layers are removed, a supporting layer of 500 nm PECVD Si_3N_4 and an active layer of 500 nm Ultra High Vacuum/Chemical Vapor Deposition (UHV/CVD) poly-SiGe are deposited, followed by an ion implant of boron with a dose of 4×10^{13} cm^{-2} (Fig. 5.6b). Then, 200 nm PECVD SiO_2 is deposited onto poly-SiGe and is patterned to form the first layer of the IR absorber. Ion implantation of boron with a dose of 5×10^{15} cm^{-2} into poly-SiGe and subsequent annealing at 650°C for 1 min in N_2 ambient in a rapid thermal annealing (RTA) reactor are performed (Fig. 5.6c). The poly-SiGe is patterned to define the active area and the support legs, which also act as electrical contacts. Then, 200 nm plasma enhanced chemical vapor deposition (PECVD) Si_3N_4 is deposited as the second layer of IR absorber (Fig. 5.6d). The contacts of the electrodes are patterned and 1.0 μm aluminum is sputtered, patterned and metallized. After patterning the stack of two PECVD Si_3N_4 films, reactive ion etching (RIE) technique is used to etch the Si_3N_4 layers until the top surface of the silicon appears (Fig. 5.6e). Finally, the remaining 30 μm silicon substrate is further etched using an induced couple plasma (ICP) system. The etchant is pure SF_6 gas. With SF_6 flow of 6 sccm at 20 Pa and a power of 600 W during etching reaction, the silicon is etched at a rate of 1.1 μm/min (Fig. 5.6f).

PolySiGe is a promising material system because it exhibits low thermal conductance, high TCR, and moderate noise levels combined with low stress, suitable for surface micromachining. Furthermore, it is a complementary metal oxide semiconductor (CMOS) frontend material used in BiCMOS processes and its process modules are available in most silicon foundries. Therefore, compared to nearly all microbolometer processes, deep-submicron lithography is used, which (among other outcomes) leads to superlative uniformity in direct material parameters (TCR, noise) and indirect process-related properties (thermal isolation). Compared to metals and monocrystalline silicon, polycrystalline silicon shows much lower thermal conductivity of 18 W/mK. By adding germanium, phonon scattering at point defects further reduces the thermal conductivity, and polySiGe with 30% Ge achieves a thermal conductivity of 3 W/mK (i.e., 25 times less than that of monocrystalline silicon and more than 50 times lower than that of aluminum) [40].

5.4 2D materials including graphene

Since the discovery and isolation of a single atomic layer of carbon for the first time [42] in the literature, research on the development of infrared detectors based on graphene [43–47] and other two dimensional (2D) materials [48–50] has grown significantly. The classification of mechanisms of photodetectors, based on 2D materials, is summarized in Fig. 5.7.

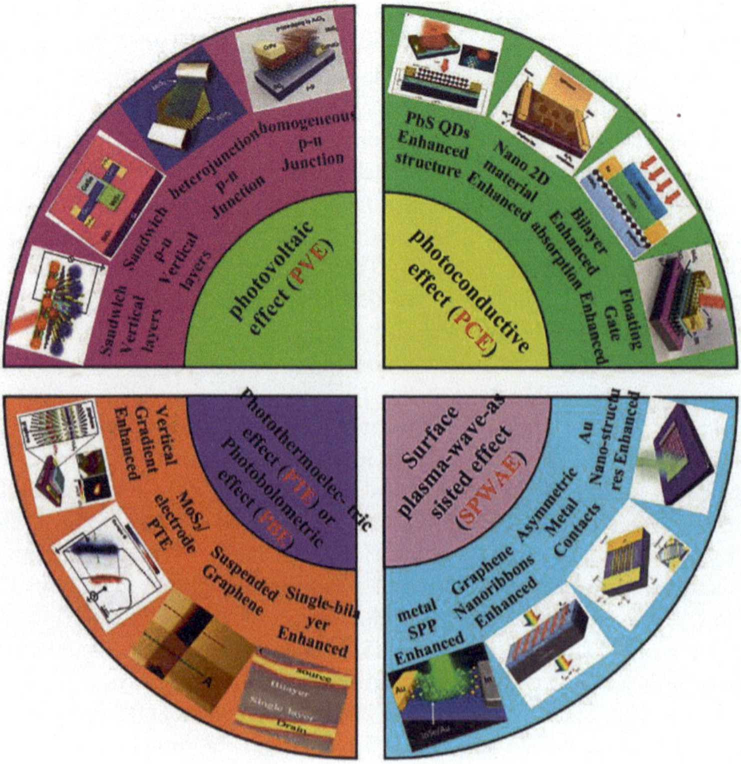

Fig. 5.7 Classification of mechanisms of photodetectors, based on 2D materials [48].

Graphene detectors have higher power signals from far infrared to mid infrared in the IR optical region. Since graphene is a single layer of carbon and has ultra-small volume, it has low heat capacity, enabling fast detectivity. The very small value of electron-phonon thermal conductivity of graphene also results in high intrinsic sensitivity for graphene-based bolometers. The low and gate-tunable resistance makes it possible to integrate the devices with a planar antenna with high coupling efficiency. Graphene is generally used as the radiation absorber and can detect either the photon energy or the radiation power by measuring the electron temperature rise due to the incoming radiation [43].

In this example from the literature [44], single layer graphene (SLG) is grown by CVD on 35-μm-thick Cu. Fig. 5.8 shows the schematic of an SLG pyroelectric bolometer. An SLG channel with source and drain contacts is fabricated on a pyroelectric substrate: 500 μm-thick z-cut $LiNbO_3$ (LN) crystal. A 10 nm-thick Al_2O_3 dielectric layer isolates the SLG from an H-shaped floating Au structure designed to overlap the oxide-coated SLG in the center, while lateral pads are placed in direct contact with the substrate. The design is such that the SLG channel conductivity can be modulated by a dual-gate capacitive structure. From the bottom, a pyroelectric polarization is generated directly by the substrate.

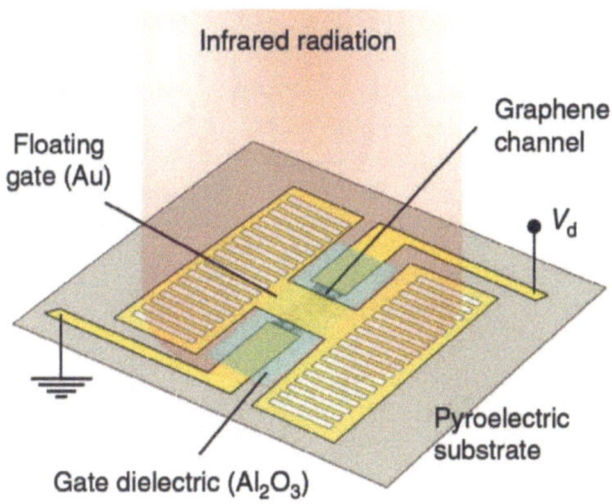

Fig. 5.8 Schematic of an individual graphene pyroelectric bolometer [44].

A schematic of the structure, band structure and recombination mechanism of ultrafast and ultrasensitive graphene-based photodetectors is presented in Table 5.2 [49].

5.5 Yttrium barium copper oxide ($YBa_2Cu_3O_{7-\delta}$)

Historically, Yttrium Barium Copper Oxide ($YBa_2Cu_3O_{7-\delta}$) was the first superconductor to exhibit a critical temperature (92 K) [51, 52] above the boiling point of nitrogen (77.36 K) [53]. The room temperature coefficient of resistivity of this material has a high and relatively constant value of TCR $\geq$ 3%/K. Amorphous YBaCuO thin films deposited on silicon wafers with a MgO buffer layer have exhibited TCRs as high as 4%/K. The responsivity of 3.8×10^5 V/W and detectivity of 1.6×10^9 cmHz$^{1/2}$ W^{-1} for a $YBa_2Cu_3O_{7-\delta}$ detector, reported in the literature, are possible if these thin films are integrated into typical air-gap microbolometer structure [54, 55].

Fig. 5.9 shows a schematic of the YBaCuO microbolometer. In this case study [55], the fabrication begins with the deposition of 1 μm thick MgO sacrificial layer on n-type silicon wafers. MgO is deposited by RF magnetron sputtering at 300 W from an 8-in. target with a mixture of gases of 90% Ar and 10% O_2 at 10mTorr pressure, yielding a deposition rate of 100 nm/h. The Si_3N_4 membrane is deposited next by radio frequency (RF) sputtering with 300 W of RF power and 10 mTorr of Ar, yielding the desired 400 nm thickness. The metallization layers are then deposited. Two metal electrode process is adopted utilizing a thin Nb layer (about 30 nm) for a low thermal conductivity lead and a thicker Au layer (300 nm) to make electrical contact to the YBaCuO and to form bond pads. Both gold and niobium layers are deposited by RF magnetron sputtering in pure Ar at 500 W. Patterning is performed with standard lithographic techniques, to leave gold only on the bond pad and contact areas and to define the Nb leads. Approximately 100 nm of YBaCuO is deposited by RF sputtering at a power of 150 W in a gas environment of 20% Ar and 80% O_2 (10 mTorr pressure). Photoresist layer is patterned to form the trench-cuts. The Si_3N_4 layer is etched with CF_4, exposing the MgO layer underneath. The wafer is then placed in a weak 1:1000

Table 5.2 Illustration of the structure, band structure, and recombination mechanism of ultrafast and ultrasensitive graphene-based photodetectors [49].

Structure	Band Structure	Recombination Mechanism
Ultra-fast (pristine) 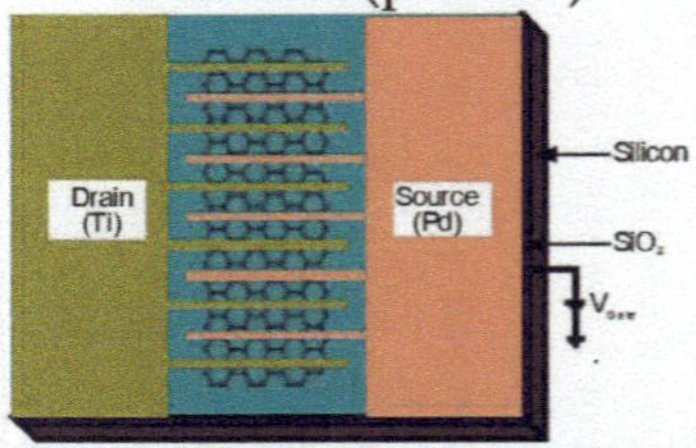	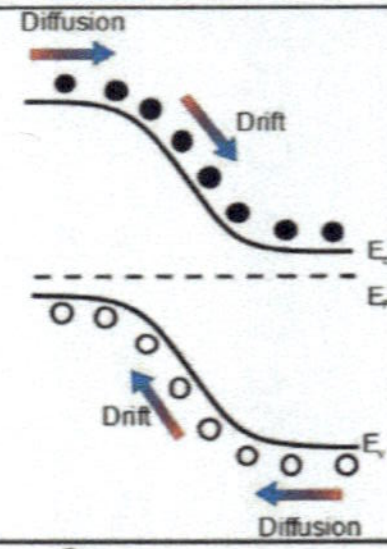	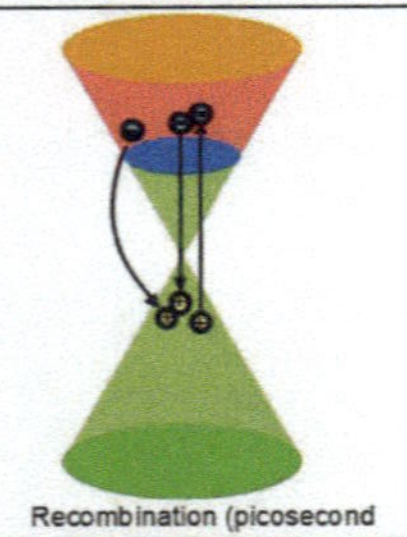
Ultra-sensitive (hybrid) 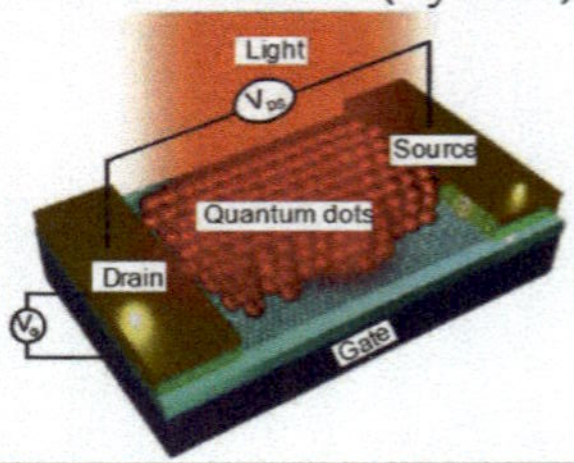	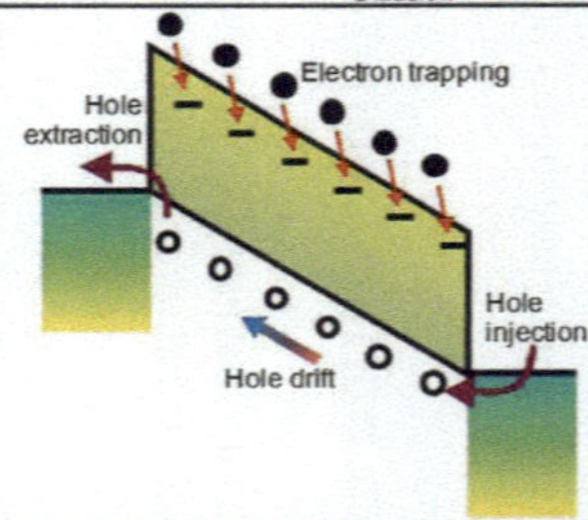	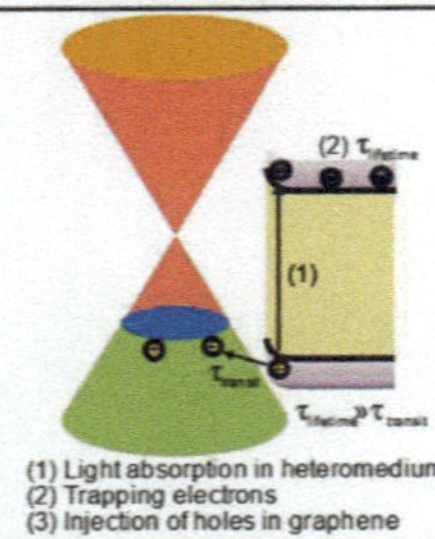

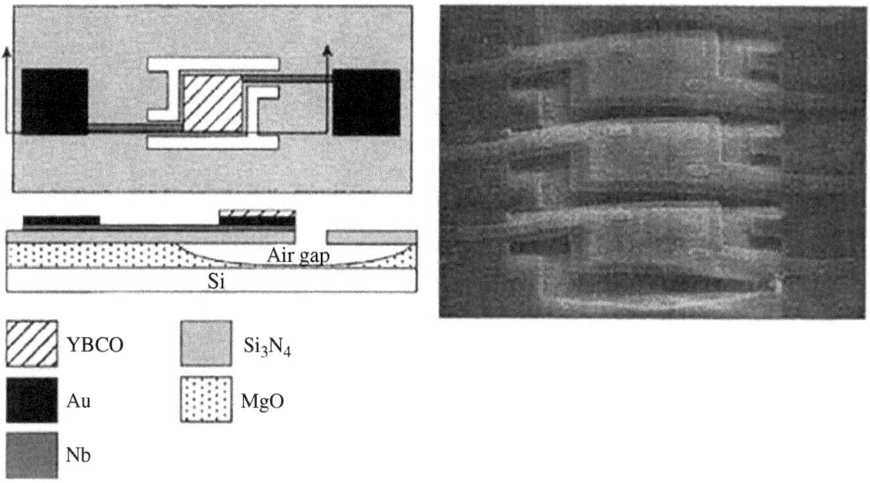

Fig. 5.9 Schematic and SEM micrograph of YBaCuO microbolometers fabricated with MgO sacrificial layer. The pixel size is 40 × 40 μm [55].

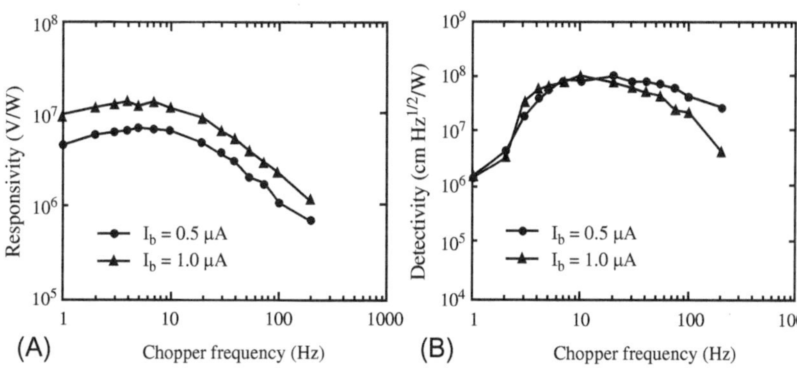

Fig. 5.10 (A) Voltage responsivity and (B) detectivity as a function of chopper frequency at different current biases for a MgO sacrificial layer micromachined YBaCuO microbolometer pixel (40 × 40 μm) [55].

acetic acid etch to isotropically remove the MgO sacrificial layer through the trench-cut openings and undercut the pixel. The photoresist trench-cut mask is retained in place for this step to protect the semiconducting YBaCuO from the acetic acid etch.

The highest measured detectivity is 10^8 cm $Hz^{1/2}$ W^{-1}, as shown in Fig. 5.10. The current independence and decline of the detectivity at lower chopper frequencies are indications of 1/*f* noise.

5.6 GaAs

An example of a Si-Ge microbolometer is presented in Fig. 5.11 [56]. Several binary, ternary, quaternary, and tertiary III–V compound semiconductors are being utilized in the manufacture of infrared detectors. An example of this includes the GaAs/AlGaAs

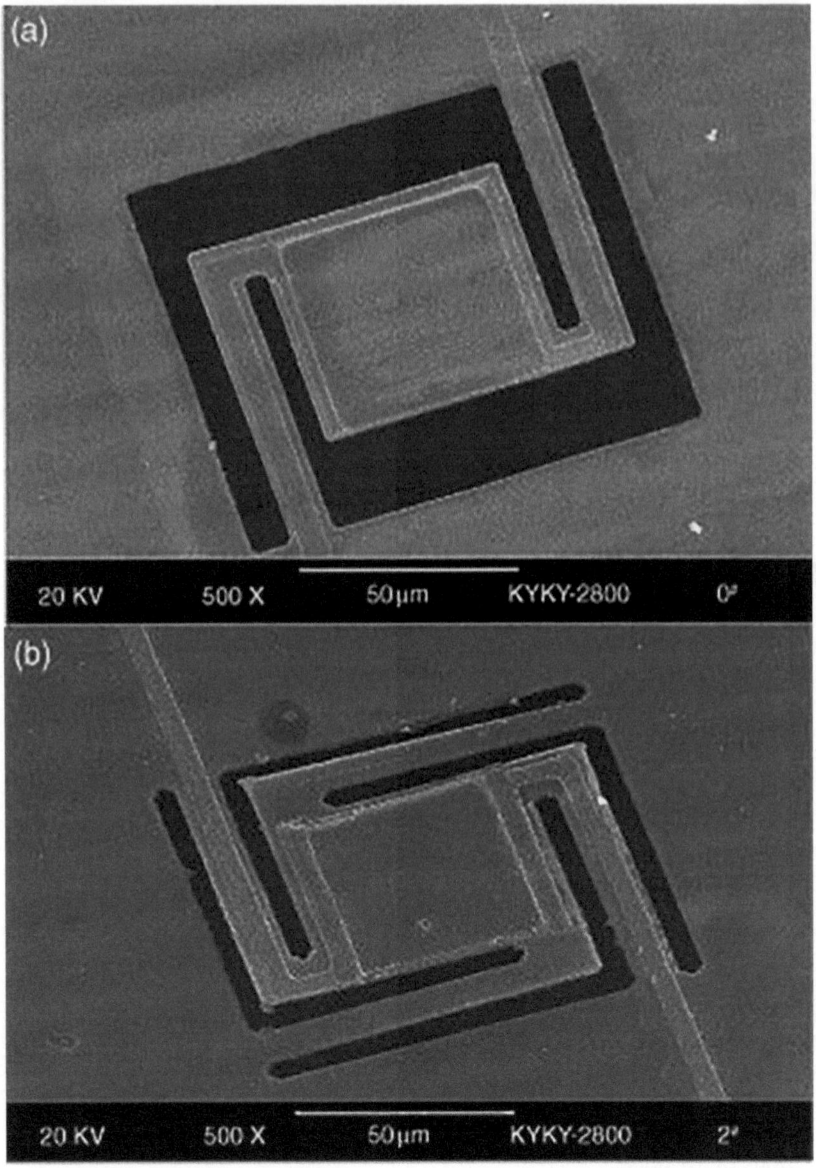

Fig. 5.11 SiGe microbolometers with (A) two suspensions and (B) four suspensions [56]. Reprinted from Sens. Actuators A, 105, Dong, L. et al., An uncooled microbolometer infrared detector based on poly-SiGe thermistor, 286–92, Copyright (2003), with permission from Elsevier.

heterostructure. A properly optimized GaAs/AlGaAs heterostructure is a competitive candidate as a microbolometer with high detectivity due to its high TCR.

Conventional microbolometers are fabricated using LTGaAs (low temperature GaAs) as a detector material between bow-tie antenna structures. A 0.75 μm layer of LTGaAs is grown epitaxially on a semi-insulating (SI) GaAs substrate at 270°C in an MBE system. An antenna structure is evaporated onto the LTGaAs layer and patterned by a bilayer photoresist liftoff process. An electron beam is used to evaporate 25 nm of Cr followed by a 150 nm layer of gold for these antennas. The LTGaAs layer is then etched into 10 × 10 μm square mesas, which provide a small defined electrical connection between the two antenna leads of each microbolometer [57].

5.7 Black silicon

Black silicon (BSi) is a surface modification of crystalline silicon. It has been used successfully in the fabrication of photodetectors [58]. The use of BSi as a thermal detector [59] has been studied in this section. BSi has a rough surface [60]. The effective optical constants of BSi are determined by using the effective medium approximation for normal incidence [61–63]. The refractive index (n) and extinction coefficient (k) vary with wavelength and temperature. Due to the high emissivity of BSi, it can be used in microbolometer detectors [60]. The average emissivity of BSi in the near IR range of 0.8–2.4 μm is ~0.854.

After several investigations, the structure of the proposed free standing BSi microbolometer comprises of the following layers (Fig. 5.12):

- Black silicon (BSi) layer—3 μm—IR absorbing element
- Silicon nitride/silicon dioxide/alumina—Isolation layer—0.004 μm
- p-Silicon or n-Silicon—0.25 μm/0.50 μm/0.75 μm—active element

In the literature, there have been several approaches to the simulation of optical properties of multilayers. In general, these models are based on the Bruggeman's Effective Medium Approximation [64]. In these simulations discussed in this section, the multilayer structure is considered to be suspended in vacuum. Normal angle of incidence has been considered. The propagation of electromagnetic radiation in various multilayers has been treated by a general transfer-matrix method.

In the literature, two approaches as examples, to modeling of the optical properties of multilayers have been proposed—Multi-Rad proposed by the Massachusetts Institute of Technology [65] and Radpro proposed by Georgia Tech [66].

A simulation tool that considers Multi-Rad has been used in this study for modeling purposes.

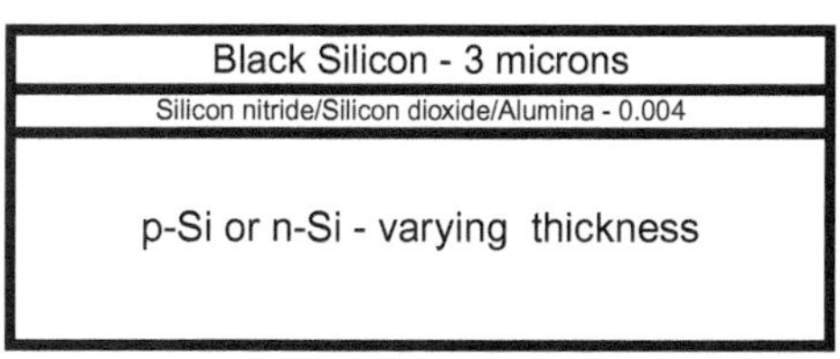

Fig. 5.12 Proposed multilayer structure of BSi based microbolometer [63].

Fig. 5.13a shows the emissivity of the microbolometer structure with Si_3N_4 layer. The average emissivity for the structure with varying thicknesses is about 0.869. The reflectance is negligible (~3%). The average transmittance is about 10%. Fig. 5.13b shows the emissivity of the microbolometer structure with SiO_2 layer. The average emissivity for the structure with varying thicknesses is about 0.870. The reflectance is negligible (~3%). The average transmittance is about 10%. Fig. 5.13c shows the emissivity of the microbolometer structure with Al_2O_3 layer. The average emissivity for the structure with varying thicknesses is about 0.868. The reflectance is negligible (~3%). The average transmittance is about 10%.

Fig. 5.14a shows the emissivity of the microbolometer structure with Al_2O_3 layer. The average emissivity for the structure with varying thicknesses is about 0.869. The reflectance is negligible (~3%). The average transmittance is about 10%. Fig. 5.14b shows the emissivity of the microbolometer structure with Si_3N_4 layer. The average emissivity for the structure with varying thicknesses is about 0.868. The reflectance is negligible (~3%). The average transmittance is about 10%. Fig. 5.14c shows the emissivity of the microbolometer structure with SiO_2 layer. The average emissivity for the structure with varying thicknesses is about 0.869. The reflectance is negligible (~3%). The average transmittance is about 10%.

In Tables 5.3 and 5.4, a summary of the total amount of IR radiation, absorbed in the 0.8–2.5 μm spectral range, by the proposed multilayer structure is presented. The total emissivity (Ɛ) of the multilayer structure is integrated with respect to wavelength

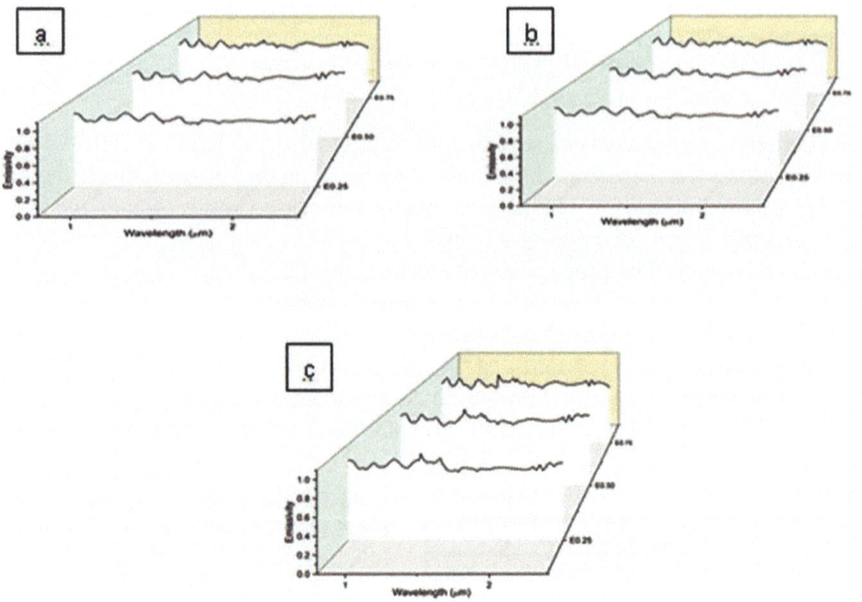

Fig. 5.13 (a) Emissivity of BSi/Si_3N_4/p-Si. (b) Emissivity of BSi/SiO_2/p-Si. (c) Emissivity of BSi/Al_2O_3/p-Si [63].

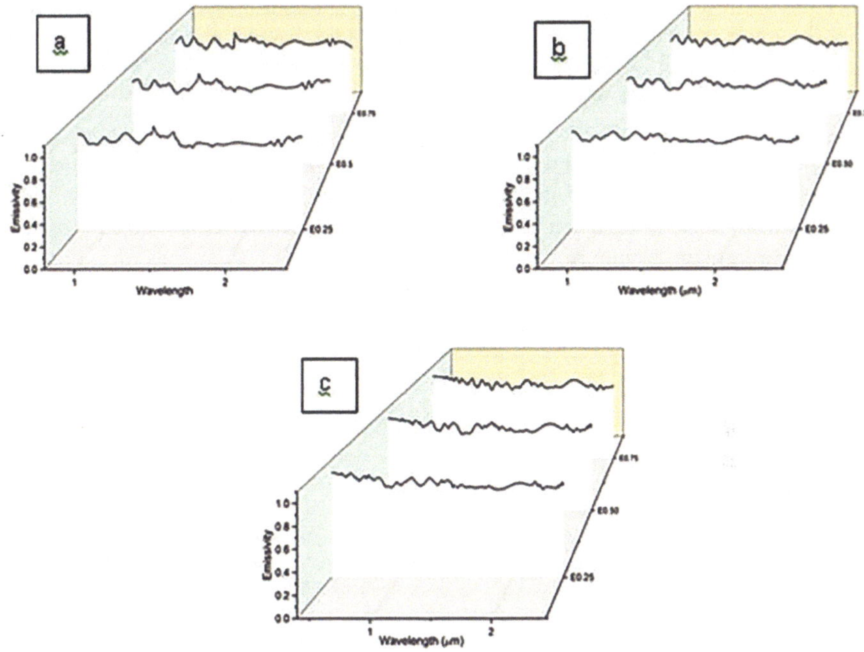

Fig. 5.14 (a) Emissivity of BSi/Al_2O_3/n-Si. (b) Emissivity of BSi/Si_3N_4/n-Si. (c) Emissivity of BSi/SiO_2/n-Si [63].

Table 5.3 Emissivity of multilayer for various thicknesses of p-Si active layer and isolation layers (Si_3N_4, SiO_2, Al_2O_3) [63].

p-Si thickness (μm)	ε (BSi/Si_3N_4/p-Si)	ε (BSi/SiO_2/p-Si)	ε (BSi/Al_2O_3/p-Si)
0.25	0.867	0.868	0.868
0.50	0.898	0.898	0.898
0.75	0.842	0.842	0.842

Table 5.4 Emissivity of multilayer for various thicknesses of n-Si active layer and isolation layers (Si_3N_4, SiO_2, Al_2O_3) [63].

n-Si thickness (μm)	ε (BSi/Si_3N_4/n-Si)	ε (BSi/SiO_2/n-Si)	ε (BSi/Al_2O_3/n-Si)
0.25	0.854	0.855	0.854
0.50	0.894	0.894	0.894
0.75	0.860	0.859	0.859

for a blackbody distribution of energy; and it is a function of the thickness of low doped Si and the isolation layers (Si_3N_4, SiO_2, Al_2O_3). From these tables, it can be noted that the emissivity for the structure with p-Si thermosensing layer has the highest total emissivity of 0.898.

The figure of merit, described in Chapter 3, has been calculated for the proposed free standing structure connected by polysilicon to the readout circuit. The figure of merit is tabulated in Tables 5.5–5.10.

From Tables 5.5 to 5.10, it can be observed that the figure of merit for the proposed BSi microbolometer structure, with n-Si thermosensing layer, has the highest

Table 5.5 BSi/Si_3N_4/p-Si [63].

p-Si thickness (μm)	0.25	0.50	0.75
R (*V*/*W*)	1.064×10^3	1.102×10^3	1.033×10^3
Power (mW)	0.056	0.094	0.122
τ (mS)	0.070	0.079	0.089
Δf (Hz)	2.287×10^3	2.002×10^3	1.781×10^3
V_n (V)	1.646×10^{-6}	1.188×10^{-6}	9.844×10^{-7}
D^* (cm $Hz^{1/2}$ W^{-1})	5.252×10^7	7.056×10^7	7.528×10^7
NEP (W $Hz^{-1/2}$)	3.237×10^{-11}	2.409×10^{-11}	2.258×10^{-11}

Table 5.6 BSi/SiO_2/p-Si [63].

p-Si thickness (μm)	0.25	0.50	0.75
R (V/W)	1.065×10^3	1.102×10^3	1.033×10^3
Power (mW)	0.056	0.094	0.122
τ (mS)	0.070	0.079	0.089
Δf (Hz)	2.287×10^3	2.003×10^3	1.781×10^3
V_n (V)	1.646×10^{-6}	1.188×10^{-6}	9.844×10^{-7}
D^* (cm $Hz^{1/2}$ W^{-1})	5.258×10^7	7.056×10^7	7.528×10^7
NEP (W $Hz^{-1/2}$)	3.233×10^{-11}	2.409×10^{-11}	2.258×10^{-11}

Table 5.7 BSi/Al_2O_3/p-Si [63].

p-Si thickness (μm)	0.25	0.50	0.75
R (V/W)	1.065×10^3	1.102×10^3	1.033×10^3
Power (mW)	0.056	0.094	0.122
τ (mS)	0.070	0.080	0.089
Δf (Hz)	2.283×10^3	2.000×10^3	1.779×10^3
V_n (V)	1.645×10^{-6}	1.187×10^{-6}	9.839×10^{-7}
D^* (cm $Hz^{1/2}$ W^{-1})	5.258×10^7	7.056×10^7	7.528×10^7
NEP (W $Hz^{-1/2}$)	3.233×10^{-11}	2.409×10^{-11}	2.258×10^{-11}

Table 5.8 BSi/Si_3N_4/n-Si [62].

n-Si thickness (μm)	0.25	0.50	0.75
R (V/W)	1.076×10^3	1.126×10^3	1.083×10^3
Power (mW)	0.056	0.094	0.122
τ (mS)	0.070	0.079	0.089
Δf (Hz)	2.287×10^3	2.002×10^3	1.781×10^3
V_n (V)	1.646×10^{-6}	1.188×10^{-6}	9.844×10^{-7}
D^* (cm $Hz^{1/2}$ W^{-1})	5.312×10^7	7.212×10^7	7.894×10^7
NEP (W $Hz^{-1/2}$)	3.200×10^{-11}	2.357×10^{-11}	2.154×10^{-11}

Table 5.9 BSi/SiO_2/n-Si [63].

n-Si thickness (μm)	0.25	0.50	0.75
R (V/W)	1.077×10^3	1.126×10^3	1.082×10^3
Power (mW)	0.056	0.094	0.122
τ (mS)	0.070	0.079	0.089
Δf (Hz)	2.287×10^3	2.003×10^3	1.781×10^3
V_n (V)	1.646×10^{-6}	1.188×10^{-6}	9.844×10^{-7}
D^* (cm $Hz^{1/2}$ W^{-1})	5.318×10^7	7.212×10^7	7.885×10^7
NEP (W $Hz^{-1/2}$)	3.197×10^{-11}	2.357×10^{-11}	2.156×10^{-11}

Table 5.10 BSi/Al_2O_3/n-Si [63].

p-Si thickness (μm)	0.25	0.50	0.75
R (V/W)	1.076×10^3	1.126×10^3	1.082×10^3
Power (mW)	0.056	0.094	0.122
τ (mS)	0.070	0.080	0.089
Δf (Hz)	2.283×10^3	2.000×10^3	1.779×10^3
V_n (V)	1.645×10^{-6}	1.187×10^{-6}	9.839×10^{-7}
D^* (cm $Hz^{1/2}$ W^{-1})	5.312×10^7	7.212×10^7	7.885×10^7
NEP (W $Hz^{-1/2}$)	3.200×10^{-11}	2.357×10^{-11}	2.156×10^{-11}

simulated responsivity and detectivity. Also, with p-Si or n-Si as the thermosensing layer, the highest simulated responsivity has been achieved for 0.50 μm thickness. Detectivity is the most important factor in the calculation of the figure of merit. It is used to compare the microbolometers for better performance. It is independent of area and is dependent on the emissivity of the microbolometer structure. The emissivity and hence the figure of merit can further be improved by adding a gold (Au) layer on top of the silicon substrate in combination with the optimized resonant cavity.

An improved structure of the microbolometer consists of two level structures with a reflective layer of gold deposited on top of the lower level of the structure. The upper level is comprised of a thermally isolated structure suspended over a silicon readout circuit on a substrate. The thermally isolated structure is the same structure as in Fig. 5.12 considering SiO_2 as the isolation layer. Polysilicon columns act as an interconnection between two levels. The bottom level consists of gold (Au) as a reflecting layer. Heavily doped n-type polysilicon (10^{22} cm^{-3}) has been used as an interconnection material as it provides an excellent Ohmic contact with the n-Si thermosensing element and it also has low thermal conductivity, a key factor to achieve high responsivity. This improved structure is shown in Fig. 5.15.

Table 5.11 shows the figures of merit calculated for p-Si and n-Si with thickness of 0.5 μm. By comparison, it can be discerned that n-Si, as active element, has higher responsivity and detectivity. The thin layer of gold and the vacuum cavity act as a resonating cavity, and the emissivity is enhanced by ~7%.

Table 5.12 shows the comparison of the figures of merit of the microbolometer structure (n-Si), considered in this study, with devices that have been presented in

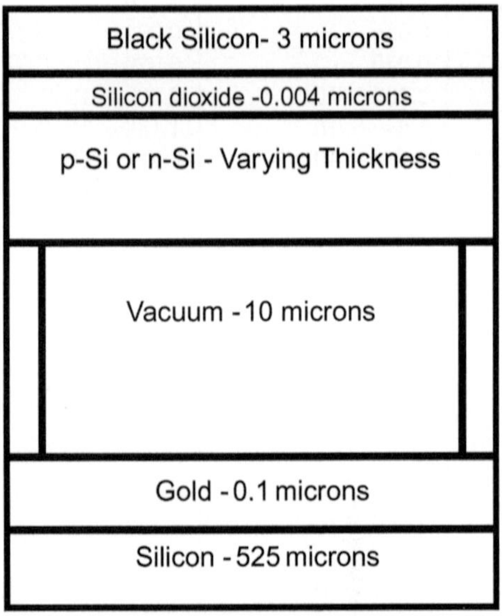

Fig. 5.15 Proposed structure of BSi based microbolometer [63].

Table 5.11 Figure of merit for microbolometer with gold resonating cavity [63].

Active element	p-Si	n-Si
R (V/W)	1.180×10^3	1.212×10^3
Power (mW)	0.094	0.094
τ (mS)	0.079	0.079
Δf (Hz)	2.003×10^3	2.003×10^3
V_n (V)	1.188×10^{-6}	1.188×10^{-6}
D^* (cm $Hz^{1/2}$ W^{-1})	7.559×10^7	7.761×10^7
NEP (W $Hz^{-1/2}$)	2.249×10^{-11}	2.190×10^{-11}

Table 5.12 Comparison of n-Si studied microbolometer with literature [63].

Active element	SiGe-Poly	Si-a	Si-Poly	n-Si
TCR (%/K)	−2	−4.0	−2.54	0.75
τ (ms)	10.35	9.70	3.79	0.079
R (V/W)	139.7×10^3	78.388×10^3	0.35	1.212×10^3
D^* (cm $Hz^{1/2}$ W^{-1})	2.31×10^9	8.67×10^8	6.04×10^8	7.761×10^7
NEP (W $Hz^{-1/2}$)	–	0.13×10^{-10}	9.25×10^{-9}	2.190×10^{-11}

the literature. As can be seen in Table 5.12, the response time of the proposed device is lower, the responsivity is comparable, the detectivity is slightly lower, and the noise equivalent power for the proposed device is comparable to devices considered in the literature.

Optical properties of the proposed structure, with varying thickness of silicon and different isolation layer materials, have been simulated in the near-infrared range of wavelengths. Results show that the use of black silicon results in increased absorption in the infrared. Highest emissivity of 0.898 is obtained for p-Si with a thickness of 0.5 μm and this is independent of the isolation layer material.

Based on a comparison of the performance characteristics of the proposed microbolometer structure with similar devices in the literature, it can be concluded that the proposed two-level structure is a viable candidate for the fabrication of uncooled microbolometers. The studied multilayer structure, with BSi as an IR absorber in combination with a resonant cavity, exhibits a high radiative emissivity. The modeled IR radiation absorptance and the numerical calculations show the possibility to consider the studied two-level structure as an alternative in the fabrication of uncooled microbolometers.

References

[1] V.R. Mehta, et al., Silicon-integrated uncooled infrared detectors: perspectives on thin films and microstructures, J. Electron. Mater. 34 (5) (2005) 484–489.

[2] F.M. Tong, et al., High T_c $YBa_2Cu_3O_{7-\delta}$ superconducting transition-edge microbolometers, Infrared Phys. Technol. 36 (1995) 1053–1058.

[3] W.F. Kosonocky, Review of Schottky-barrier imager technology, SPIE advent technologies series, in: R. Feinberg (Ed.), Vol. AT 2: Current Overviews in Optical Science and Engineering II, SPIE, Bellingham, WA, 1990, pp. 446–470.

[4] M.P. Lepselter, A.T. Fiory, N.M. Ravindra, Platinum and rhodium silicide-germanide optoelectronics, J. Electron. Mater. 37 (4) (2008) 403–416.

[5] J.M. Densmore, et al., High-speed two-camera imaging pyrometer for mapping fireball temperatures, Appl. Optics 50 (33) (2011) 6267–6271.

[6] W.F. Kosonocky, et al., Multi-wavelength imaging pyrometer, in: Proceedings of SPIE, vol. 2225, 1994, pp. 26–43.

[7] G. Roelkens, et al., Silicon-based photonic integration beyond the telecommunication wavelength range, IEEE J. Sel. Top. Quantum Electron. 20 (4) (2014), 8201511. 11 pages.

[8] B.E. Lemoff, MAUI: enabling fiber-to-the-processor with parallel multiwavelength optical interconnects, J. Lightwave Technol. 22 (9) (2004) 2043–2054.
[9] H.C. Liu, B.F. Levine, J.Y. Andersson (Eds.), Quantum Well Intersubband Transition Physics and Devices, Kluwer Academic, Dordrecht, 1994.
[10] H. Schneider, et al., High-speed infrared detection by uncooled photovoltaic quantum well infrared photodetectors, Appl. Phys. Lett. 70 (1997) 1602.
[11] E. Peytavit, Room temperature terahertz microbolometers, in: 2005 Joint 30th International Conference on Infrared and Millimeter Waves & 13th International Conference on Terahertz Electronics, IEEE, 2005, pp. 257–258.
[12] A.J. Miller, A. Luukanen, E.N. Grossman, Micromachined antenna-coupled uncooled microbolometers for terahertz imaging arrays, in: Proceedings of SPIE 5411, Terahertz for Military and Security Applications II, 2004.
[13] C. Greboval, et al., Infrared Narrow Band Gap Nanocrystals: Recent Progresses Relative to Imaging and Active Detection, 2020.
[14] J.C. Bean, Silicon-based semiconductor heterostructures: column IV bandgap engineering, Proc. IEEE 80 (4) (1992) 571–587.
[15] G.-L. Tan, L. Liu, W. Wu, Mid-IR bandgap engineering of $Cd_xPb_{1-x}S$ nanocrystals by mechanochemical reaction, AIP Adv. 4 (2014), 067107.
[16] I.M. Baker, II-VI Narrow bandgap semiconductors: optoelectronics, in: S. Kasap, P. Capper (Eds.), Springer Handbook of Electronic and Photonic Materials, Springer Handbooks, Springer, Cham, 2017, https://doi.org/10.1007/978-3-319-.
[17] M. Walther, G. Weimann, Infrared imaging with InAs/GaSb type-II superlattices, Phys. Status Solidi A 203 (14) (2006) 3545–3549.
[18] R. Guntupalli, R. Allen, Evaluation of InGaAs camera for scientific near infrared imaging applications, in: R.E. Longshore, A. Sood (Eds.), Infrared and Photoelectronic Imagers and Detector Devices II Proceedings of SPIE, vol. 6294, 2006, p. 629401, https://doi.org/10.1117/12.674137.
[19] M. Vollmer, K.-P. Möllmann, Infrared Thermal Imaging Fundamentals, Research and Applications, second ed., Wiley-VCH Verlag GmbH & Co. KGaA, 2018.
[20] A.M. Itsuno, Bandgap-Engineered HgCdTe Infrared Detector Structures for Reduced Cooling Requirements, PhD Dissertation, Electrical Engineering, The University of Michigan, 2012.
[21] B.B. Lahiri, et al., Medical applications of infrared thermography: a review, Infrared Phys. Technol. 55 (2012) 221–235.
[22] R. Berz, H. Sauer, The medical use of infrared-thermography—history and recent applications, in: Thermografie-Kolloquium 2007—Vortrag 04, 2007, pp. 1–12. www.ndt.net/search/docs.php3?MainSource=61.
[23] E.F.J. Ring, K. Ammer, Infrared thermal imaging in medicine, Physiol. Meas. 33 (3) (2012) R33–R46, https://doi.org/10.1088/0967-3334/33/3/R33.
[24] A.A. Gurjarpadhye, et al., Infrared imaging tools for diagnostic applications in dermatology, SM J. Clin. Med. Imaging 1 (1) (2015) 1–5.
[25] S. Fussell, Infrared Cameras can Spot a Fever, But May Not Slow Covid 19, 2020. https://www.wired.com/story/infrared-cameras-spot-fever-not-slow-covid-19/. (Accessed 16 June 2020).
[26] The Evolution of Thermal Imaging Cameras, Palmer Wahl Instrumentation Group, 170 years of Continued Innovation, 2007.
[27] A. Rogalski, Next decade in infrared detectors, in: Proceedings of SPIE 10433, Electro-Optical and Infrared Systems: Technology and Applications XIV, 104330L (9th October 2017), Event: SPIE Security+Defense, 2017, Warsaw, Poland, 2017, https://doi.org/10.1117/12.2300779.

[28] A. Rogalski, History of infrared detectors, Opto-Electron. Rev. 20 (3) (2012) 279–308.
[29] A. Rogalski, Infrared detectors for the future, Acta Phys. Pol. A 116 (3) (2009) 389–406.
[30] A. Rogalski, Infrared detectors: an overview, Infrared Phys. Technol. 43 (2002) 187–210.
[31] Z. Zou, et al., Object Detection in 20 Years: A Survey, arXiv, 2019. 1905.05055v2[cs.CV].
[32] N.M. Ravindra, S. Auluck, V.K. Srivastava, Temperature dependence of the energy gap in PbS, PbSe and PbTe, Phys. Status Solidi A 52 (1979) k151.
[33] W.D. Lawson, S. Nielsen, E.H. Putley, A.S. Young, Preparation and properties of HgTe and mixed crystals of HgTe-CdTe, J. Phys. Chem. Solid 9 (1959) 325–329.
[34] A. Rogalski, HgCdTe infrared detector material: history, status and outlook, Rep. Prog. Phys. 68 (2005) 2267–2336.
[35] F.M. Tong, Y. Haoxin, Y. Xiuzhen, N.M. Ravindra, HgCdTe photodiodes—a device study, Infrared Phys. 33 (6) (1992) 511–522.
[36] M. Kopytko, P. Martyniuk, HgCdTe Mid- and Long-Wave Barrier Infrared Detectors for Higher Operating Temperature Condition, Intech Open Science, 2016, https://doi.org/10.5772/63943.
[37] R.E. Longshore, MCT properties, growth methods and characterization, Chapter 7, in: M. Henini, M. Razeghi (Eds.), Handbook of Infrared Detection Technologies, first ed., Elsevier Science, 2002, pp. 234–266.
[38] A. Rogalski, Infrared Detectors, second ed., CRC Press, 2010.
[39] R. Xu, et al., Boron-doped nanocrystalline silicon germanium thin films for uncooled infrared bolometer applications, Infrared Phys. Technol. 58 (2013) 32–35.
[40] C.V. Hoof, P.J.J. De Moor, PolySiGe uncooled microbolometers for thermal IR detection, Chapter 12, in: Handbook of Infra-Red Detection Technologies, 2002, pp. 449–479, https://doi.org/10.1016/B978-185617388-9/50012-5.
[41] L. Dong, R.-F. Yue, L.-T. Liu, Characterization of uncooled poly SiGe microbolometer for infrared detection, Chin. Phys. Lett. 20 (5) (2003) 770–773.
[42] K.S. Novoselov, et al., Electric field effect in atomically thin carbon films, Science 306 (2004) 666–669.
[43] X. Du, et al., Graphene-based bolometers, Graphene 2D Mater. 1 (2014) 1–22. De Gruyter OPEN.
[44] U. Sassi, et al., Graphene-based, mid-infrared, room-temperature pyroelectric bolometers with ultrahigh temperature coefficient of resistance, Nat. Commun. 8 (2017) 14311.
[45] A. Blaikie, D. Miller, B.J. Aleman, A fast and sensitive room-temperature graphene nanomechanical bolometer, Nat. Commun. 10 (2019) 4726.
[46] A. Rogalski, Graphene-based materials in the infrared and terahertz detector families: a tutorial, Adv. Opt. Photon. 11 (2) (2019) 314–379.
[47] D.K. Efetov, Fast thermal relaxation in cavity-coupled graphene bolometers with a Johnson noise read-out, Nat. Nanotechnol. 13 (2018) 797–801.
[48] J. Wang, et al., Design strategies for two-dimensional material photodetectors to enhance device performance, InfoMat 1 (2019) 33–53.
[49] A. Rogalski, 2D materials in infrared detector family, Proceedings MDPI 27 (2019) 33. 1–4.
[50] Y. Lin, Infrared Detectors Based on Two-Dimensional Materials and Heterostructures, PhD Dissertation, Massachusetts Institute of Technology, 2019.
[51] B.A. Howe, Crystal Structure and Superconductivity of $YBa_2Cu_3O_{7-x}$, Thesis, Master of Science in Physics, Minnesota State University, Mankato, 2014.
[52] S.M. Wentworth, D.P. Neikirk, The transition-edge microbolometer (TREMBOL), in: SPIE Vol. 1292 Superconductivity Applications in Infrared and Microwave Devices, 1990, pp. 148–154.

[53] The Element Nitrogen, It's Elemental, Jlab Science Education, 2020. https://education.jlab.org/itselemental/ele007.html. (Accessed 18 June 2020).

[54] A.J. Kreisler, V.S. Jagtap, A.F. Dégardin, Infrared response in the 95 to 300 K temperature range of detectors based on oxygen-depleted Y-Ba-Cu-O thin films, Phys. Procedia 36 (2012) 223–228.

[55] D.P. Butler, Z.G. Butler, R. Sobolewski, Yttrium barium copper oxide as an infrared radiation sensing material, chapter 4, in: Handbook of Advanced Electronic and Photonic Materials and Devices, vol. 3, 2001, pp. 169–195.

[56] L. Dong, et al., An uncooled microbolometer infrareddetector based on poly-SiGe thermistor, Sens. Actuators A 105 (2003) 286–292.

[57] J.M. Lewis, D.P. Neikirk, S.M. Wentworth, Low growth temperature GaAs microbolometers, in: R.J. Temkin (Ed.), 15th International Conference on Infrared and Millimeter Waves, 1990, pp. 398–400.

[58] S. Huang, et al., Black silicon photodetector with excellent comprehensive properties by rapid thermal annealing and hydrogenated surface passivation, Adv. Opt. Mater. 8 (7) (2020), https://doi.org/10.1002/adom.201901808.

[59] J.E. Carey, J.W. Sickler, IR Detectors: Black Silicon Sees Further into the IR, Laser Focus World, 2009.

[60] N.M. Ravindra, S.R. Marthi, S. Sekhri, Black Silicon: Processing, Properties, and Applications, Momentum Press, 2016.

[61] S.R. Marthi, S. Sekhri, N.M. Ravindra, Optical properties of black silicon: an analysis, JOM 67 (2015) 2154–2159.

[62] N.M. Ravindra, S. Marthi, S. Sekhri, Modeling of optical properties of black silicon/crystalline silicon, J. Sci. Ind. Metrol. 1 (1) (2015) 1–7.

[63] S.R. Marthi, Black Silicon: Optical Properties and Applications, PhD Dissertation, Interdisciplinary Program in Materials Science & Engineering, New Jersey Institute of Technology, 2017.

[64] D.A.G. Bruggeman, Berechnung verschiedener physikalischer Konstanten von heterogenen Substanzen, Ann. Phys. 24 (1935) 636–679.

[65] J. Hebb, K.F. Jensen, "Multi-Rad" User's Manual, SEMATECH, Austin, Texas, 1997.

[66] B.J. Lee, Z. Zhang, Rad-Pro User Manual Version 1.1, Georgia Institute of Technology, 2005.

Optical and thermal detector fundamentals, microbolometer, and readout integrated circuits

6

6.1 Introduction

The electromagnetic spectrum range, 0.2–1000 μm, is considered to be optical radiation; it includes visible radiation (0.4–0.75 μm) and infrared radiation (0.75–1000 μm). The human eye can detect wavelengths in the visible range. Infrared radiation is the second most intense source of radiation on earth; this is the reason for the efforts, during the last century, toward the study and development of infrared radiation detectors. There is significant interest in sensing infrared radiation for a variety of applications, both in the civilian sector as well as in the defense sector.

An infrared detector is an optoelectronic device that reacts to infrared radiation, transducing the incident infrared radiation into an electrical signal. Electromagnetic radiation interacts with matter in various ways that determine the physical mechanisms involved in the detection process and can be classified as due to: photon effect, thermal effect, and wave interaction effect. This radiation-matter interaction depends on the infrared absorber material and the infrared spectral range. Infrared radiation detectors can be classified into two main types: photon detectors and thermal detectors [1].

Infrared detectors can be further subclassified into several types; these types of detectors, along with their important performance characteristics, are summarized in Fig. 6.1.

Thermal detectors respond to the heating effect caused by the absorbed photon radiation that causes changes in the temperature of the detector and in its electrical properties. The detection process of thermal detectors is divided in two steps. First, the radiation is absorbed generating phonons and causes changes in the lattice temperature of the absorber material. Second, the change in temperature induces changes in a measurable parameter of the detector active element. The response of the thermal detector is proportional to the energy absorbed [1]. Thermal detectors, most used by the infrared industry, are bolometers, thermopiles, and pyroelectrics. Bolometers are detectors in which the absorption of radiation causes changes in the electrical resistance of its thermal sensing materials (for example, VO_x, a-Si). They exhibit high responsivity and bias is required (DC response). Thermoelectric detectors (thermopiles) are formed by the junction of two materials (Bi/Sb, Al/Poly-Si) with high Seebeck coefficients. They exhibit low responsivity and no bias is required (DC response). Pyroelectric detectors are made using dielectric materials ($LiTaO_3$, $Ba_xSr_{1-x}TiO_3$) that modify their polarization under the influence of changes in temperature and need

Microbolometers. https://doi.org/10.1016/B978-0-08-102812-4.00006-1

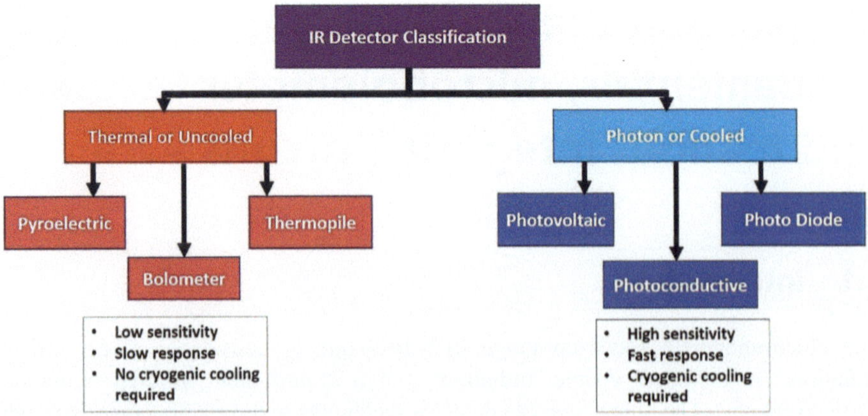

Fig. 6.1 Types of infrared detectors.

a chopper system in order to generate a modulation in their polarization. They exhibit moderately high responsivity, and no bias is required (current response).

Photon detectors are devices that detect optical radiation by direct interaction of individual photons with active charge carriers within the atomic lattice of the detector material. The transition of the active charges from the valence band to the conduction band depends on the energy band gap of the material and causes changes in the physical parameters such as resistance, capacitance, voltage, or current [2]. The photodetector response is proportional to the number of photons absorbed [3]. Among the many types of photon detectors, the photoconductive and the photovoltaic devices are the most used. Photoconductive detectors use the increase in electrical conductivity resulting from an increase in the number of free carriers generated when photons are absorbed (generation of current), whereas in photovoltaic detectors, current is generated as a result of the absorption of photons due to a voltage difference across a p-n junction (generation of electron-hole pair across the junction).

6.2 Photon detectors versus thermal detectors

A wide variety of photon detectors and thermal detectors have been developed by the infrared detection/imaging industry; each detection mechanism has advantages and disadvantages that determine the most appropriate approach depending on the infrared spectrum range for the detection, application, and measurement conditions.

The time response of the thermal detector is a two-step process; it is slower than the corresponding time response of the photon detector, in which the absorption of a photon and the generation of an electron-hole pair is faster. Photon detectors are more sensitive than thermal detectors, but they can only operate at specific and narrow wavelength range in comparison with thermal detectors that can respond continuously over a broad range of wavelengths. Thermal detectors have the advantage to operate at room temperature, in contrast with photon detectors that need to be

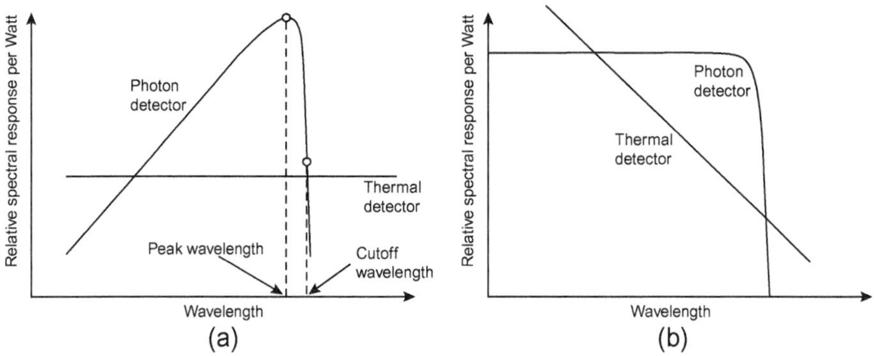

Fig. 6.2 Comparison of the relative spectral response of a photon detector and thermal detector for (A) constant incident radiant power and (B) photon flux [6].

cooled, as a direct result of the thermal generation-recombination of carriers that increases the effect due to noise. In photon detectors, in order to reduce the thermally induced generation of charge carriers, the temperature of the detector must be maintained at cryogenic levels of 77 K or below [4], by using cooling systems such as the Stirling cooling and Peliter cooling. This increases the cost, complexity, weight, size, and power consumption of the infrared detection imaging systems [5]. The low cost and portability are the main reasons why uncooled thermal detectors are considered as an alternative to replace cooled photon detectors. Additionally, some applications do not necessarily require the sensitivity that is provided by the cooled photon detectors. A schematic of the relative performance of the photon detector with thermal detector is shown in Fig. 6.2 [6].

6.3 Thermal infrared detector

Thermal detectors have been utilized extensively for applications in infrared detection, where it is impractical to use cooled infrared detectors (for example, military systems, space applications, and portable electronics where system weight is a primary concern). There are several types of thermal detectors. These are based on the nature of the interaction between heat and the sensing element. For example, thermal expansion is utilized in Golay cells [7]. Thermopiles and thermocouples are based on the thermoelectric effect [8]. The change in resistance due to the incident thermal radiation is the working principle of a bolometer [9]. Pyroelectric sensors work on the principle of heat induced change in polarization that results in electrical potential [10]. Among the thermal detectors, the most commonly used is the bolometer. Advantages of bolometer include operation at the desired temperature including room temperature, flat response over the infrared spectrum, and conservative time response; this allows us to consider bolometer infrared detectors as the ideal choice in the development of uncooled infrared detectors and imaging systems.

6.4 Bolometer-fundamentals

A bolometer is an infrared thermal detector, used in the infrared spectral range, that transduces the optical radiation absorbed into an electrical readable signal. A typical bolometer has two main components—an optical radiation absorber material coupled to an active element material. The absorbed optical radiation is converted into heat in the lattice by the absorber material, which produces changes in a measurable physical property (electrical resistivity) of the active element material (thermosensing material) and this change is transduced into an electrical output signal [11]. This information on change in resistance is electrically transferred to the read-out integrated circuit (ROIC) for further processing, and the end result is a thermal output/image.

The schematic of each pixel in the microbolometer array, along with the functionality of each layer, is shown in Fig. 6.3 [12].

An example of the details of a structure of a vanadium oxide based uncooled microbolometer is shown in Fig. 6.4 [13]. In this structure, vanadium oxide is utilized as the thermal sensitive layer. The absorption of the incident infrared radiation is enhanced by the ($\lambda/4$) cavity structure [13,14]. The gap between the silicon nitride membrane and the bottom reflector is 2 μm. This design (see Fig. 6.4) corresponds to a wavelength range of 8 to 14 μm for the operation of the microbolometer. The combination of TiN film as the absorber and the ($\lambda/4$) cavity structure leads to a perfect absorber.

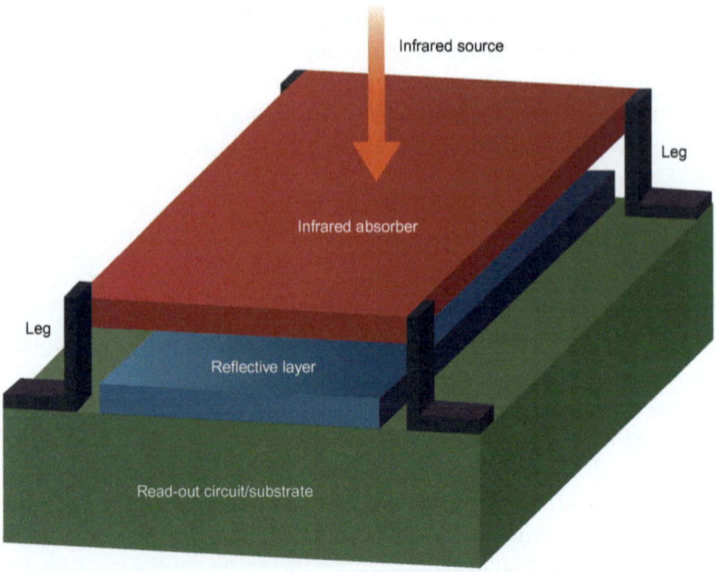

Fig. 6.3 Schematic of each pixel in a microbolometer array; an infrared-absorbing surface is elevated above the substrate and is thermally isolated from the adjacent pixels. The low mass of the absorber layer makes the layer highly sensitive to the change in temperature from absorption of heat. Read-out circuits are typically in the base layer. This base layer may be coated with a reflective layer to reflect the transmitted infrared photons and thus increase the absorption and hence the sensitivity of the pixel [12].

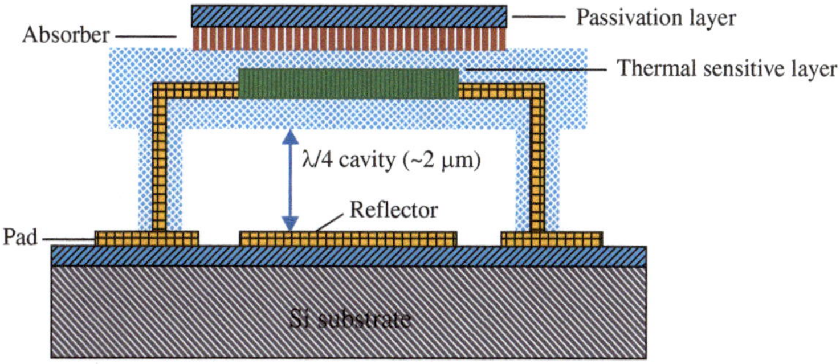

Fig. 6.4 An example of the structure of a vanadium oxide microbolometer [13].

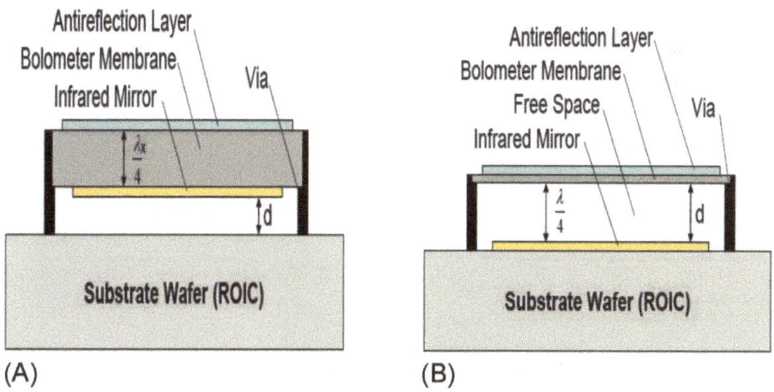

Fig. 6.5 Cross-sectional image of two uncooled microbolometer designs with resonant optical cavities for high absorption of the incident infrared radiation [15].

The top layer in the structure is a passivation layer of silicon nitride. The metal pads are connected through gold wire [13].

In another approach [15], Niklaus et al. have reported a detailed performance model and tradeoffs of uncooled infrared bolometer arrays of two different designs, based on uncooled microbolometer arrays. The resonant optical cavities in the microbolometer in Fig. 6.5A emit infrared radiation from both the upper and lower surfaces of the membranes of the bolometer. In this design, only the upper surface of the bolometer membrane absorbs the incident infrared radiation and the mirror layer at the lower surface of the bolometer reflects the incoming infrared radiation. In the microbolometer design, as shown in Fig. 6.5B, only the upper surface of the bolometer acts as the radiating surface; the infrared mirror reflects the infrared radiation that is emitted from the lower surface of the bolometer to the substrate. Concurrently, both the upper and lower surfaces of the bolometer membrane absorb the infrared radiation.

While there have been several approaches to the design of the uncooled microbolometer in the industry, an example of Mitsubishi's dual-diode sensing microbolometer is presented in Fig. 6.6 [12,16]. This structure leads to reduced diode area and thus results in significant increase in the number of diodes in a pixel and hence an enhanced infrared responsivity of the pixel.

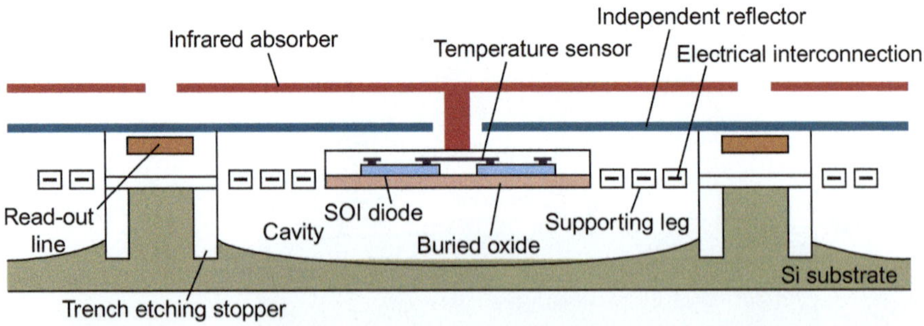

Fig. 6.6 Mitsubishi's dual-diode sensing microbolometer has a three-layer structure. The top layer absorbs thermal radiation and transmits heat to the bottom heat-sensing layer through the vertical pillar. The middle layer reflects any IR light that reaches it back to the absorbing layer to increase its sensitivity. The bottom layer paired SOI diodes sense the temperature of the absorber through the pillar [12,16].

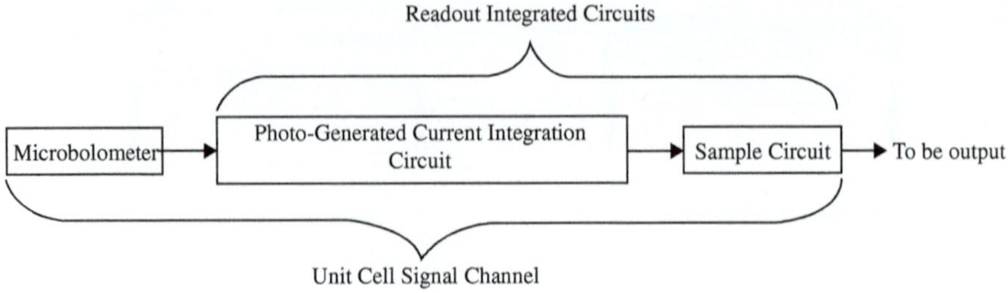

Fig. 6.7 The infrared response signal channel for a typical unit cell microbolometer.

The microbolometer infrared focal plane array (IRFPA) comprises of microbolometer array and read out integrated circuits (ROICs). The variation in resistance of each of the detector in the array is caused by the incident infrared radiation from a target that can be transformed to voltage or current signals to be output by the ROICs to form infrared image of the target. A schematic of the infrared response signal channel for a typical unit cell microbolometer is shown in Fig. 6.7.

In commercial uncooled infrared bolometer arrays, the typical microbolometer pixel pitch ranges from 12 to 37.5 μm [17]. In recent years, there has been significant developments in pixel technology [17–21]. The evolution in pixel size over the last 30 years is summarized in Fig. 6.8 [17].

The fundamental component of an infrared imaging system is the infrared imager array. The physical distance between the centers of two adjacent pixels in the imager array is known as the pitch. In determining the trade-off between resolution, sensitivity and size of the pixel, the following should be duly noted: smaller pixels will facilitate the manufacture of more devices (pixels) per wafer, resulting in improved yield and reduced cost per pixel as well as in the cost of the lens [17]. By addressing packaging technology at the wafer level, manufacturing costs can be reduced significantly [17]. It must however be noted that since larger pixels collect more incident radiation, they will have less inherent noise and will therefore yield better quality image [22]. The resolution of an imager is determined by its ability to react to the incident

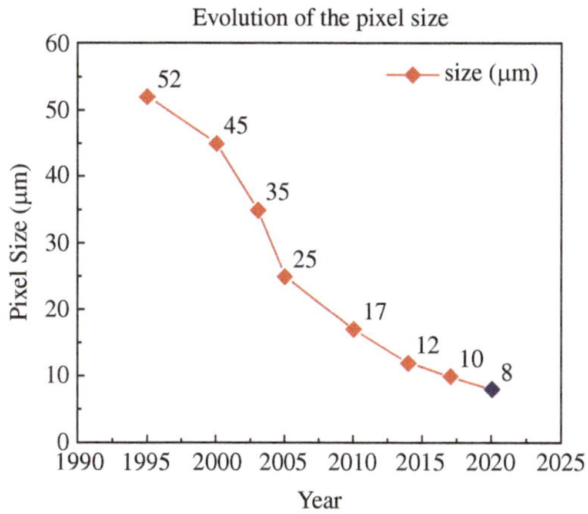

Fig. 6.8 Evolution of pixel size (μm). Redrawn based on Pulsfort U., Challenges and trends on microbolometer design, Stemmer Imaging Technology Forum (2017).

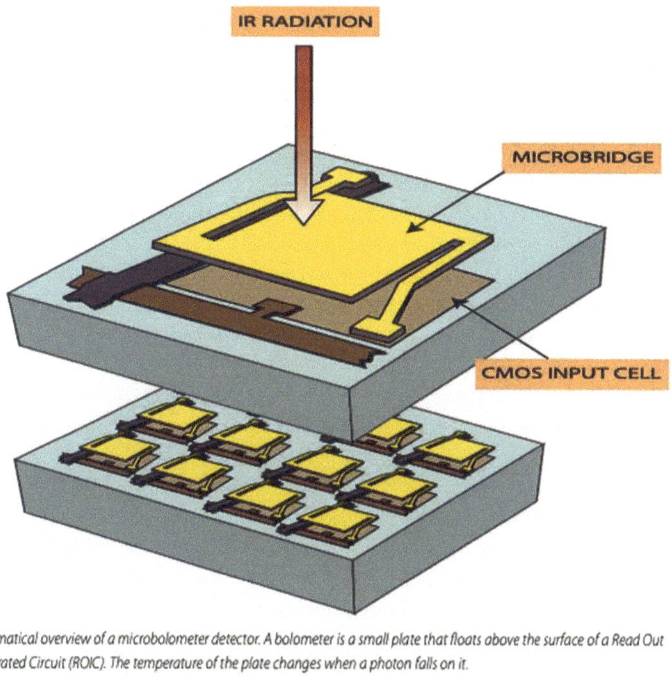

Fig. 6.9 Schematical overview of a microbolometer detector (http://www.flirmedia.com/MMC/CVS/Appl_Stories/AS_0015_EN.pdf) [22].
Reproduced with permission from FLIR.

radiation, which is proportional to the number of pixels. Thus, smaller pixels, with smaller pitch, will lead to larger number of pixels per unit area of the imager, leading to a higher resolution for the imager. A schematic of the various elements of a detector is presented in Fig. 6.9 [22].

The IR radiation that is absorbed by each pixel can result in a fraction of nW. The resulting radiation signal current can also be as low as a few nanoamperes. The design and implementation of ROICs for microbolometer arrays is challenging. Furthermore, the various sources of noise at these minute signal compounds the challenges involved in the design of microbolometer ROICs. The sources of noise in a microbolometer include the following [23]:

- Thermal noise in resistances;
- flicker noise, a type of electronic noise with a 1/*f* spectral density; the effects of flicker noise are usually observed at low frequencies in electronic components;
- burst noise;
- thermal fluctuations in the temperature of the bolometer;
- noise associated with the incident radiation.

The various sources of noise in an uncooled vanadium oxide based microbolometer is summarized in Table 6.1 [24].

Table 6.1 Sources of noise in uncooled vanadium oxide based microbolometers [24].

Source		3D noise designation	Sources	Comment
	• **Pixel Noise**			
1	• Rapidly varying	σ_{tvh}	AKA temporal or white noise Only source of noise in ideal system Less important for systems with human observers than spatial noise	Detector; electronics
	• **Slowly varying**		AKA spatial or fixed pattern noise	
2	• High spatial frequency	σ_{vh}	Especially important for systems with human observers	Detector; electronics
3	• Low spatial frequency		Can often be removed or ignored by system	Temp. drifts in optics/ housing; shutter internal to housing; Imperfect gain or offset
	• **Row Noise**			
4	• Rapidly varying	σ_{tv}		Electronics
5	• Slowly varying	σ_{v}		Electronics; imperfect gain or offset correction

Table 6.1 Continued

Source		3D noise designation	Sources	Comment
	• **Column noise**			
6	• Rapidly varying	σ_{th}		Electronics
7	• Slowly varying	σ_h		Electronics; imperfect gain or offset correction
	• **Frame bounce**		Important for radiometric systems; less important for other systems unless magnitude is very large	
8	• Rapidly varying	σ_t		Electronics, rapid scene temperature variations
9	• Slowly varying			Electronics

Niklaus et al. [15] have performed a detailed analysis of the noise equivalent temperature difference (*NETD*) of uncooled infrared bolometer arrays. Their model suggests that an *NETD* of ~150 mK can be attained with these arrays operating in vacuum with pressures of ~6 mbar. This facilitates to simplify wafer-level vacuum packaging as well as long-term reliability, leading to reduced costs of infrared imaging systems [15]. Typically, cooled photon-based infrared detectors can approach an *NETD* value of 10 mK while *NETD* of uncooled microbolometers is in the range of 20–200 mK [25,26].

The ROIC is essential to carefully apply noise cancellation techniques to be insensitive to process and supply voltage variations [27]. The IR detector consists of the optical system followed by the electrical system. The optical system comprises of a group of lenses that focus the incident IR power into the pixels. The lenses are made of IR transparent materials. The transmittance spectra of various optical materials are shown in Fig. 6.10 [28]. The electrical system is composed of array of microbolometers, decoder, analog to digital converters (ADCs), and digital multiplexer. The array consists of $M \times N$ pixels. Each pixel is a microbolometer that is capable of absorbing the IR radiation and has a ROIC to convert the radiation into electrical signal in itself. The ROIC is attached to the microbolometer on its bottom using flip chip technology or monolithic technology. Each row or column is activated one at a time by using row or column decoder.

In Table 6.2, a summary of the representative commercially available uncooled infrared bolometer array, along with their specifications, is presented [29].

The primary parameters of infrared systems are the detector size d and the F-number ($F/\#$) of the lens [29,30]. Due to their dependence on $F\lambda/d$ [29,31], these

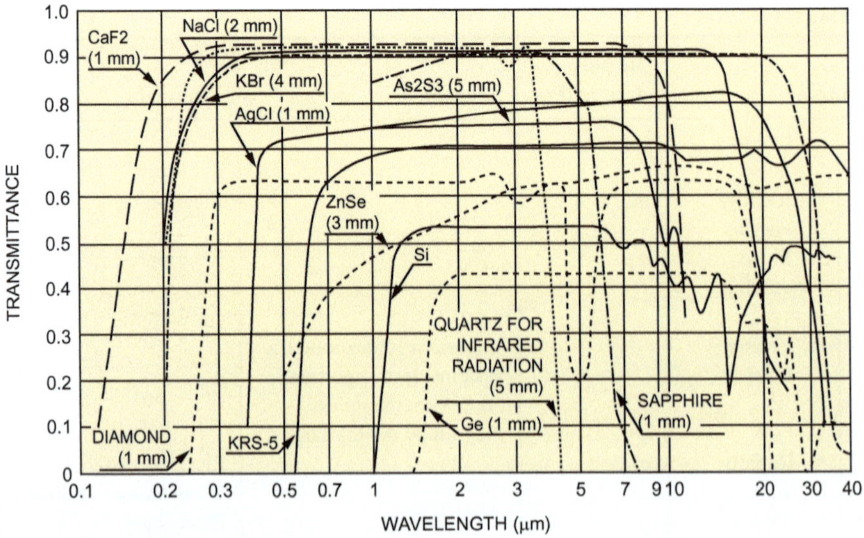

Fig. 6.10 Transmittance of various materials in the infrared region [28].

Table 6.2 A summary of representative commercially available uncooled infrared bolometer arrays [29].

Company	Bolometer type	Array format	Pixel pitch (μm)	Detector *NEDT* (mK) (f/1, 20–60 Hz)	Time constant (ms)
L-3 (USA)	VO_X bolometer	320 × 240	37.5	50	
www.l-3com.com	a-Si bolometer	160 × 120–640 × 480	30	50	
	a-Si/a-SiGe bolometer	320 × 240–1024 × 768	17	30–50	
BAE (USA)	VO_X bolometer	320 × 240	28	<35	<10–15
http://www.baesystems.com	VO_X bolometer (standard design)	640 × 480	12	<50	<10–15
	VO_X bolometer (standard design)	1024 × 768	17		
DRS (USA)	VO_X bolometer (standard design)	320 × 240	25	≤40	≤18
www.drsinfrared.com	VO_X bolometer (umbrella design)	320 × 240	17	≤40	≤14
	VO_X bolometer (umbrella design)	640 × 480, 1024 × 768	17	<40	≤14

Table 6.2 Continued

Company	Bolometer type	Array format	Pixel pitch (μm)	Detector *NEDT* (mK) (f/1, 20–60 Hz)	Time constant (ms)
Raytheon (USA) http://www.raytheon.com/capabilities/products/uncooled/	VO_x bolometer	320 × 240, 640 × 480	25	30–40	
	VO_x bolometer (umbrella design)	320 × 240, 640 × 480	17	50	
	VO_x bolometer (umbrella design)	1024 × 480, 2048 × 1536	17		
ULIS (France) www.ulis-ir.com	a-Si bolometer	80 × 80	34	<100	
	a-Si bolometer	160 × 120	25	<60	<10
		320 × 240	12	<60	<10
		384 × 240	17	<55	<10
		640 × 480, 1024 × 768	17	<50	<10
SCD (Israel) www.scd.co.il	VO_x bolometer	384 × 288	25	<20	22
	VO_x bolometer	384 × 288	25	<35	16
	VO_x bolometer	640 × 480	17	<35	16
	VO_x bolometer	1024 × 768	17	<35	14
FLIR Systems http://www.flir.com	VO_x bolometer	640 × 512	17	<60	<12
	VO_x bolometer	336 × 256	17	<50	<15
NEC (Japan) http://www.nec.com	VO_x bolometer	320 × 240	23.5	<75	
	VO_x bolometer	640 × 480	23.5	<75	
	VO_x bolometer	640 × 480	12	60	
	VO_x bolometer	320 × 240	23.5	NEP < 100 pW[a]	

[a] At 4 THz.

two primary parameters have a major influence on both detection/identification range and the noise equivalent difference temperature (*NEDT*) [29]. The equations for *Range* and *NEDT* are as follows [29]:

$$Range = [D\Delta x/M\lambda][F\lambda/d], \tag{6.1}$$

$$NEDT \approx \{2/[C\lambda(\eta\varphi_B{}^{2\pi}\tau_{int})]\}[F\lambda/d], \tag{6.2}$$

where λ is the wavelength, D is the aperture, M is the number of pixels required to identify a target Δx, C is the scene contrast, η is the detector collection efficiency, $\varphi_B^{2\pi}$ is the background flux into a 2π FOV (field of view), and τ_{int} is the integration time.

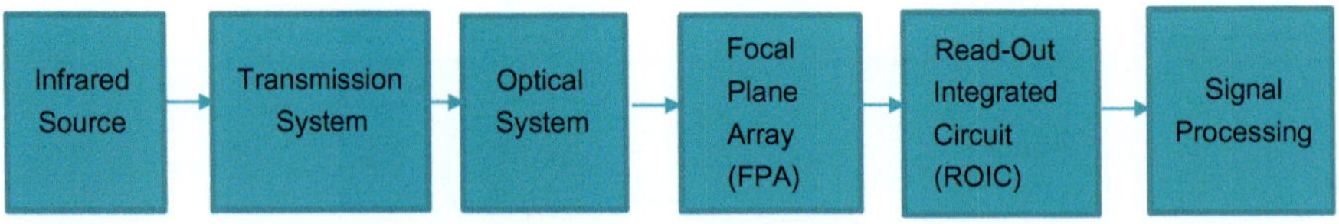

Fig. 6.11 Sequence of conversion of the incident infrared radiation into IR video image. Redrawn based on Khoshakhlagh A., Design of a Readout Integrated Circuit (ROIC) for Infrared Imaging Applications, Thesis, Master of Science in Electrical Engineering, University of New Mexico, http://digitalrepository.unm.edu/ece_etds/135 (2011).

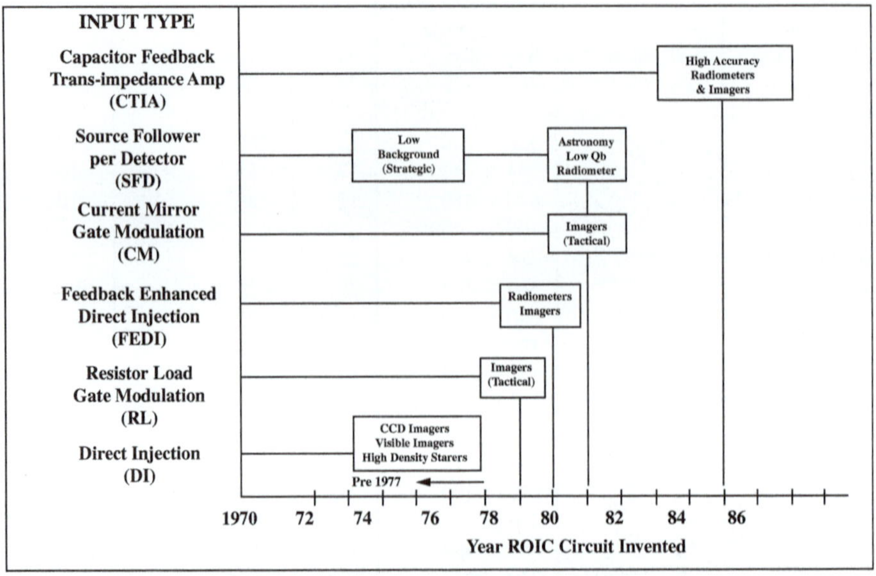

Fig. 6.12 Timeframe of infrared circuits and applications [36].

The various steps involved in the sequence of conversion of the incident infrared radiation into a video signal is described in a schematic in Fig. 6.11 [32].

6.5 Read-out integrated circuits

The ability to integrate a large number of detectors in the form of two-dimensional sensor arrays is made possible through read-out integrated circuits (ROICs). ROICs provide the signal conversion and multiplexing capabilities that are required to handle the large numbers of detector pixel elements [33]. The history of the transition from the digital imaging technology of charge-coupled devices (CCDs) [34,35] to the present day ROICs is summarized in Fig. 6.12 [36]. Factors such as the improvements in silicon processing technology, device, and wafer scaling and enhanced performance requirements of the readout circuitry have been responsible for this evolution [36].

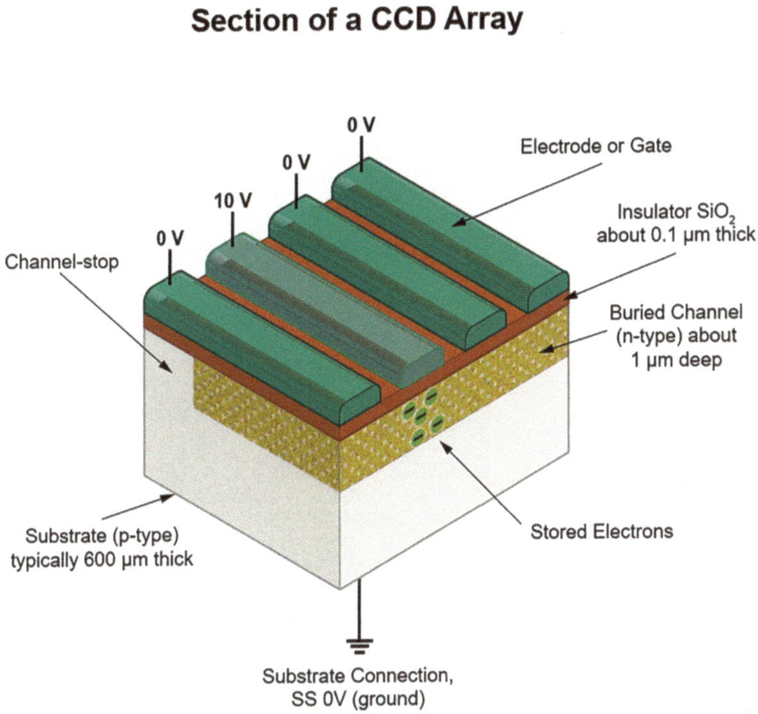

Fig. 6.13 Schematic of the section of a CCD array [35].

Willard S. Boyle and George E. Smith were awarded the 2009 Nobel Prize in Physics for their invention of the CCD sensor. They shared the prize with Charles K. Kao for his contributions to transmission of light in fibers for optical communication [37].

A schematic of the section of the CCD array is presented in Fig. 6.13 [35]. CCDs operate on the principle of interaction of the incident photon with the photoreceptor, generally an epitaxial layer of p + (boron) doped silicon on p++ silicon substrate, and the resulting output is stored within the sensor as an electrical charge and is subsequently converted to voltage, buffered, and transmitted as an analog signal upon closure of the shutter [38]. CMOS (complementary metal oxide semiconductor) image sensors consist of each pixel having a photoreceptor performing all the required electronic functions, including charge to voltage conversion, amplification, noise correction, and digitization, facilitating the sensor to output digital data directly [38]. While CCDs offer higher quantum efficiency and lower noise compared to CMOS technology, they require more power. The fundamental advantage of CMOS technology is that its output is digital [39]. CCDs continue to be utilized in astronomy, digital photography, electron microscopy, medical fluoroscopy, and optical and UV spectroscopy [39].

An example of the hybrid structure of the focal plane array (FPA) and the ROIC is shown in Fig. 6.14 [32,40]. In Fig. 6.15, the structures of hybrid (a, b) and monolithic array (c, d) technologies are illustrated [41]. In flip-chip technology (see Fig. 6.15a),

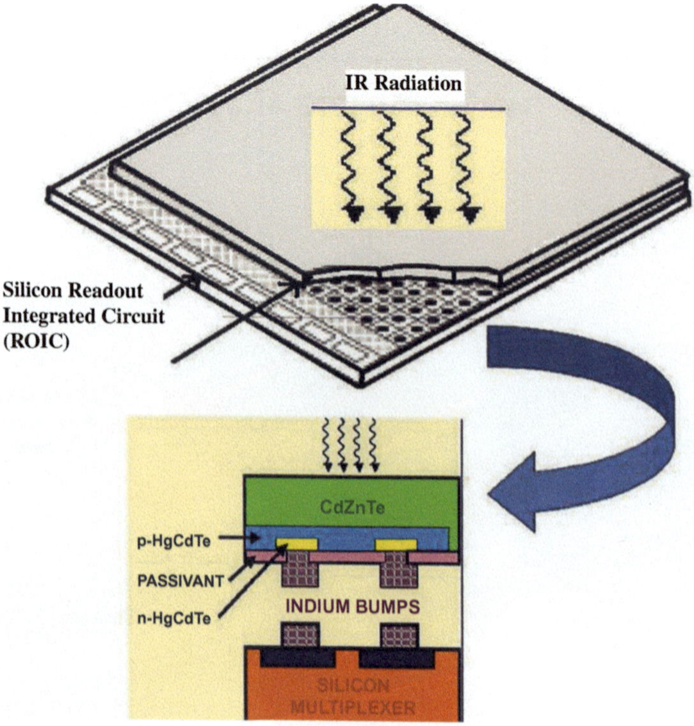

Fig. 6.14 Hybrid structure of FPA and ROIC [32,40].
From Recent infrared detector technologies, applications, trends and development of HgCdTe based cooled infrared focal plane arrays and their characterization; R.K. Bhan, V. Dhar; Opto-Electronics Rev. 27(2), 2019, 174–193.

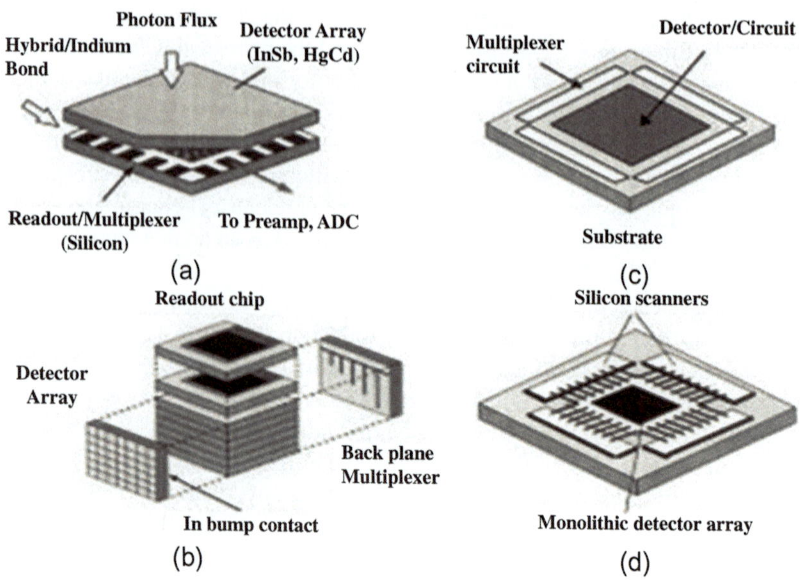

Fig. 6.15 Schematic of hybrid array structures (a), (b): (a) flip-chip array technology [41,42]; (b) *Z*-plane technology [41,43]; (c) structure of monolithic array technology [41,43]; (d) structure of pseudomonolithic array technology [41,43].

the aligned pixel of the IR detector and the silicon readout chip are held together by an indium bump. In Z-plane technology (see Fig. 6.15b), the readout devices are stacked one on top of the other followed by mounting the detector array to the third dimensional plate on the edge. In hybrid technology, flip-chip technology is the most used structure in the semiconductor industry. While the Z-plane technology offers the benefit of integrating all the electronic functions and circuitry in the readout chip, the thickness of the readout chip limits the image resolution. In general, the uniformity of indium bumps, chip alignment, thermal expansion, and potential mechanical damage to the detectors are some of the key elements that need to be considered in implementing hybrid array technology. Monolithic array technology, described in Fig. 6.15c, has the inherent advantage of permitting the fabrication of the IR detector and the readout circuits within the same silicon substrate. The compatibility of the detector materials with the silicon substrate is the limitation in this technology. The via-hole technique utilized in pseudomonolothic array integration technology, described in Fig. 6.15d, permits the integration of the readout chip and the detector chip. However, in this method, the surfaces of both the readout chip and the IR detector chip must be optically flat for integration.

An infrared imager consists of two key components—the microbolometer and the readout integrated circuit (ROIC) [44]. Fundamentally, the microbolometer is a thermal infrared detector. In order to enhance its sensitivity, it is critical that the microbolometer be thermally isolated from its surroundings. The support structure of the microbolometer must have an extremely low thermal conductance [45]. A resistor with a large temperature coefficient of resistance (TCR) is the active element, i.e., sensor, in the microbolometer. In Fig. 6.16, a basic thermal block diagram of a bolometer is presented. In this figure, P is the incident infrared power, C is the heat capacity of the pixel, T is the temperature of the pixel, G is the thermal conductance, α is the absorption coefficient, T_0 is the heat sink temperature, and t is the time.

The thermal balance for such a bolometer is given by

$$\alpha P = C(dT/dt) + G(T - T_0). \tag{6.3}$$

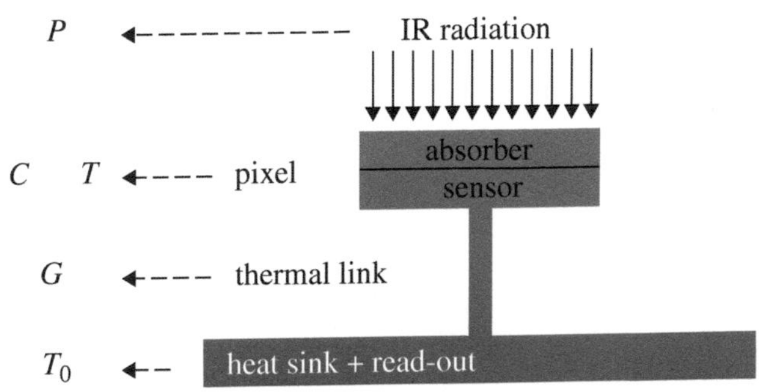

Fig. 6.16 A basic thermal block diagram of a bolometer [45].

From Eq. (6.3), it can be noted that in steady-state, the temperature increase of the bolometer is proportional to the absorbed power and is inversely proportional to the thermal conductivity [45]. The solution to differential equation in Eq. (6.3) leads to the thermal time constant $\tau = (C/G)$ for the new thermal equilibrium [45]. By decreasing the thermal conductance G and optimizing the absorption efficiency, the response from the bolometer is maximum. MEMS (microelectromechanical structures) technology, in the form of bulk micromachining and surface micromachining techniques, has been the enabler to address the stringent requirements for achieving the thermal isolation of the sensor in the manufacture of microbolometers. The basic structure of an uncooled microbolometer consists of a thermally isolated resistor coupled with a thin dielectric membrane which also acts as the absorber of the incident infrared radiation [46].

The ROIC is the interface between the microbolometer array and the thermal image display [44]. Generally, the pixel pitch of IR FPAs is reduced with increasing array size and resolution. Moreover, the total power dissipation of IR FPAs is limited by the image system. These two major factors often put constraints on circuit design, space, and complexity. Thus, the design of IR FPA readout electronics requires a tradeoff between circuit performance and complexity.

The fundamentals of ROICs and their applications for uncooled microbolometers have been discussed in several books, papers, and theses in the literature. Among these, case studies presented in three sources [32,41,44] are considered in detail in this chapter. An example of a basic configuration of the readout circuit for the microbolometer FPA is shown in Fig. 6.17 [44]. The microbolometer, R, is biased with a constant voltage V through a load resistor R_L. The interaction of the infrared radiation with the microbolometer, R, results in a small change in voltage, ΔV. The preamplifier

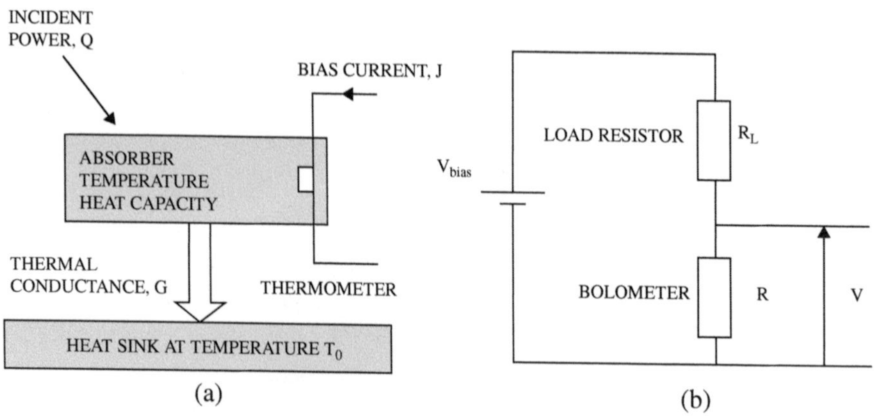

Fig. 6.17 (A) Schematic illustration of bolometer operation and (B) bolometer bias circuit. Reproduced with permission from Fig. 2 in R.K. Bhan, R.S. Saxena, C.R. Jalwania, and S.K. Lomash, Uncooled Infrared Microbolometer Arrays and their Characterisation Techniques, Defence Sci. J. 59(6), 2009, 580–589.

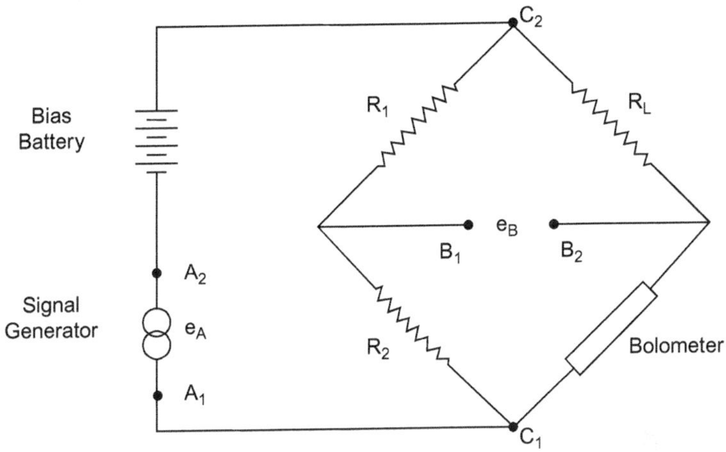

Fig. 6.18 The basic elements of a bridge circuit that permits one to measure the responsivity of a bolometer by purely electrical means.
Reproduced with permission from Fig. 13 in R. Clark Jones, The general theory of bolometer performance, J. Opt. Soc. Am. 43(1) 1–14, (1953). https://doi.org/10.1364/JOSA.43.000001.

amplifies this voltage ΔV_d. However, in this approach, the large voltage V_d is also amplified. This results in overriding the IR signal [44].

The Wheatstone bridge structure, shown in Fig. 6.18 [44], is utilized in most practical circuits. The bridge is balanced when $(R_L/R_{Bolometer}) = (R_1/R_2)$. $R_{Bolometer}$ represents the resistance of the microbolometer at room temperature. This approach has the advantage of removal of the influence of bias voltage; the power supply induced fluctuations and noise are eliminated. In the implementation of the uncooled microbolometer for infrared imaging, one of the major challenges is self-heating. The readout process of the microbolometer requires that the microbolometer element be electrically biased. Thus, the microbolometer element is subjected to heating from the incident radiation as well as the current flowing in the microbolometer readout that leads to Joule heating. The thermal conductance of the microbolometer is significantly reduced due to the micromachined structure. This makes it difficult for the dissipation of heat generated by self-heating through thermal conduction to the substrate. Thus, the change in temperature of the microbolometer, induced by self-heating, is significantly greater than that due to the incident infrared radiation [44]. Therefore, readout circuits must compensate for self-heating effect.

An example of a schematic circuit of the 128 × 128 pixel FPA is shown in Fig. 6.19 [46]. The basic functionalities of the readout electronics for microbolometer focal plane arrays include the following: bias for the microbolometer, front-end amplification, signal conditioning, pixel addressing, digitization, and error correction circuits [44].

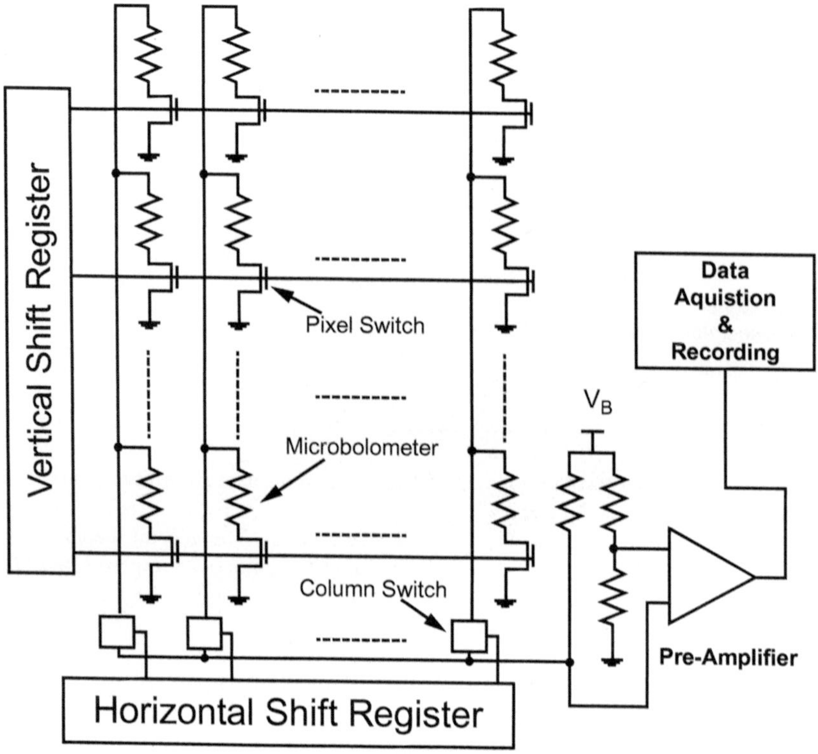

Fig. 6.19 Schematic circuit diagram of the 128 × 128 microbolometer focal plane array with integrated readout electronics [46].

One of the commonly used methods for compensation and cancellation of self-heating is to utilize a "blind" (dummy) microbolometer [44,47,48], as shown in Fig. 6.20. In this scheme, the detector bolometer and the blind bolometer are identical except that the blind bolometer is blocked from IR radiation and is therefore subject only to self-heating. This approach permits the subtraction of the signal of the blind bolometer from the output of the detector bolometer. The challenge in this methodology is to be able to shield the blind bolometer from the incident infrared radiation [44].

Another example of self-heating compensation circuit, based on the use of dummy microbolometer, is illustrated in Fig. 6.21 [44,49,50]. In this approach, the dummy microbolometer has a higher thermal conductivity than the detector microbolometer but has the same thermal mass as the detector microbolometer. The microbolometer is pulse biased, and no IR shielding is required.

A detailed analysis of CMOS readout techniques for IR imaging systems has been presented by Hsieh et al. [41]. Recently, Park et al. have reported their studies on "An uncooled microbolometer infrared imager with a shutter-based

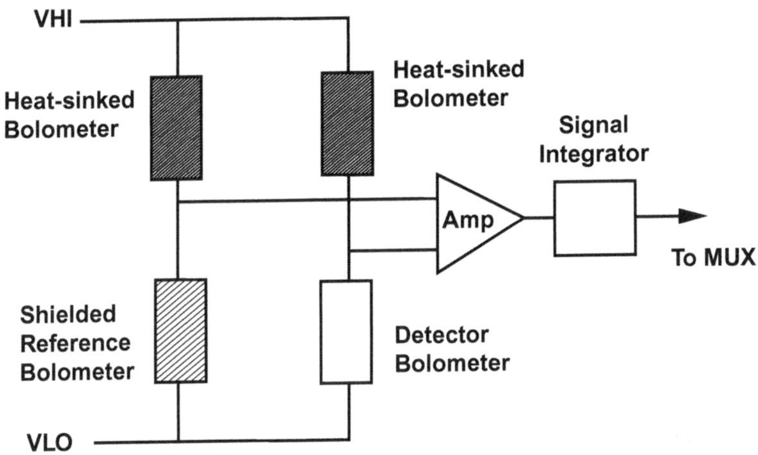

Fig. 6.20 Front-end readout circuit with self-heating cancellation using blind bolometer [44,47,48].
Reproduced with permission from Fig. 3 in Howard P.E., Advances in microbolometer focal plane technology at Boeing, Proc. SPIE 3379, 47–57 (1998).

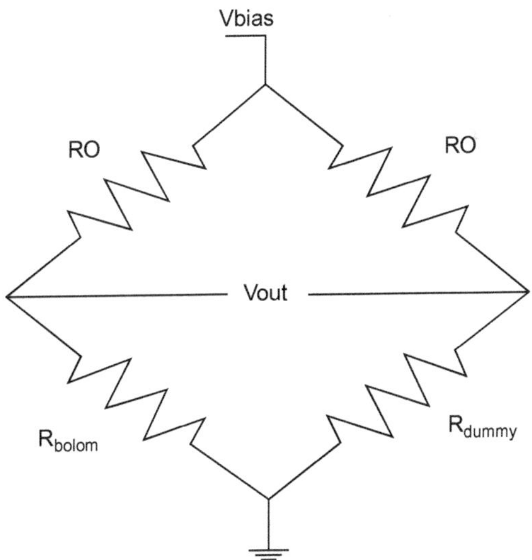

Fig. 6.21 Self-heating compensating circuits [44,49,50].

successive-approximation calibration loop" [51]. Their proposed approach, as shown in Figs. 6.22 and 6.23, has been shown to compensate effectively for fixed pattern noise that is caused by the nonuniformity of the uncooled microbolometers.

The summary of the design characteristics of the imager is presented in Table 6.3 [51].

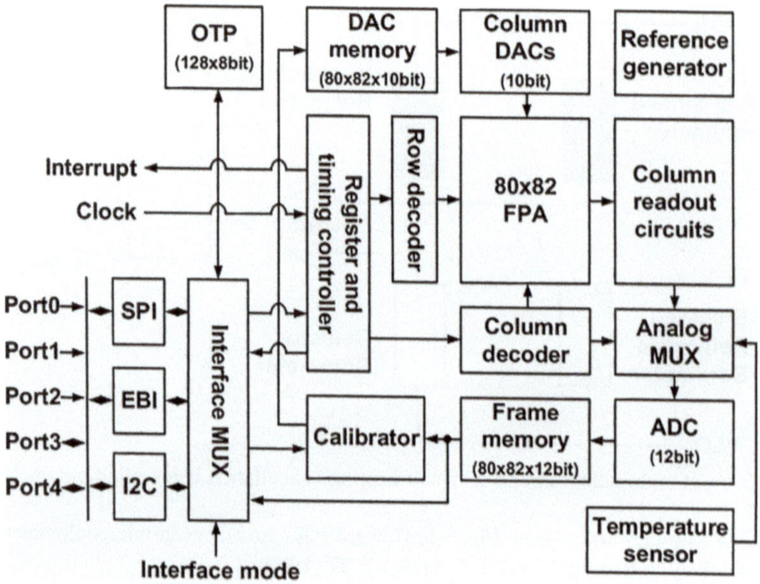

Fig. 6.22 Block diagram of the infrared imager [51].

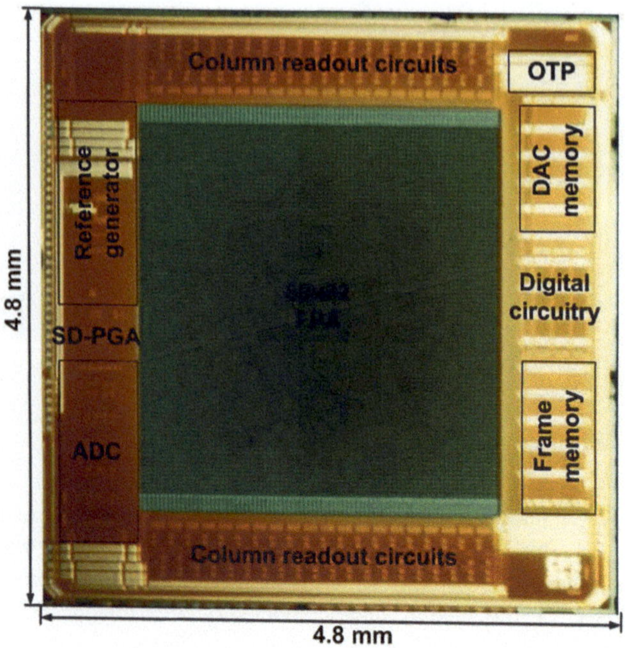

Fig. 6.23 Photograph of the imager [51].

Table 6.3 Summary of design characteristics of imager [51].

Parameter	Values
Technology	0.18 μm 1P5M CMOS
Supply voltage	3.3 V (analog and I/O), 1.8 V (digital)
Active pixels	80 × 82
Pixel size	35 μm × 35 μm
Spectral response	8–14 μm
NETD	188 mK
Frame-rate	1–120 fps
Clock frequency	1 MHz (analog), Max. 32 MHz (digital)
Interface	EBI, SPI, I2C
Memory	SRAM 9.8 kbyte (for frame memory) SRAM 8.2 kbyte (for DAC memory)
Operating temperature	−20°C ~ 60°C
Storage temperature	−50°C ~ 150°C
Chip area	4.8 × 4.8 mm^2
Power dissipation	50 mW (typical) 2 μA (power down)

References

[1] P.W. Kruse, The Photon Detection Process in Optical and Infrared Detectors, second ed., Springer-Verlag, New York Heidelberg Berlin, 1980, ISBN: 0-387-10176-4.

[2] A. Rogalski, Infrared detectors: an overview, Infrared Phys. Technol. 43 (2002) 187–210.

[3] A. Rogalski, Infrared Photon Detectors, *SPIE The International Society of Optical Engineering*, SPIE Optical Engineering Press, 1995, ISBN: 978-1420076714.

[4] R.B. Darling, S. Inawaga, Structure, properties, and MEMS and microelectronic applications of vanadium oxides, Sadhana 34 (2009) 531–542.

[5] G. Hyseni, et al., Infrared thermal detectors parameters: semiconductor bolometers versus pyroelectrics, WSEAS Trans. Syst. Control. 9 (2010) 238–241.

[6] A. Rogalski, M. Kopytko, P. Martyniuk, Antimonide-Based Infrared Detectors: A New Perspective, *SPIE The International Society of Optical Engineering Press*, 2018. Rogalski A., Graphene-Based Materials in the Infrared and Terahertz Detector Families: A Tutorial, *Adv. Opt. Photon.* 11(2), 314–379 (2019).

[7] M. Schossig, V. Norkus, G. Gerlach, A pneumatic infrared detector with capacitive readout circuit, in: *SENSOR+TEST Conferences* 2011—*IRS*^{2} *Proceedings*, 2011, pp. 109–114.

[8] N.M. Ravindra, B. Jariwala, A. Banobre, A.A. Maske, Thermoelectrics—Fundamentals, Materials Selection, Properties, and Performance, SpringerBriefs in Materials, 2019.

[9] M.I. Pech-Canul, N.M. Ravindra, Semiconductors—Synthesis, Properties and Applications, Springer, 2019.

[10] C.R. Bowen, et al., Pyroelectric materials and devices for energy harvesting applications, Energ. Environ. Sci. (2014), https://doi.org/10.1039/c4ee01759e.

[11] R.K. Bhan, et al., Uncooled infrared microbolometer arrays and their characterization techniques, Def. Sci. J. 59 (2009) 580–589.

[12] C. Hanson, Photonic Frontiers: Room Temperature IR Imaging: Microbolometer Arrays Enable Uncooled Infrared Camera, *Laser Focus World*, 2012.
[13] B. Wang, et al., Nanostructured vanadium oxide thin film with high TCR at room temperature for microbolometer, Infrared Phys. Technol. 57 (2013) 8–13.
[14] K.C. Liddard, Application of interferometric enhancement to self-absorbing thin film thermal IR detectors, Infrared Phys. 34 (4) (1993) 379–387.
[15] F. Niklaus, Uncooled infrared bolometer arrays operating in a low to medium atmosphere: performance model and tradeoffs, Proc. SPIE (2007).
[16] A. Takamuro, et al., Development of new SOI diode structure for beyond 17 μm pixel pitch SOI diode uncooled IRFPAs, Proc. SPIE (2011) 80121E, https://doi.org/10.1117/12.884556.
[17] U. Pulsfort, Challenges and trends on microbolometer design, in: Stemmer Imaging Technology Forum, 2017.
[18] C. Johnston, Trends in Thermal Imaging; Small-pixel LWIR Microbolometer Sensors and Cameras, *Techbriefs*, 2018.
[19] S. Tohyama, et al., Uncooled infrared detectors toward small pixel pitch with newly proposed pixel structure, Opt. Eng. 52 (12) (2013). 123105-1-123105-11.
[20] M. Kimata, Uncooled infrared focal plane arrays, EEJ Trans. Electr. Electron. Eng. 13 (2018) 4–12.
[21] K.-M. Muckensturm, et al., Measurement results of a 12 μm pixel size microbolometer array based on a novel thermally isolating structure using a 17 μm ROIC, SPIE Defense+Security, 2016, Baltimore, Maryland, in: Proceedings Volume 9819, Infrared Technology and Applications XLII, 2016, p. 98191N, https://doi.org/10.1117/12.2223608.
[22] FLIR Technical Note, http://www.flirmedia.com/MMC/CVS/Appl_Stories/AS_0015_EN.pdf.
[23] C. Ringqvist, et al., FLIR Project—Modeling Noise in Bolometer Signal, KTH Royal Institute of Technology, Stockholm, Sweden, 2018.
[24] M. Kohin, N. Butler, Performance limits of uncooled VO_x microbolometer focal plane arrays, Proc. SPIE (2004), https://doi.org/10.1117/12.542482.
[25] opgal, Intro to IR (Part 2): Cooled vs. Uncooled Cameras, Sensitivity, Resolution, Frame Rate, 2018. https://www.opgal.com/blog/thermal-cameras/intro-to-ir-part-2-cooled-vs-uncooled-cameras-sensitivity-resolution-frame-rate/. (Accessed 28 May 2020).
[26] W. Radford, et al., Microbolometer uncooled infrared camera with 20mK NETD, Proc. SPIE (1998), https://doi.org/10.1117/12.317595.
[27] A.H. Ahmed, M. Aboudina, S.E.D. Habib, Offset calibration technique for capacitive transimpedance amplifier used in uncooled infrared detection, Microelectron. J. 53 (2016) 56–64.
[28] G. Carugno, Infrared emission: A new possible approach to particle detectors, ICFA Instrumentation Bulletin Volume 17 (1998). http://www.slac.stanford.edu/pubslicfa/.
[29] A. Rogalski, History of infrared detectors, Opto-Electronics Review 20 (3) (2012) 279–308.
[30] G.C. Holst, T.S. Lomheim, CMOS/CCD Sensors and Camera Systems, JCD Publishing and SPIE Press, Winter Park, 2007.
[31] M.A. Kinch, State-of-the-Art Infrared Detector Technology, SPIE Press, Bellingham, 2014.
[32] A. Khoshakhlagh, Design of a Readout Integrated Circuit (ROIC) for Infrared Imaging Applications, *Thesis*, Master of Science in Electrical Engineering, University of New Mexico, 2011. http://digitalrepository.unm.edu/ece_etds/135.

[33] J.D. Vincent, S.E. Hodges, J. Vampola, M. Stegall, G. Pierce, Readout integrated circuits, chapter 7, in: Fundamentals of Infrared and Visible Detector Operation and Testing, second ed., Wiley, 2015.
[34] W.S. Boyle, G.E. Smith, Charge-coupled semiconductor devices, Bell Syst. Tech. J. (1970) 587–593.
[35] Teledyne Imaging, 50 Years of the Charge Coupled Device (CCD), 2020. www.teledyneimaging.com. (Accessed 23 May 2020).
[36] M.J. Hewitt, J.L. Vampola, S.H. Black, C.J. Nielsen, Infrared readout electronics: a historical perspective, Proc. SPIE (1994), https://doi.org/10.1117/12.178474.
[37] bell-labs, 2009 Nobel Prize in Physics: The Charge Coupled Device (CCD), 2020. https://www.bell-labs.com/about/recognition/2009-charge-coupled-device-ccd/. (Accessed 24 May 2020).
[38] P. Dempster, CCD and CMOS Sensors, Photonics Tech Briefs Magazine, 2014.
[39] newworldencyclopedia, Charge-Coupled Device, in New World Encyclopedia, 2020. https://www.newworldencyclopedia.org/entry/Charge-coupled_device. (Accessed 24 May 2020).
[40] P. Zarkesh-Ha, S. Krishna, Design of a Readout Integrated Circuit (ROIC) customized for a dual-band quantum dot infrared Focal Plane Array (FPA), in: Application to the UNM Research Allocation Committee Discipline-Specific Large Grant Competition, 2008.
[41] C. Hsieh, C.Y. Wu, et al., Focal-plane-arrays and CMOS readout techniques of infrared imaging systems, EEE Trans. Circuits Syst. Video Technol. 7 (4) (1997) 594–605. and references therein.
[42] E.R. Fossum, B. Pain, Infrared readout electronics for space science sensors: state of the art and future directions, Proc. SPIE 2020 (1993) 262–285.
[43] D.A. Scribner, M.R. Kruer, J.M. Killiany, Infrared focal-plane array technology, Proc. IEEE 79 (1) (1991) 66–85.
[44] Q. Xinbo, Readout Electronics for Microbolometer Infrared Focal Plane Array, PhD Thesis, Department of Electrical & Computer Engineering, National University of Singapore, 2004.
[45] C.V. Hoof, P.D. Moor, PolySiGe uncooled microbolometers for thermal IR detection, in: M. Henini, M. Razeghi (Eds.), Handbook of Infra-Red Detection Technologies, Elsevier, 2002, pp. 449–479.
[46] X. He, et al., Performance of microbolometer focal plane arrays under varying pressure, IEEE Electron Device Lett. 21 (5) (2000) 233–235.
[47] P.E. Howard, Advances in microbolometer focal plane technology at boeing, Proc. SPIE 3379 (1998) 47–57.
[48] T.K. Kaisha, Non-Contacting Thermometer, European Patent, No. EP 0798545A1, 1997.
[49] X. Gu, G. Karunasiri, J. Yu, G. Chen, U. Sridhar, W.J. Zeng, On-chip compensation of self-heating effects in microbolometer infrared detector arrays, Sensors Actuators A 69 (1998) 92–96.
[50] M.V.S. Ramakrishna, G. Karunasiri, P. Neuzil, U. Sridhar, W.J. Zeng, Highly sensitive infrared temperature sensor using self-heating compensated microbolometers, Sensors Actuators A 79 (2000) 122–127.
[51] Y. Park, et al., An uncooled microbolometer infrared imager with a shutter-based successive-approximation calibration loop, IEEE Trans. Very Large Scale Integr. VLSI Syst. 26 (1) (2018) 122–132.

Methods of calibration

7

7.1 Introduction

The evolution in infrared thermography is based on the development of small handheld and portable devices that are power-efficient and consist of low-cost IR sensors. As the terminology implies, uncooled microbolometer focal plane arrays (FPAs) operate at room temperature without the need for any cooling system. A combination of a low-cost fabrication approach with a reasonable sensitivity is the reason why uncooled microbolometers are widely used in thermal monitoring and control-related applications, such as in fire prevention, surveillance, and control of industrial processes as well.

Microbolometers are thermal infrared detectors. Thermopiles and pyroelectric detectors are other types of thermal infrared detectors, but advanced developments of microbolometer elements and their associated packaging has led to the manufacture of more efficient and accurate microbolometer FPAs, reducing the costs of the infrared camera [1].

The working principle of the microbolometer is based on the temperature dependence of an active physical property (in this case, its electrical resistance) of the thermosensing sensor material. The absorbed infrared radiation changes the temperature of the microbolometer element, thereby changing its electrical resistance. The sensitivity of the microbolometer is quantified by its temperature coefficient of resistance (TCR). Each pixel of a microbolometer sensor array acts as a single sensor element. A read out integrated circuit (ROIC) converts the change in electrical resistance into an output voltage that is proportional to the absorbed infrared radiation.

The internal thermal energy created by the microbolometer element and the ROIC generates errors in the conversion of the absorbed energy from the radiation source. Therefore, a system calibration is required in order to reduce the noise in the image. The raw electrical output signal data comprises of the signal voltages of the entire FPA and contains temperature information from the source scene and also from the interior of the camera.

An uncooled IR camera consists of the following main components: infrared optics, microbolometer sensor FPAs, and processor-based camera electronics. Fig. 7.1 shows the signal flow in an uncooled infrared camera [2].

In general, there are various sources of thermal radiation in an infrared camera. A schematic describing these sources is presented in Fig. 7.2 [3].

7.2 Radiometric calibration methods

The thermal calibration of uncooled thermal infrared cameras for absolute temperature measurements is a time-consuming process. It is a complex and lengthy process that contributes significantly to the manufacturing costs of an infrared camera. The

Microbolometers. https://doi.org/10.1016/B978-0-08-102812-4.00012-7

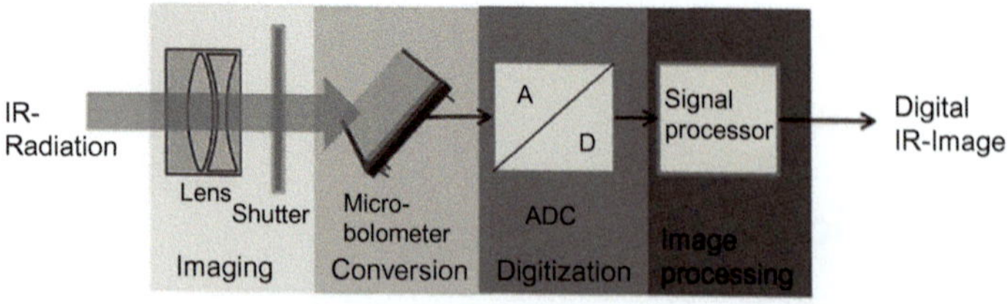

Fig. 7.1 Signal flow in uncooled IR camera [2].

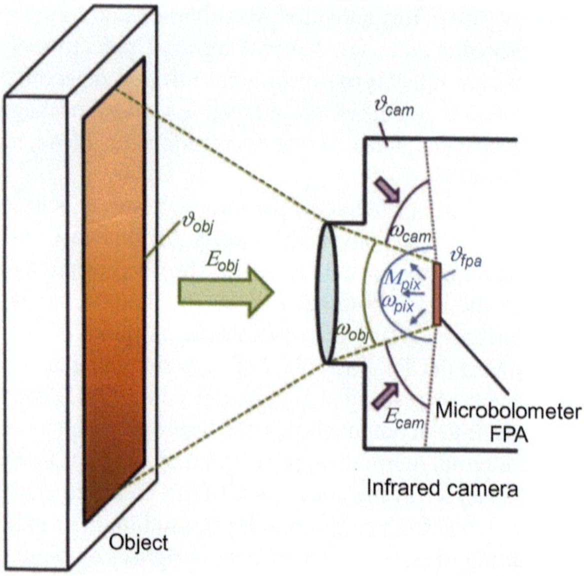

Fig. 7.2 Radiation components in an infrared camera. E_{obj} radiation derived from the object depending on its temperature ϑ_{obj} covering the projected solid angle ω_{obj}, E_{cam} radiation from the camera interior related to the camera temperature ϑ_{cam} and the solid angle ω_{cam}, as well as the radiant exitance M_{pix} of the sensor depending on the sensor temperature ϑ_{fpa} emitted into the projected solid angle ω_{pix} [3].

thermal calibration process takes into consideration the electronic and optical properties of an infrared camera and its components in order to provide an accurate representation of a thermal scenery.

Obtaining quantitative and qualitative data from uncooled thermal infrared imagers requires a precise radiometric calibration to relate the output signal from the camera to radiance or brightness of the radiation source temperature. Uncooled microbolometer detectors do not use any cooling system to stabilize the FPA temperature and need to be calibrated continuously. The output of uncooled imaging cameras corresponds to changes in scene temperature and FPA temperature.

In order to measure the temperature of thermal sources with accuracy, by utilizing uncooled FPAs based on microbolometer-based infrared cameras, one has to deal with a disturbing radiation flux due to the ambient temperature derived from the interior of the camera enclosure. In the FPA without any temperature stabilization system, its responsivity is affected by the temperature changes of the camera. Heat dissipated by the microbolometer element, biasing voltage variation, and changes in the temperature of the observed scene produce a signal output noise that affects the responsivity and the noise equivalent detector temperature (NEDT) of the detector. This noise degrades the temperature resolution, the accuracy, and image quality of the thermal camera. A schematic of the sequence of calibration of a microbolometer infrared camera is presented in Fig. 7.3 [2].

Advancements in MEMS-based microbolometer thermal infrared detectors have led to the manufacture of infrared cameras that operate without any active temperature stabilization system. The response of these infrared cameras varies with their own temperature, and thermal calibration is required to compensate for internal thermal errors. In microbolometer infrared cameras that lack thermal stabilization, the correction for the FPA temperature dependence can be implemented using the sequence of steps as described in Fig. 7.4 [4].

A number of methods have been developed in the infrared imaging industry for radiometric calibration such as a polynomial curve fit [5] to relate the variation in the output digital number to the variation in the FPA temperature, alterations to camera parameters [6–9], Kalman filters to update the gain, a per-pixel calibration at every FPA temperature to be experienced during operation [10], corrections based on measurements of the camera response over a wide range of temperatures [11, 12], an autoregressive moving average to compensate for dynamic changes in the camera's FPA temperature [13], use of a mechanical shutter [14], and use of one or multiples of external blackbody sources.

The two most used methods of updating the thermal calibration of uncooled infrared cameras is to periodically calibrate them in the presence of one or multiples of external blackbody sources and the use of a mechanical shutter [14].

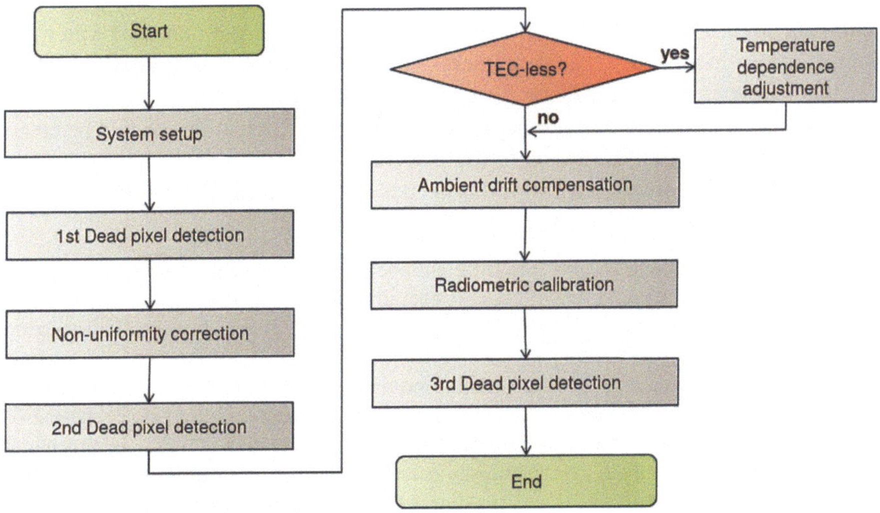

Fig. 7.3 Sequence of calibration [2].

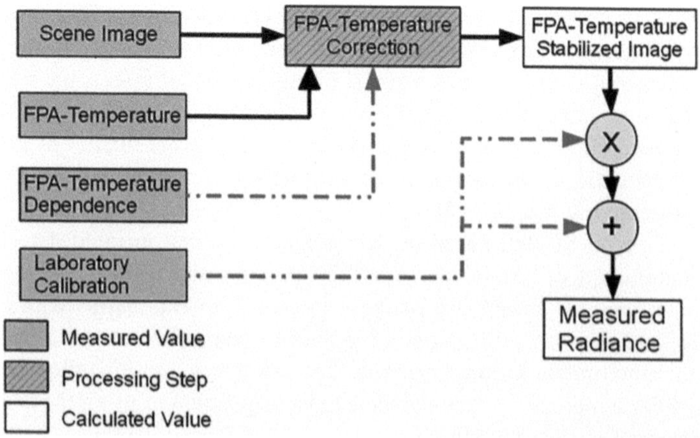

Fig. 7.4 Flow diagram for stabilization and radiometric calibration using only the FPA temperature [4].

One method to maintain a highly accurate calibration is to expose the infrared camera to one or more external blackbody sources during deployment, and each measured reading should be close to one another in time to each scene of the image, to create a relationship between the digital number and the radiometric value, and dividing the nonlinear characteristic to piecewise linear sections. The level of correction depends on the number of sources used; while one single blackbody image source enables a one-point offset correction, images from two or more blackbody sources allow for correcting both offset and gain. This technique is very commonly used for long-term deployment of infrared vision and thermal cameras [15, 16]. An alternative approach is to incorporate the FPA temperature into the calibration algorithm.

In general, infrared cameras that require cooling utilize Stirling engine or thermoelectric cooler. In uncooled thermal cameras, the microbolometer is not stabilized to a constant temperature. This leads to fluctuations in the sensor temperature along with the temperature of the camera. These sensor temperature fluctuations should be taken into account during the calibration of the camera [17].

There are several other corrections that need to be incorporated in the calibration of the infrared camera [17, 18]. These include the following:

- Nonuiformity correction (NUC): The different operating points of the individual pixels of a microbolometer imager leads to NUC. Generally, this is corrected by implementing a smoothing process with a goal to equalize their performance. This is performed by utilizing procedures such as polynomial fit.
- Defective pixel correction: This correction is required when the pixels do not work or their performance parameters vary greatly from the mean. This is a characteristic of the sensor array and is generally specified by the manufacturer. The correction of these pixels is based on their location and their interpolation is based on the data obtained from neighboring pixels. The main objective of this correction is to have a high-quality visual image rather than a high-quality radiometric value. Such corrections are rigorously required for cameras

that utilize larger number of pixels and are designed to yield high spatial and temporal temperature resolution.

- Shutter correction: This refers to the correction required due to the radiance of the camera interior that also varies with sensor temperature. Current uncooled thermal cameras perform an automatic shutter correction based on the time or change in sensor temperature.
- Radiometric calibration: This refers to establishing the relationship between the response of the sensor and the temperature of the object. It is possible to approximate the sensor output signal with a Planck curve. The physical principles governing the laws of Blackbody Radiation and emissivity models are implemented in such corrections.
- Temperature dependence correction: This refers to the effect of the sensor temperature on the response of the sensor. A linear correction that considers the signal from the object and the signal from the camera is typically used to perform this type of correction. This correction is a function of the camera temperature. Such correction procedures may require the use of dynamic models to reflect the nature of the temporal dependence of temperature measurements.

In the current approach to the manufacture of uncooled thermal cameras, NUC, defective pixel correction, and shutter correction are corrected by the firmware that is included in the system. Radiometric calibration in devices is corrected by the digital acquisition system. However, temperature dependence correction is not usually performed, which leads to errors in the temperature measurements that are not acceptable for many applications, such as in some applications that relate to agriculture and the environment. Scientists [18] have found that the digital response of the camera is affected by the camera temperature in a nonlinear manner. Thus, linear and nonlinear classical models such as the polynomial models are compared to calibrate the thermal camera. Also, artificial neural networks models are implemented to calibrate the camera due to their ability to address and solve highly nonlinear problems [19]. Recent applications of uncooled microbolometer cameras in autonomous vehicles, drones, medical imaging, and 3-D Printing may require the implementation of more stringent and rigorous correction and calibration schemes due to the constant change in the target scenery as well as the associated temperature.

Other sources of inaccuracies in thermal measurements using uncooled cameras include the following [20]: size-of-source effect, distance effect, and environmental effects. It may be noted that these are also present in cooled cameras. An example of a flowchart describing the algorithm of NUC with the temperature drift compensation is shown in Fig. 7.5 [21].

An illustration of a case study of radiometric cloud imaging with an uncooled microbolometer thermal infrared camera is shown in Fig. 7.6. In this figure, the principal components of an infrared cloud imager (ICI) is shown [15].

Fig. 7.7 shows the radiometric calibration method based on multiple blackbody sources in a laboratory environment with the use of an infrared camera; in this case, three black bodies are set at different temperatures along with a data acquisition card and an inline computer [21].

Fig. 7.8 shows the results after the calibration considering two sources of blackbodies [21], where U is the response of the detector to the radiation flux Φ from a black body, and $U(\Phi_1)$ and $U(\Phi_2)$ are the mean value responses of a single detector to the radiation flux coming from a black body at a temperature of T_1 and T_2, respectively.

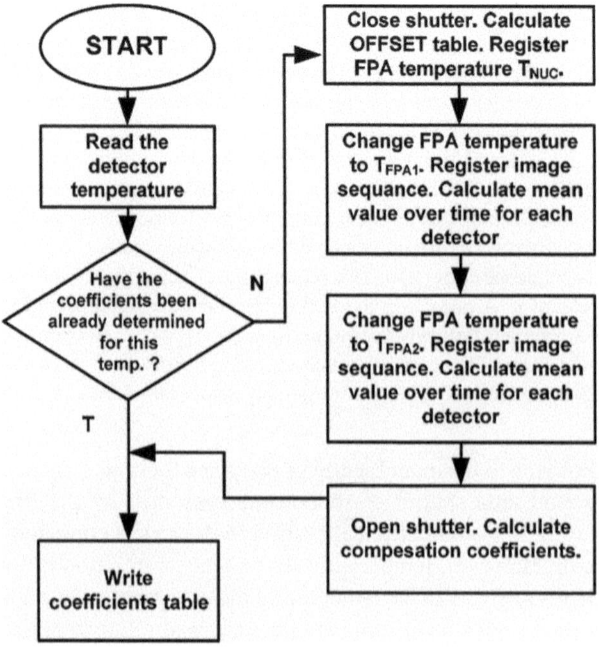

Fig. 7.5 Example of a flowchart describing the algorithm of NUC with the temperature drift compensation [21].

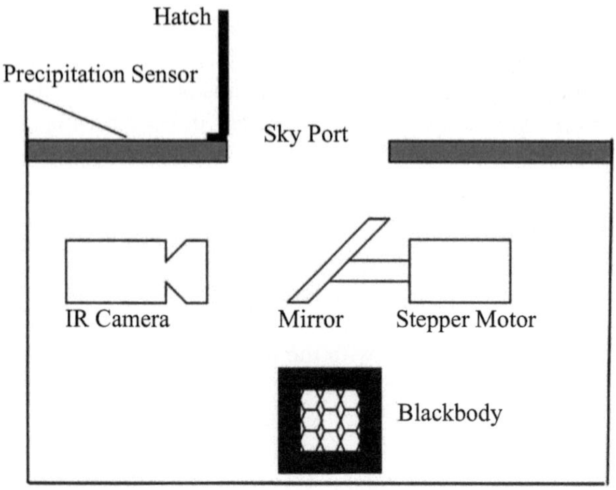

Fig. 7.6 Principal components of the ICI optical system. The IR camera alternately views the sky and two blackbody calibration sources (only one shown for convenience) [15].

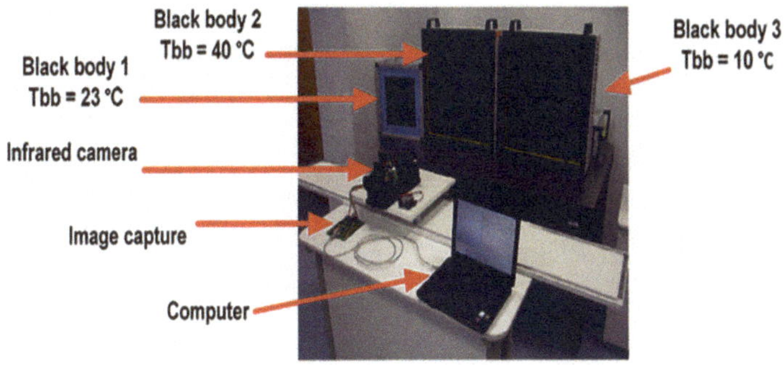

Fig. 7.7 Radiometric calibration using multiple black-body sources [21].

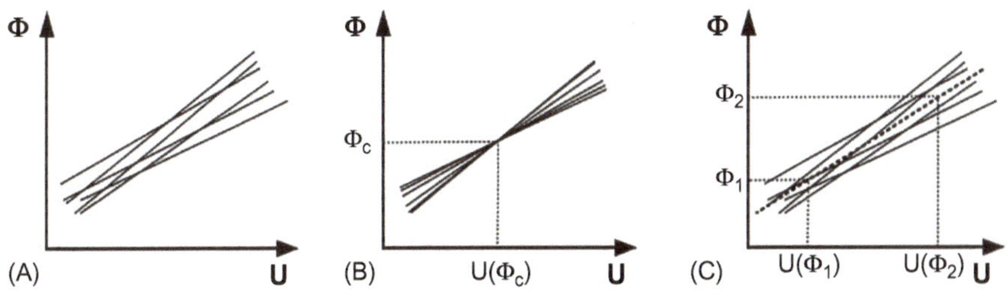

Fig. 7.8 The characteristics of detectors: before the Nonuniformity correction (A), after the one-point correction (B), and during the two-point correction (C) [21].

The first image depicts a spread in the microbolometer characteristics before the implementation of NUC. The second image illustrates the microbolometer characteristics after the one-point correction. In this case, only a shift in the characteristics is corrected. This correction is executed inside the camera by the use of a mechanical shutter. The last image presents the microbolometer characteristics during the two-point correction with the use of two reference black bodies. The two-point correction allows the correction of the shift and the slope of the microbolometer characteristics. The calculated detector coefficients are stored in a nonvolatile memory in the camera and the process of correcting the output signal is performed by the camera in real time [21].

The second method most used in the calibration is the shutter-based FPA temperature compensation. Radiometric shutter corrections are based on consecutive images of a constant scene as the camera temperature changes to derive a FPA temperature dependent calibration.

The standard approach for calibration involves the determination of this radiation flux by using an optical shutter occasionally and subtracting the derived correction signal from the measurement signal. In order to compensate for disturbing influences derived from the changing ambient conditions, radiometric cameras use an optical shutter for online recalibration. New calibration methods are needed to be developed in order to eliminate the mechanical components such as the optical shutter to continue with the miniaturization of infrared cameras. These efforts must also be accompanied by a detailed understanding and assessment of sensor temperature stabilization. Infrared cameras should consist only of infrared optics, a sensor array, and a digital signal processing (DSP) unit. For acceptable measurements, uncertainty values in the received flux without using an optical shutter (shutterless) and the disturbing influences of changing thermal conditions have to be treated based on temperature measurements of the camera interior.

Most microbolometer cameras have an internal shutter, located between the optical lens and the detector. This shutter is normally used to perform a NUC during operation. The shutter-based FPA temperature calibration uses a post-NUC image of this shutter and the temperature of the shutter to perform a secondary correction for the FPA temperature dependence in the camera [21]. A schematic of the correction procedure for shutter- and thermoelectric coolerless IR cameras is presented in Fig. 7.9 [3].

The shutter-based method for calibration is established on the assumption that a shutter image can be mathematically modified to account for the camera's lens and a pseudo external blackbody image can be formed. The use of the internal shutter as an equivalent external blackbody requires careful characterization of the shutter

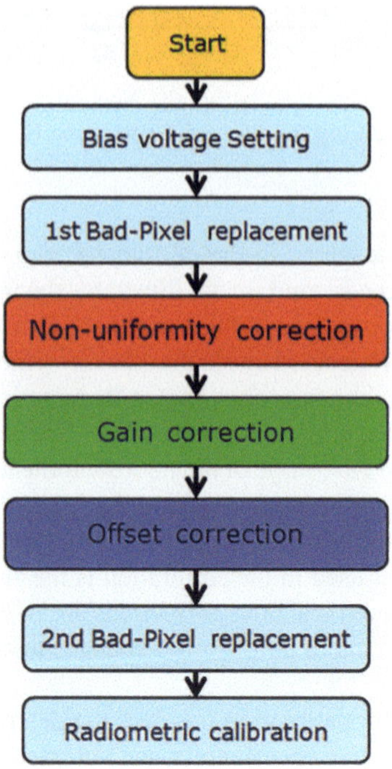

Fig. 7.9 Correction procedure for shutter- and thermoelectric cooler-less (TEC-less) IR cameras [3].

and optical lens. The radiometry of this situation is depicted for a FPA exposed to a blackbody source in Fig. 7.10A and the shutter in Fig. 7.10B [22, 23].

These two methods allow the reduction of the FPA temperature dependence in microbolometer based infrared cameras through a direct FPA temperature correction with multiple blackbody sources and the use of the internal camera shutter as an equivalent external blackbody source. These calibration methods significantly improve the camera calibration and reduce the errors correlated with the camera-lens temperature.

The purpose of the radiometric calibration is to relate the infrared camera output to scene radiance or scene temperature, depending on the camera application (vision or thermal). Fig. 7.11 shows the thermal correction of an image after the application of the two-point correction method, using multiple blackbody sources [17]. In a detailed report authored by the scientists at the National Institute of Standards and Technology (NIST), the calibration and measurement procedures for a high magnification thermal camera are discussed in detail [24]. In particular, their study [24] focuses on approaches to uncertainty evaluation in thermographic measurements. Methods for the measurement and reduction of sources of uncertainty in infrared thermography are also discussed.

It must be emphasized that, in many cases, the applications of microbolometers determine their calibration. For example, in noncontact monitoring of semiconductor processes or semiconductor circuits, the repeatable and reproducible measurement of absolute temperature as well as its spatial and temporal distribution is critical.

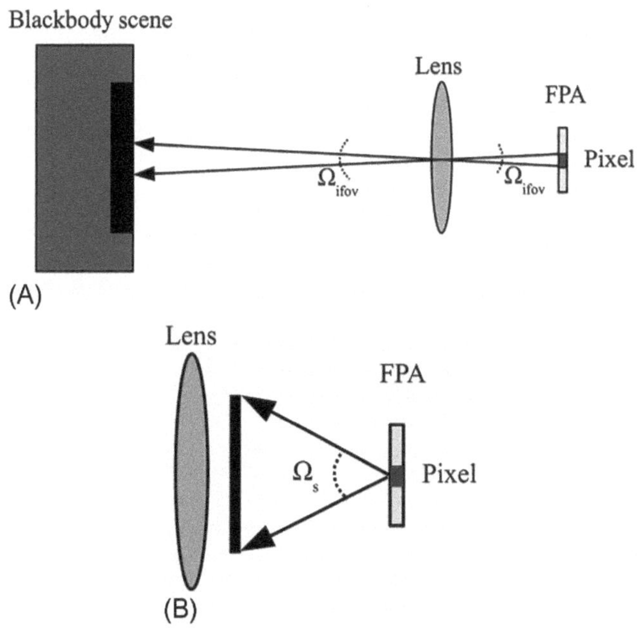

Fig. 7.10 (A) Geometric optical layout of the camera FPA viewing an external blackbody with instantaneous field-of-view (ifov) solid angle Ω_{ifov} determined by the pixel area and the lens focal length [22, 23]. (B) Geometric optical layout of the camera FPA viewing the shutter with solid angle Ω_s determined by the shutter area and shutter-to-FPA distance [22, 23].

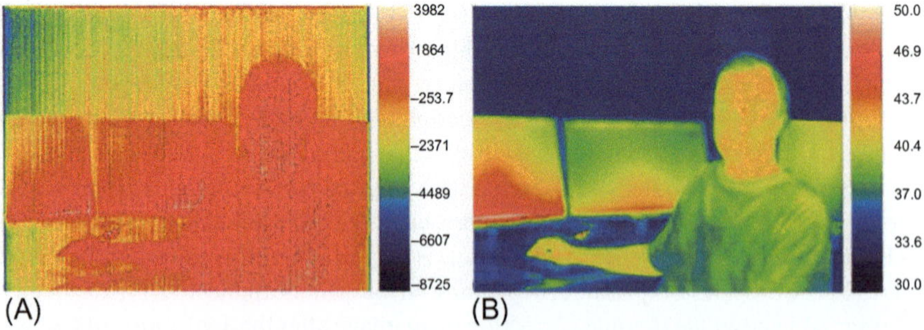

Fig. 7.11 Effect of two-point correction. (A) Raw image (color bar chart in *grey* values) and (B) thermal image (color bar chart in °C) [17].

Therefore, emphasis is placed on the calibration of the microbolometer with respect to accuracy of measurement of the absolute temperature. In situations like this, high-resolution thermal imaging cameras, equipped with microscope lenses, can be used [25]. While the distance between the camera and the object to be measured can be 10 cm, these cameras have been reported to provide a spatial resolution of up to 28 μm [25]. There are several governmental and private laboratories that provide calibration services for infrared cameras. These include NIST [26], Applied Technical Services [27], Colbert Infrared Services [28], Electro Optical Industries [29], Infrared Cameras Inc. [30], Innovative Imaging & Research [31], ired Remote Sensing [32], Optris Infrared Sensing [33], and TRANSCAT [34], in addition to infrared camera manufacturers such as DIAS Infrared Systems [35] and FLIR [36].

An example of a calibration set-up of infrared imagers, at NIST, is shown in Fig. 7.12. As can be seen in this figure, the calibration of infrared imagers is a very

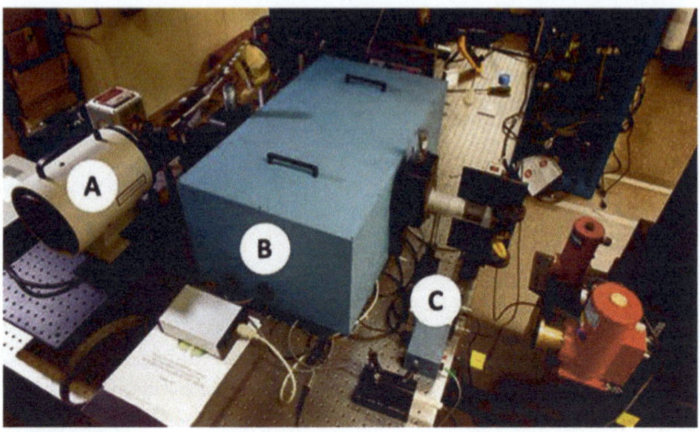

Fig. 7.12 Calibration setup: (A) infrared source, (B) monochromator, and (c) output beam and device under test *(red)* next to NIST standard detector [26].

complex process. It involves the use of several standards in the form of infrared sources such as black bodies, infrared optics including filters and lenses [37], a temperature controlled chamber as well as mathematical models for radiation. Thus, the design and construction of optical components that can be utilized as standards in the calibration of uncooled microbolometers is a critical step. This step must be implemented conjointly with a good understanding of emissivity, its dependence of wavelength, temperature, and surface morphology.

References

[1] H. Budzier, G. Gerlach, Thermal Infrared Sensors, Wiley, Chichester, 2011.

[2] H. Budzier, G. Gerlach, Calibration of infrared cameras with microbolometers, in: AMA Conferences 2015—Sensor 2015 and IRS2, 2015, pp. 889–894.

[3] A. Tempelhahn, H. Budzier, V. Krause, G. Gerlach, Shutter-less calibration of uncooled infrared cameras, J. Sens. Sens. Syst. 5 (2016) 9–16.

[4] P.W. Nugent, J.A. Shaw, N.J. Pust, Correcting for focal-plane-array temperature dependence in microbolometer infrared cameras lacking thermal stabilization, Opt. Eng. 52 (6) (2013). 061304-1 to 061304-8.

[5] P.W. Kruse, Uncooled Thermal Imaging, Arrays, Systems, and Applications, SPIE Press, Bellingham, Washington, 2001.

[6] N.R. Butler, Methods and Apparatus for Compensating a Radiation Sensor for Temperature Variations of the Sensor, US Patent No.6730909 B2, 2004.

[7] S.M. Anderson, T.J. McManus, Microbolometer Focal Plane Array With Temperature Compensated Bias, US Patent No. 7105818B2, 2006.

[8] M.A. Wand, Ambient Temperature Micro-Bolometer Control Calibration and Operation, Patent No. US 6267501 B1, 2001.

[9] B. Hornback, Imaging Device With Multiple Fields of View Incorporating Memory-Based Temperature Compensation of an Uncooled Focal Plane Array, US Patent No. 7235785 B2, 2007.

[10] S.N. Torres, J.E. Pezoa, M.M. Hayat, Scene-based non-uniformity correction for focal plane arrays using the method of the inverse covariance form, Appl. Optics 42 (29) (2003) 5872–5881.

[11] P.E. Howard, Infrared Sensor Temperature Compensated Response and Offset Correction, US Patent No. 6433333 B1, 2002.

[12] C.S. Kaufman, R.S. Carson, W.B. Hornback, Method and Apparatus for Temperature Compensation of an Uncooled Focal Plane Array, US Patent No. 6476392 B1, 2002.

[13] A. Wolf, R. Redlich, M. Figueroa, J.E. Pezoa, On-line nonuniformity and temperature compensation of uncooled IR FPAs using embedded digital hardware, Proc. SPIE 8868 (2003) 88680H.

[14] S. Vidas, P. Moghadam, Ad hoc radiometric calibration of a thermal infrared camera, in: 2013 International Conference on Digital Image Computing: Techniques and Applications (DICTA), 2013.

[15] J. Shaw, P. Nugent, N. Pust, B. Thurairajah, K. Mizutani, Radiometric cloud imaging with an uncooled microbolometer thermal infrared camera, Opt. Express 13 (15) (2005) 5807–5817.

[16] J.A. Shaw, L.S. Fedor, Improved calibration of infrared radiometers for cloud temperature remote sensing, Opt. Eng. 32 (5) (2014) 10002–11010.

[17] H. Budzier, G. Gerlach, Calibration of uncooled thermal infrared cameras, J. Sens. Sens. Syst. 4 (2015) 187–197.
[18] K. Ribeiro-Gomes, D. Hernández-López, J.F. Ortega, R. Ballesteros, T. Poblete, M.A. Moreno, Uncooled thermal camera calibration and optimization of the photogrammetry process for UAV applications in agriculture, Sensors 17 (2017) 2173. 23 pages.
[19] C.M. Bishop, Neural networks for pattern recognition, *J. Am. Stat. Assoc.* 92 (1995) 482.
[20] H. Mcevoy, R. Simpson, G. Machin, Quantitative InfraRed Thermography Review of current thermal imaging temperature calibration and evaluation facilities, practices and procedures, across EURAMET, in: Proceedings of the 11th International Conference on Quantitative InfraRed Thermography (QIRT 2012), Naples, Italy, 11–14 June, 2012.
[21] M. Krupiński, G. Bieszczad, T. Sosnowski, H. Madura, S. Gogler, Non-uniformity correction in microbolometer array with temperature influence compensation, Metrol. Meas. Syst. 21 (4) (2014) 709–718.
[22] P.W. Nugent, J.A. Shaw, N.J. Pust, Radiometric calibration of infrared imagers using an internal shutter as an equivalent external black body, Opt. Eng. 53 (12) (2014). 123106–1 to 123106–7.
[23] P. Nugent, J.A. Shaw, Calibration of uncooled LWIR microbolometer imagers to enable long-term field Deployment, Proc. SPIE 9071 (2014) 90710V. 1–10.
[24] B. Lane, E.P. Whitenton, Calibration and Measurement Procedures for a High Magnification Thermal Camera, NISTIR 8098, National Institute of Standards and Technology, US Department of Commerce, 2015. https://doi.org/10.6028/NIST.IR.8098.
[25] Microscope Lens for Temperature Monitoring of Ultra-Small Targets, 2020. https://www.ien.eu/article/microscope-lens-for-temperature-monitoring-of-ultra-small targets/. (Accessed 15 October 2020).
[26] A New, World-Class Facility for Infrared Calibrations, National Institute of Standards and Technology, 2020. https://www.nist.gov/news-events/news/2015/10/new-world-class-facility-infrared-calibrations. (Accessed 15 October 2020).
[27] https://atslab.com/calibrations/infrared-calibration/, Accessed 15 October 2020.
[28] https://colbertinfrared.com/service/camera-calibration/, Accessed 15 October 2020.
[29] https://www.electro-optical.com/Products/Test-and-Measurement/Blackbodies/CN-37-Thermal-cameras-reference-source-for-real-time, Accessed 15 October 2020.
[30] https://infraredcameras.com/infrared-camera-calibration/, Accessed 15 October 2020.
[31] https://www.i2rcorp.com/main-business-lines/camera-calibration-services, Accessed 15 October 2020.
[32] https://ired.co.uk/thermal-camera-calibration/, Accessed 15 October 2020.
[33] https://www.optris.com/calibration-service, Accessed 15 October 2020.
[34] https://www.transcat.com/calibration-services/instruments/thermal-imager-calibration-services?stt=adword_google_*&
gclid=CjwKCAjwiaX8BRBZEiwAQQxGx5SQz1ts39y
KWl3bXQ1DeqOKXIpRHPhxxSti8N5rTNeQfO-XQkc7-RoC4lEQAvD_BwE,
Accessed 15 October 2020.
[35] https://www.dias-infrared.com/, Accessed 15 October 2020.
[36] https://www.flir.com/support-center/Instruments/service/service-and-support/, Accessed 15 October 2020.
[37] S. Papini, I. Klapp, N. Sochen, Calibrating uncooled micro-bolometer FPA By lenses, in: Light, Energy and the Environment, OSA Technical Digest (online), Optical Society of America, 2016. paper EW3A.6 https://www.osapublishing.org/abstract.cfm?URI=EE-2016-EW3A.6. (Accessed 15 October 2020).

Types of microbolometers

8

8.1 Microbolometer

Uncooled microbolometers have been utilized extensively for applications in infrared detection, where it is impractical to use cooled infrared detectors (for example, military systems, space applications, portable electronics—applications in which the imaging system characteristics such as the total system weight and power requirements are low and the system performance parameters such as response time and noise equivalent temperature are not critical). Among the thermal detectors, the most used in a variety of applications is the uncooled microbolometer. Advantages of uncooled microbolometers include operation at room temperature, flat response over the infrared spectrum, low cost, and conservative time response; this allows users to consider these infrared detectors as the ideal choice in the development of uncooled infrared imaging systems.

A microbolometer is a thermal infrared detector that absorbs infrared radiation, transducing the incident infrared radiation into a readable electrical signal. The physical mechanism involved in the infrared detection process can be classified as a thermal effect that leads to change in the electrical resistance of the detector element. These devices exhibit high responsivity and voltage bias is required (DC response). A wide variety of microbolometer detectors have been developed by the infrared detection/imaging industry; each configuration has advantages and disadvantages that determine the most appropriate approach depending on the infrared spectral range for the detection, application, desired sensitivity/resolution, fabrication cost, and measurement conditions.

Microbolometer detectors respond to the heating effect caused by the absorbed photon radiation that causes changes in the temperature of the detector and in its electrical properties. The process of detection of microbolometer detectors occurs in two steps. First, the incident infrared radiation is absorbed generating phonons; this causes changes in the lattice temperature of the absorber material. Second, the change in temperature induces changes in a measurable parameter of the detector active element. Microbolometers are detectors in which the absorption of radiation causes changes in the electrical resistance of its thermal sensing materials (for example, VO_x, a-Si—at room temperature). The response of the microbolometer detector is proportional to the energy absorbed [1]. In order to maximize the responsivity of the microbolometer detector, its thermal conductance must be minimized. This requires that the length of the thermal link be maximized and the cross-sectional area be minimized.

A summary of the thermal conductivity of various materials of interest to bolometers is presented in Table 8.1 [2]. The thermal conductivity of vanadium dioxide has been reported to be 0.2 W/mK [3]. However, higher values of the thermal conductivity have been revealed for bulk single crystals of VO_2 [4]. A value of 5 W/mK has been used for the thermal conductivity of VO_x in the literature [5].

Microbolometers. https://doi.org/10.1016/B978-0-08-102812-4.00014-0

Table 8.1 Thermal conductivity of various materials of interest to microbolometers [2] and references therein.

Material	Thermal conductivity (W/cm K)
Si (Bulk)	~1.3
SiO_2	0.014
Si_3N_4	~0.032–0.0385
Bi (Thin film)	~0.0015–0.018
Bi (Bulk film)	~0.0792
Au	~3.15
Cr	~0.019

8.2 Microbolometer-principle of operation

A microbolometer is an infrared detector that is generally fabricated by micromachining technology. A microbolometer consists of a thin multilayer structure thermally isolated and connected to a substrate through support structures that act as heat links and electrical connectors.

A typical microbolometer has two main components: an optical radiation infrared absorber material coupled to an active element material. The absorbed incident infrared radiation is converted into heat in the lattice by the absorber material, which produces changes in the electrical resistivity of the active element material (thermosensing material) and the change is transduced into an electrical output signal [6]. The information on change in electrical resistance is electrically transferred to the read-out integrated circuit (ROIC) for further processing, producing, thereby, a thermal output/image. Fig. 8.1 shows the schematic block diagram of a microbolometer detector [6].

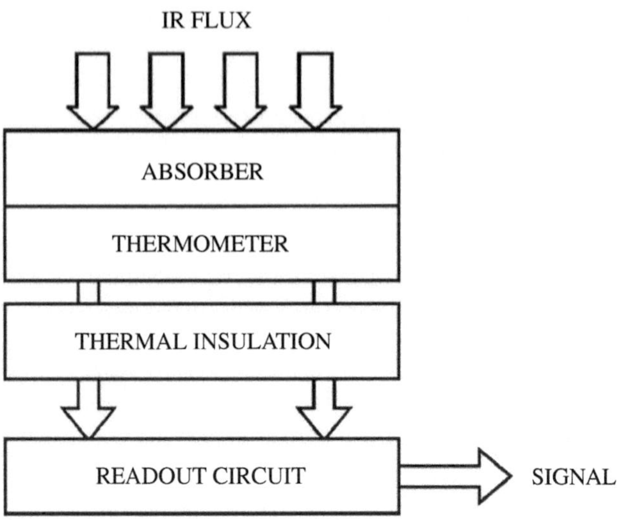

Fig. 8.1 Schematic block diagram of a microbolometer detector [6].

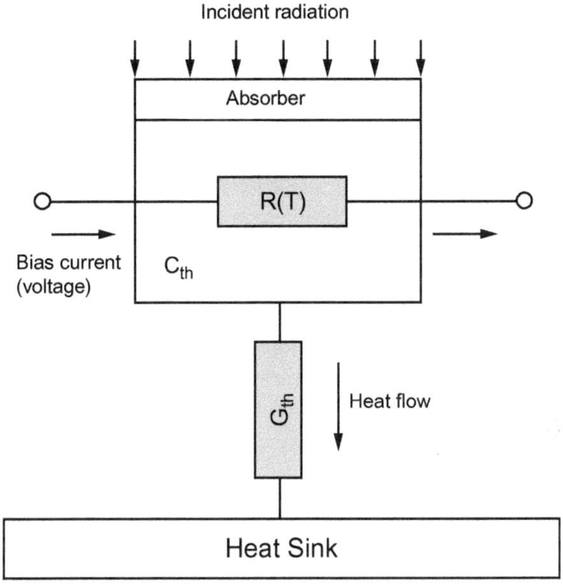

Fig. 8.2 Schematic diagram of a thermal detector operation [7].

Table 8.2 Summary of TCR of materials of interest to bolometers [2].

Material	TCR (K^{-1})
YBCO	~+0.5–1 (Transition at 88 K)
VO_x	~−0.015–0.028 (Room Temp)
Ag	~+0.037 (Room Temp)
Ni	~+0.005 (Room Temp)
Au	~+0.0036 (Room Temp)
Bi	~−0.003 (Room Temp)
Cr	~−0.000522 (Room Temp)

The operation of a microbolometer is based on the change in electrical resistance of a thermosensing material as a function of temperature. The optical radiation absorber and the thermosensing material are connected thermally and electrically through a support structure to the substrate that acts as a heat sink. The basic representation of the microbolometer detector operation is shown in Fig. 8.2. The microbolometer detector, with thermal capacitance $C_{th,}$ is connected by a thermal link, of thermal conductance G_{th}, to a heat sink, assuming that there is no other path for heat loss [7].

The infrared absorber materials of interest to applications in microbolometers should have large temperature coefficient of resistance $\{TCR = (1/R)(dR/dT)\}$, where R is the resistance and T is the temperature. A summary of the values of TCR of materials of interest to bolometers is presented in Table 8.2 [2].

The sensitivity of the uncooled microbolometer is maximized by providing high thermal isolation.

8.3 Microbolometer structural design

In a standard infrared imaging system, single IR detector elements (pixels) are laid out in a two-dimensional array, called focal plane array (FPA). The thermal isolation and efficient radiation absorption of single detector elements (microbolometers) are the key factors for high performance infrared focal plane array systems [8].

The performance of a microbolometer is determined by the material characteristics and its structural design parameters (structure geometry and dimensions) [9]. In order to obtain a high performance microbolometer, some conditions are required in the design of the multilayer structure. These include the following:

- high temperature coefficient of resistance, TCR, of thermosensing material;
- efficient absorption of IR radiation by the absorber layer;
- low thermal conductance with surroundings;
- low thermal capacitance of pixel;
- high thermal isolation;
- fast response;
- low noise.

Efficient infrared radiation absorption by the absorber layer, in combination with a thermosensing material with high TCR, provides a high responsivity of the microbolometer. A high TCR of the thermosensing material is key to the performance of the detector. In order to obtain a high TCR, a low doping concentration of the thermosensing material, which leads to a high electrical resistivity, is required. This increases the Johnson noise and is undesirable for the read-out integrated circuit. Consequently, a trade-off must be found between the TCR and the electrical resistivity of the detector.

One of the most important microbolometer design parameters is a low thermal conductance between the bolometer and its surroundings (important at low frequencies); this is the reason why conventional bolometers operate in a vacuum package to reduce the thermal conductance between the bolometers and their surroundings through the surrounding gas. The vacuum environment makes possible that the principal heat loss occurs by the conduction path through the support structure of the suspended membrane to the substrate. Microbolometers need to be packaged in vacuum with vacuum levels <0.01 mbar [10]. The support structure provides three functions: mechanical support, a thermally conducting path, and an electrically conducting path. The thermal conductance through the support arms can be minimal, reducing the cross-sectional area, increasing the length and using materials with low thermal conductivity. The optimum thermal isolation of the microbolometer structure can be obtained in devices with the microbolometer multilayer structure suspended over a cavity [11]. In order to enhance the IR radiation absorption in the microbolometer upper level membrane, a resonant optical cavity (Fabry-Perot) structure that is optimized for the targeted wavelength interval is required [12].

In order to minimize the thermal capacitance (important at high frequencies) per unit of area of the microbolometer, it is needed to reduce the thermal mass of the multilayer structure while maintaining a high absorption. The lower the thermal capacitance of the suspended structure, the lower is the thermal time constant response required by the imaging system [13].

The most important set of parameters of a thermal infrared sensor, for use in an imaging system, are high responsivity and short thermal time constant response. High responsivity is achieved by increased absorption and a short thermal time response requires small heat capacity of the structure and a high thermal conductance to the substrate. A desired high responsivity and fast thermal time constant response in a microbolometer cannot be achieved; a trade-off is needed depending on the microbolometer application requirements [14].

8.4 Classification of microbolometers

Recent advancements in microelectromechanical systems (MEMS) technology facilitate the fabrication of sensitive microbolometer detectors on thermally isolated hanging membranes. Microbolometers can be classified into two types, depending on the location of the readout integrated circuit (ROIC)—it can be placed underneath or besides the microbolometer membrane.

In the first approach, the suspended membrane is on top of the substrate surface of a wafer and the ROIC can be efficiently placed underneath the infrared detector structure. This approach is a cost-efficient method and it is typically used in the fabrication of Focal Plane Arrays (FPAs). Figs. 8.3–8.5 [6, 15, 16] illustrate various configurations of microbolometer single element, in which the membrane is suspended over a cavity on the substrate.

Fig. 8.3 SEM micrograph of a single pixel 50 × 50 μm n-well microbolometer with a fill factor of 40% realized in MOSIS AMI 0.8 micron CMOS process [15].

Fig. 8.4 SEM micrograph of a Ti-microbolometer element [6].

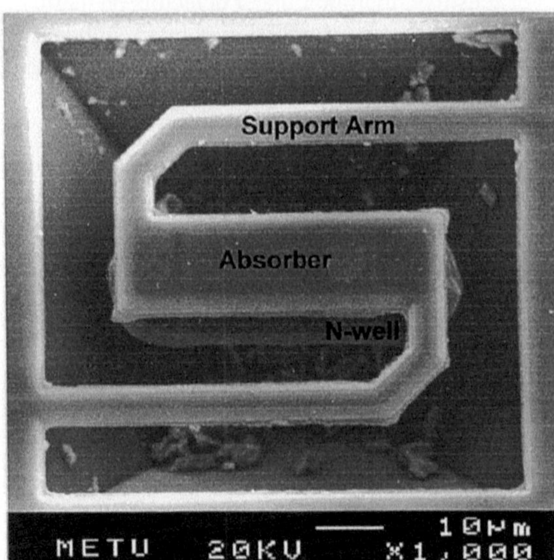

Fig. 8.5 SEM micrograph of postprocessed single pixel n-well microbolometer with 74 × 74 μm pixel size and 15% fill factor [16].

In the second approach, the suspended membrane is formed on the substrate surface of a wafer; in this case, the ROIC cannot be placed underneath the microbolometer membrane and it is fabricated besides the microbolometer structures. This second approach is also a cost-efficient method, but it is not used in the fabrication of high-end FPAs as a result of a low fill factor. Figs. 8.6–8.8 [17–19] show different configurations of microbolometer single element, with the suspended membrane, which is supported by two arms and a gap between the membrane and the substrate.

Fig. 8.6 SEM image of fabricated microbolometer [17].

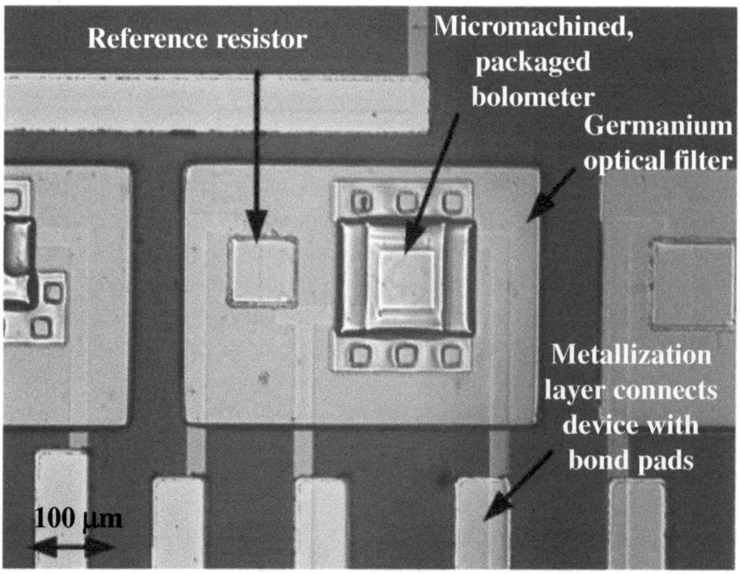

Fig. 8.7 Confocal microscope image of the fabricated microbolometer showing micromachined microbolometer and reference resistor covered with the germanium long-pass filter [18].

Source: Moinuddin Ahmed, Donald P. Butler, Zeynep Celik-Butler, Device-level vacuum packaged uncooled microbolometer on a polyimide substrate, Infrared Phys. Technol. 79 (2016) 50–57, ISSN 1350–4495, https://doi.org/10.1016/j.infrared.2016.09.011.

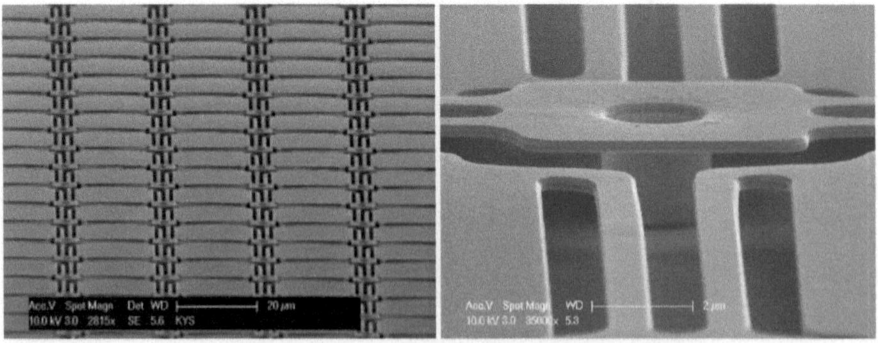

Fig. 8.8 SEM image of released microbolometers with 25 μm pixel size [19].
Source: Jung, H., Kim, T.H., Kim, G. et al. Design and characteristics of a-Si-based microbolometers with shared-anchor structure in vacuum packaged systems. Microsyst Technol 20, 899–905 (2014). https://doi.org/10.1007/s00542-014-2073-3.

8.5 Methods of microbolometer fabrication

There are three main manufacturing techniques for the fabrication of microbolometers: monolithic integration, bulk micromachining, and 3D integration process.

Monolithic integration is the most commonly used approach for the manufacture of uncooled infrared microbolometers [20–23]. In monolithic integration, the readout integrated circuit (ROIC) is premanufactured, and the microbolometer infrared absorber and thermosensing element materials are subsequently deposited and patterned on the ROIC wafer. A sacrificial layer of silicon nitride or polyimide is used as the sacrificial layer. In the final step, the sacrificial layer is removed and the suspended thermally isolated bolometer membrane is obtained.

In monolithic integration, the electronics for the signal read-out can be efficiently placed underneath the microbolometer membranes. This process is a cost-effective and well-established post-CMOS process. One disadvantage of the monolithic integration process is the potential damage to the ROIC at high temperatures; the deposition process for the temperature sensing microbolometer material is limited to about 450 °C and does not allow the deposition of monocrystalline materials that require process temperatures higher than this temperature of 450 °C [20]. An example of an integration process of uncooled microbolometer arrays is shown in Fig. 8.9 [20].

Bulk micromachining is a second approach for the manufacture of infrared microbolometers. In bulk micromachining, the microbolometers are formed in the substrate surface of a wafer. Subsequently, the substrate is etched underneath the microbolometer membranes to thermally separate them from the rest of the substrate. The advantage of bulk micromachining is that the ROIC and the microbolometers can typically be manufactured in a standard CMOS line and the free-etching of the microbolometers can then be performed in a single post-CMOS process step. In bulk-micromachining process, the electronics for the ROIC cannot be placed underneath the microbolometer membrane; it has to be placed beside the microbolometers,

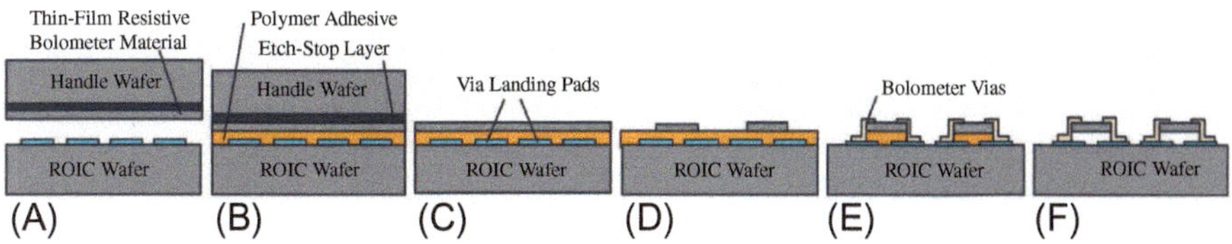

Fig. 8.9 Integration of uncooled infrared bolometer arrays: (A) separate fabrication of ROIC wafer and handle wafer with resistive bolometer material; (B) adhesive wafer bonding; (C) thinning of handle wafer; (D) bolometer definition; (E) via formation; and (F) sacrificial etching of polymer adhesive [20].
From Frank Niklaus, Adit Decharat, Christer Jansson, Göran Stemme, Performance model for uncooled infrared bolometer arrays and performance predictions of bolometers operating at atmospheric pressure, Infrared Phys. Technol. 51(3) (2008) 168–177,I SSN 1350–4495, https://doi.org/10.1016/j.infrared.2007.08.001.

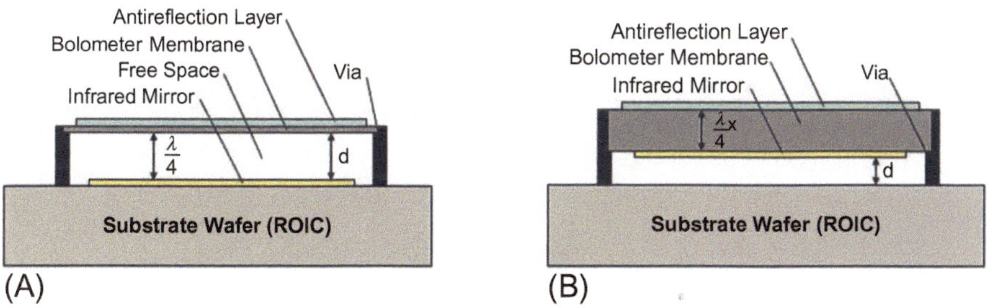

Fig. 8.10 Cross-sectional image of two bolometer designs with resonant optical cavities for high absorption of the incident infrared radiation [20].
From Frank Niklaus, Adit Decharat, Christer Jansson, Göran Stemme, Performance model for uncooled infrared bolometer arrays and performance predictions of bolometers operating at atmospheric pressure, Infrared Phys. Technol. 51(3) (2008) 168–177, ISSN 1350–4495, https://doi.org/10.1016/j.infrared.2007.08.001.

reducing the microbolometer fill factor [20]. Fig. 8.10 shows cross-sectional image of two bolometer designs with resonant optical cavities for high absorption of the incident infrared radiation [20].

Heterogeneous three-dimensional (3D) bolometer integration has been proposed for the integration of high-performance monocrystalline temperature sensing bolometer materials on ROICs [22–24]. In 3D microbolometer integration, the microbolometer materials are deposited on a separate handle wafer. The advantage of the 3D microbolometer integration is that it allows the use of high performance monocrystalline temperature sensing microbolometer materials on top of standard ROICs [20]. Example of an infrared imager from Fraunhofer Institute for Microelectronic Circuits and Systems (IMS) and its principle construction is shown in Fig. 8.11 [10]. In this figure, QVGA represents Quarter Video Graphics Array (Resolution = 320x240 pixels; Standard VGA = 640x480 pixels.)

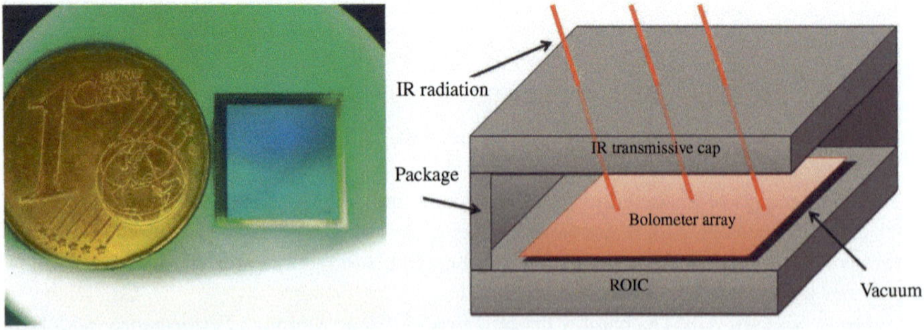

Fig. 8.11 (Left) Fraunhofer IMS infrared imager in QVGA size; (right) principle construction of an infrared imager [10].
From Michael Elßner, Holger Vogt, Reliability of microbolometer thermal imager sensors using chip-scale packaging, Procedia Eng. 120 (2015) 1191–1196.

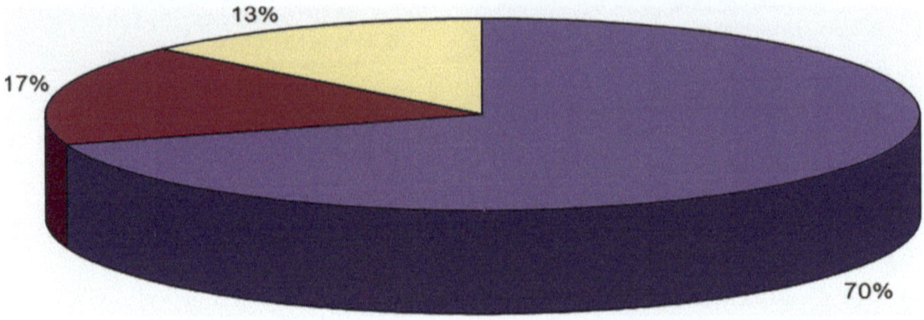

Fig. 8.12 Estimated market share of materials used in the manufacture of microbolometers—vanadium oxide: 70%, amorphous silicon: 17%, and barium strontium titanate: 13% [32].

8.6 Microbolometer types

At the pixel level, amorphous silicon [25, 26], barium strontium titanate [27–29], and vanadium oxide [30, 31] are the most commonly used materials [32] for the manufacture of uncooled microbolometers. An illustration of the market share of these materials is presented in Fig. 8.12 [32]. As can be seen in Fig. 8.12, vanadium oxide continues to account for the major share of the market.

Examples of (a) amorphous silicon-based two-color microbolometer, (b) barium strontium titanate (BST) pixel, and (c) vanadium oxide microbolometer are presented in Figs. 8.13, 8.14, and 8.15, respectively. In Chapters 5 and 6, the details of amorphous silicon and vanadium oxide microbolometers, respectively, are described. The historical origins and the details of the Honeywell approach based on vanadium oxide microbolometer and the barium strontium titanate microbolometer invention by Texas Instruments are described in Chapters 12 and 13, respectively.

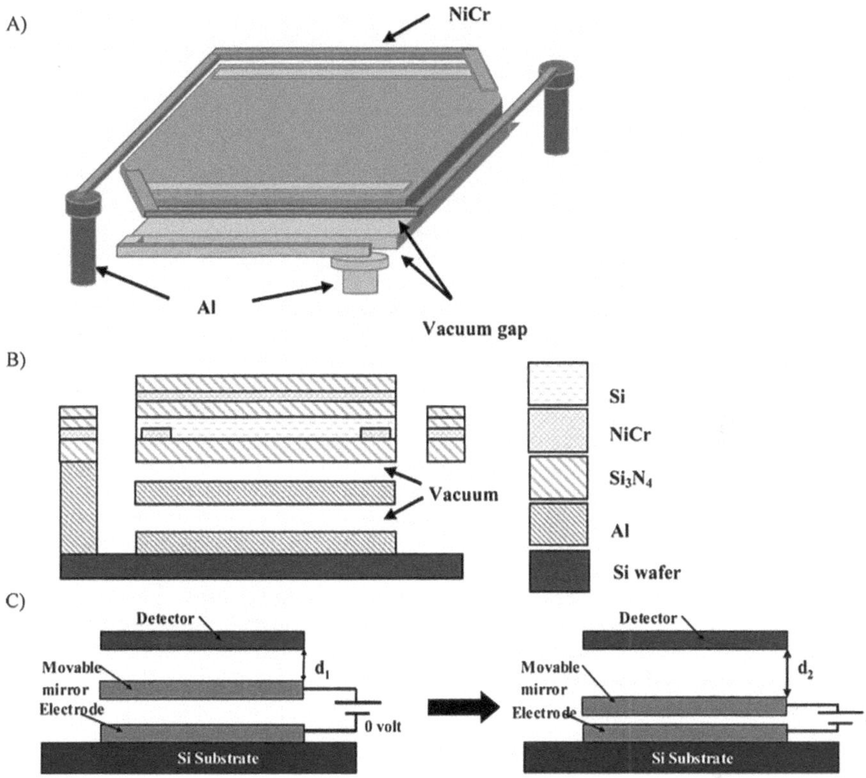

Fig. 8.13 Two-color amorphous silicon microbolometer geometry: (A) top view, (B) side view, and (C) operational view. The schematics show the device geometry consisting of movable Al-mirror, single NiCr absorber, Si N membrane and passivation layer, and NiCr electrode arms, and contacts to the suspended high TCR amorphous silicon. The air microcavity is located between the underlying Al mirror and the high TCR amorphous silicon. The movable micromirror creates a tunable resonant cavity between the detector and the mirror. The detector can be tuned to two spectral bands by changing the cavity depth by applying electrostatic voltage as indicated in the figure [25].

In recent years, with the significant progress in 2D materials, there has been considerable interest in the use of these materials [33], including graphene [33, 34] for the fabrication of detectors in the infrared [33] and terahertz frequencies [35]. While work in this area of research is in its infancy and much needs to be done to develop uncooled microbolometers based on 2D materials, an example of a graphene-based room-temperature nanomechanical bolometer has been reported in the literature [36]. It must be noted here that since the discovery of graphene, ~700 2D materials have been predicted to be stable while several of them are yet to be synthesized [35]. The spectral responsivity of graphene photodetectors, compared with commercial detectors, is illustrated in Fig. 8.16 [33].

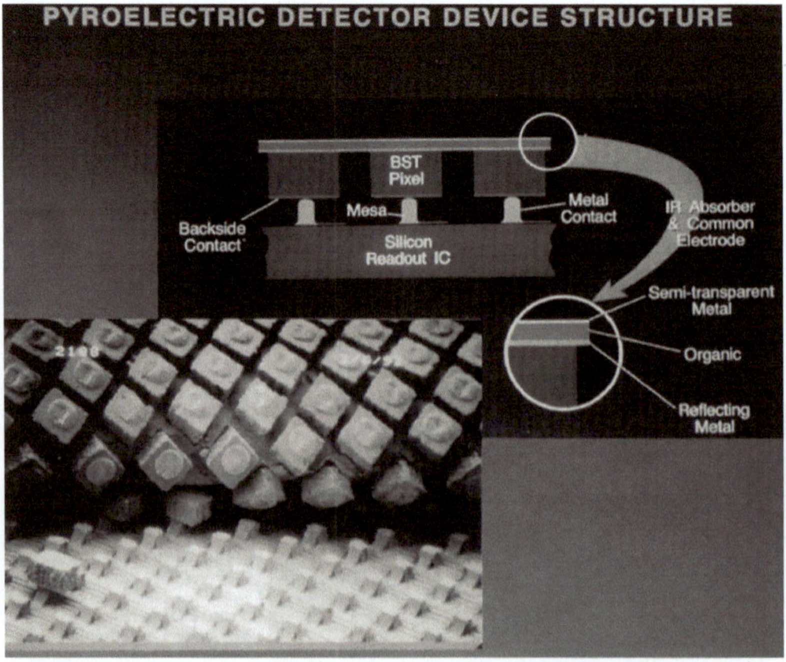

Fig. 8.14 Array structure (a peel off) showing the BST pixels, and the mesa for thermal isolation. The inset depicts vertical integration (hybrid process) of optical coating, BST, interconnect, and polymer mesa on the Si substrate [25]. Note: The 1:3 connectivity pattern of BST pixels. According to this notation, the BST elements are connected in one dimension, e.g., along the z-axis in the elevation plane and totally disconnected in the target, i.e., x y-plane. The second phase (xenon or air) surrounding the pixels is three dimensionally connected [28].

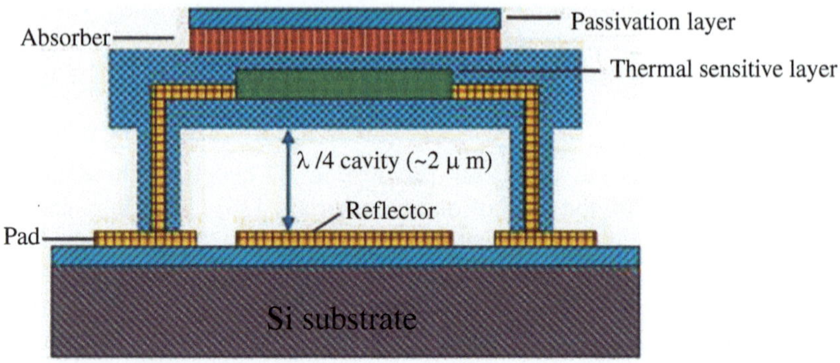

Fig. 8.15 The structure of the designed microbolometer. $\lambda/4$ cavity structure is adopted to improve infrared absorption [30].
From Bin Wang, Jianjun Lai, Hui Li, Haoming Hu, Sihai Chen, Nanostructured vanadium oxide thin film with high TCR at room temperature for microbolometer, Infrared Phys. Technol. 57 (2013) 8–13, ISSN 1350–4495, https://doi.org/10.1016/j.infrared.2012.10.006.

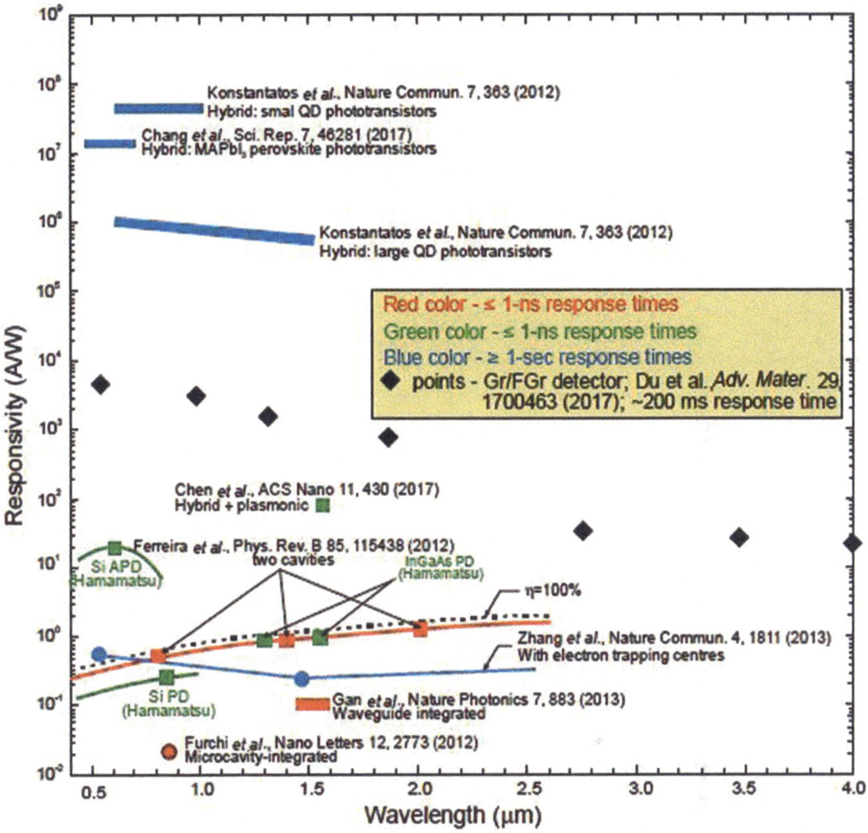

Fig. 8.16 Spectral responsivity of graphene photodetectors compared with commercial ones [33].

References

[1] P.W. Kruse, The Photon Detection Process in Optical and Infrared Detectors, Springer-Verlag, New York: Heidelberg Berlin, 1980.

[2] S.J. Yoo, Micromachined Wavelength Selective Microbolometer Sensors Operating at Room Temperature, Dissertation, PhD, University of Texas, Austin, 2000.

[3] S. Lee, et al., Anomalously low electronic thermal conductivity in metallic vanadium dioxide, Science 355 (2017) 371–374.

[4] L. Chen, et al., Enhancement of thermal conductivity across the metal-insulator transition in vanadium dioxide, Appl. Phys. Lett. 113 (2018) 061902. 1–5.

[5] J.-L. Lai, C.-J. Liao, G.-D.J. Su, Using an SU-8 photoresist structure and cytochrome C thin film sensing material for a microbolometer, Sensors 12 (2012) 16390–16403.

[6] R.K. Bhan, R.S. Saxena, C.R. Jalwania, S.K. Lomash, Uncooled infrared microbolometer arrays and their characterization techniques, Def. Sci. J. 59 (2009) 580–589.

[7] A. Kosarev, A. Torres, M. Moreno, R. Ambrosio, Thin film uncooled microbolometers based on plasma deposited materials, Can. J. Phys. 92 (2014) 570–575.
[8] M. Alhussein, S.I. Haider, Simulation and analysis of uncooled microbolometer for serial readout architecture, J. Sens. 1 (2016) 13.
[9] S. Gupta, R.K. Bhan, A. Katiyar, R. Muraldharan, Design optimization of pixel structure for α-si based uncooled infrared detector, Def. Sci. J. 63 (2013) 581–588.
[10] M. Elβner, H. Vogt, Reliability of microbolometer thermal imager sensors using chip-scale packaging, Procedia Eng. 120 (2015) 1191–1196.
[11] M. Moreno, A. Torres, R. Jiménez, Microbolometers based on amorphous silicon–germanium films with embedded nanocrystals, IEEE Trans. Electron Devices 62 (2015) 2120–2127.
[12] Q. Cheng, S. Paradis, T. Bui, M. Almasri, Design of dual-band uncooled infrared microbolometer, IEEE Sensors J. 11 (2011) 167–175.
[13] F.I.R. Sizov, Region challenges: photon or thermal detectors? Outlook and means, Semicond. Phys. Quantum Electron. Optoelectron. 15 (2012) 183–199.
[14] F. Niklaus, MEMS-based uncooled infrared bolometer arrays—a review, Proc. SPIE 6836 (2007) 68360D. 1–15.
[15] D.S. Tezcan, F. Koçer, T. Akin, An uncooled microbolometer infrared detector in any standard CMOS technology, in: 10th Int. Conf. on Solid-State Sensors & Actuators (TRANSDUCERS'99), June 7–10, Sendai, Japan, 1999, pp. 610–613.
[16] D.S. Tezcan, T. Akin, A low-cost Uncooled infrared microbolometer detector in standard CMOS technology, IEEE Trans. Electron Devices 50 (2) (2003) 494–502.
[17] S. Moon, Novel infrared absorbing material coupled uncooled microbolometer, Proc. SPIE 5074 (2004) 658–660.
[18] M. Ahmed, D.P. Butler, Z. Celik-Butler, Device-level vacuum packaged uncooled microbolometer on a polyimide susbtrate, Infrared Phys. Technol. 79 (2016) 50–57.
[19] H. Jung, T.H. Kim, G. Kim, et al., Design and characteristics of a-Si based microbolometers with shared-anchor structure in vacuum packaged systems, Microsyst. Technol. 20 (2014) 899–905.
[20] F. Niklaus, A. Decharat, C. Jansson, G. Stermme, Performance model for uncooled infrared bolometer arrays and performance predictions of bolometers operating at atmospheric pressure, Infrared Phys. Technol. 51 (3) (2008) 168–177.
[21] S. Eminoglu, D.S. Tezcan, M.Y. Tanrikulu, T. Akin, Low-cost uncooled infrared detectors in CMOS process, Sensors Actuators A 109 (2003) 102–113.
[22] F. Niklaus, E. Kälvesten, G. Stemme, A New concept for CMOS-compatible fabrication of uncooled infrared focal plane arrays using wafer-scale device transfer bonding, Proc. SPIE 4369 (2001) 397–404.
[23] F. Niklaus, E. Kälvesten, G. Stemme, Wafer-Level membrane transfers bonding of polycrystalline silicon bolometers for use in infrared focal plane arrays, J. Micromech. Microeng. 11 (2001) 509–513.
[24] J.E. Källhammer, H. Pettersson, D. Eriksson, S. Junique, S. Savage, C.J. Vieider, J.Y. Andersson, F. Niklaus, G. Stemme, Fulfilling the pedestrian protection directive using a long-wavelength infrared camera designed to meet the performance and cost targets, Proc. SPIE 6198 (2006) 74–84.
[25] M. Almasri, B. Xu, J. Castracane, Amorphous silicon two-color microbolometer for uncooled IR detection, IEEE Sensors J. 6 (2) (2006) 293–300.
[26] J.L. Tissot, et al., High performance uncooled microbolometer amorphous silicon IRFPA With 17 μm pixel-pitch, in: Sensor+Test Conference 2011—IRS^2 Proceedings, 2011, pp. 36–41.

[27] C. Mao, et al., Pyroelectric response mechanism of barium strontium titanate ceramics in dielectric mode: the underlying essence of the enhancing effect of direct current bias field, Appl. Phys. Lett. 102 (2013) 242911. 1–3.
[28] A. Amin, The role of connectivity in thermal imaging, J. Electroceram. 8 (2002) 99–106.
[29] M. Helmick, L. Nye, E. Zuniga, More power in less space: a thermal enhancement solution for thin packages, Tex. Instrum. Tech. J. 11 (4) (1994). 5–57 to 5–67.
[30] B. Wang, J. Lai, H. Li, H. Hu, S. Chen, Nanostructured vanadium oxide thin film with high TCR at room temperature for microbolometer, Infrared Physics & Technology 57 (2013).
[31] M. Abdel-Rahman, et al., Fabrication and design of vanadium oxide microbolometer, in: Proceedings of the 6th International Advances in Applied Physics and Materials Science Congress & Exhibition, AIP Conf. Proc. 1809, 020001-1–020001-12, 2017, https://doi.org/10.1063/1.4975416. Published by AIP Publishing, 020001–1 to 020001–12.
[32] Uncooled Detectors for Thermal Imaging Cameras—Making the Right Detector Choice, Technical Note, FLIRmedia, Accessed 24 June 2020.
[33] A. Rogalski, 2D materials in infrared detector family, Proceedings 27 (2019) 33, https://doi.org/10.3390/proceedings2019027033. 4 pages.
[34] M. Pech-Canul, N.M. Ravindra, Semiconductors—Synthesis, Properties and Applications, Springer, 2019.
[35] M.S. Vitiello, Nanodevices at terahertz frequency based on 2D materials, J. Phys. Mater. 3 (2020), 014008. https://doi.org/10.1088/2515-7639/ab5843.
[36] A. Blaikie, D. Miller, B.J. Alemán, A fast and sensitive room-temperature graphene nanomechanical bolometer, Nat. Commun. 10 (2019) 4726. 8 pages.

Further reading

B. Backer, M. Kohin, A. Leary, R. Blackwell, R. Rumbaugh, Advances in uncooled technology at BAE systems, Proc. SPIE 5074 (2003) 548–556.
D. Murphy, M. Ray, J. Wyles, J. Asbrock, C. Hewitt, R. Wyles, E. Gordon, T. Sessler, A. Kennedy, S. Baur, D. Van Lue, Performance improvements for VOx microbolometer FPAs, Proc. SPIE 5406 (2004) 531–540.
J.J. Yon, L. Biancardini, J.L. Tissot, L. Letellier, Infrared microbolometer sensors and their application in automotive safety, in: Proceedings AMAA, 2003, pp. 1–15.
P. Eriksson, J.Y. Andersson, G. Stemme, Thermal characterization of surface-micromachined silicon nitride membranes for thermal infrared detectors, J. Microelectron. Syst. 6 (1) (1997) 55–61.

Terahertz microbolometers 9

9.1 Introduction

The terahertz band covers the frequency range of 333 THz to 300 GHz (0.9 to 1000 μm wavelength), i.e., the transition range from the photonics (infrared region) to the microwave frequency bands. However, the term terahertz makes reference to the frequency range of 10 THz to 300 GHz (30 to 1000 μm), also known as the far-infrared (Far-IR) spectral range. Fig. 9.1 presents a schematic of the electromagnetic spectrum and the terahertz band [1]. During the last 2 decades, due to the unique spectral characteristics of terahertz (THz) radiation, there has been an increasing interest in the terahertz region to develop sources, detectors, and imaging systems.

Terahertz electromagnetic radiation is reflected by metals and penetrates dry, nonpolar, low-conductivity materials, while it is strongly absorbed by water. Terahertz radiation is a nonionizing low energy photon radiation that does not penetrate deeply into the skin, and as a result, is an excellent candidate to be considered as a new imaging tool that is complementary to X-ray or infrared thermography [2]. A disadvantage of the detection of terahertz radiation is the ability to resolve and measure the characteristics of the low energy photons, from the background, and hence requires cooling systems; for this reason, the terahertz detection systems have historically been expensive and bulky. In actuality, the terahertz band of the electromagnetic spectrum presents interesting and significant challenges to the electronic and photonic technologies in the terahertz radiation detection and processing of the signal; for this reason, the terahertz band is known as the "Terahertz gap" [3].

Terahertz radiation imaging for security screening [4], medical imaging (tomography) [5, 6], and space astronomy [7] are some of the most important areas of its applications. The absorption of terahertz radiation by water is the mechanism used by medical imaging to detect and diagnose possible cancer related tissues that have more blood vessels and more water content than healthy tissues. Terahertz radiation is routinely utilized in security screening; it allows the detection of concealed objects such as guns and knives, as well as the detection of explosives. In recent years, the technological innovations and progress in nanotechnology, photonics, and semiconductors [8, 9] has led to the development of compact and portable: (a) THz sources with high output power, efficiency, and tunability and (b) THz detectors and imagers with high sensitivity and quantum efficiency. These developments have led to enhanced applications of THz technology in a wide variety of sectors of the global economy. These applications include the following: astronomy, biological, chemical, forensic, health, materials, medical and pharmaceutical sciences, global environmental monitoring, homeland security, information and communications technology, nondestructive evaluation, quality control of food and agricultural products, satellite communications [10, 11], ultrafast computing and related applications [12].

Microbolometers. https://doi.org/10.1016/B978-0-08-102812-4.00015-2

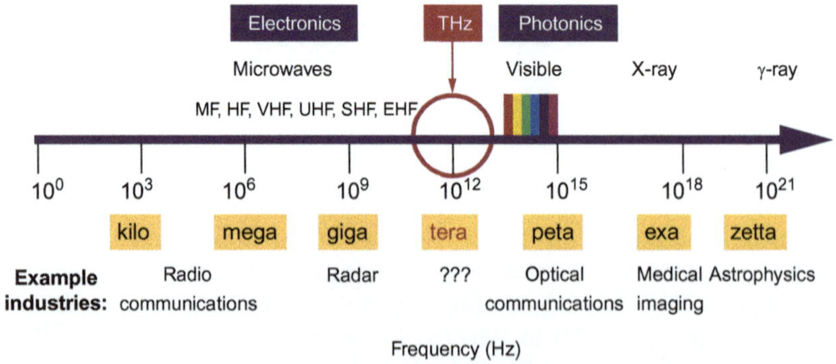

Fig. 9.1 Electromagnetic spectrum showing the location of the terahertz band [1].

In nature, the blackbody radiation consists of terahertz radiation that is associated with temperatures $>\sim 10$ K [13]. Although this radiation represents a very weak thermal emission, it presents a critical opportunity in observational astronomy for characterizing the cold 10–20 K dust in the interstellar medium in the Milky Way galaxy and in distant starburst galaxies [13, 14]. Since the invention of the femtosecond lasers, they have become extremely useful for the generation of coherent THz radiation [15]. Over the years, various sources of terahertz radiation have been developed in the industry. They can be classified into two major categories: electronic and photonic sources [16]. Electron beam and solid sources, and frequency multipliers are most widely used terahertz electronic sources [16]. Terahertz semiconductor and gas lasers, and terahertz optoelectronic sources are the commonly used terahertz photonic sources [16]. These sources include the following [13, 14, 16, 17]:

- backward wave oscillator,
- far infrared laser,
- free electron laser,
- gyrotron,
- photomixing sources,
- quantum cascade laser,
- resonant tunneling diode (RTD),
- single-cycle sources that are used in THz time domain spectroscopy (TDS), including optical rectification, photoconductive, photo-Dember, and surface field emitters,
- synchrotron light sources.

A schematic of a THz-TDS system is presented in Fig. 9.2.

It has been 25 years since Hu and Nuss demonstrated the first imaging system based on optoelectronic terahertz time-domain spectroscopy [18]. The first demonstration of free space transmitters and detectors, in the terahertz range of frequencies, was reported by Smith, Auston, and Nuss [19]. High performance terahertz radiation sources have been made possible by extending the high-frequency limit of radio frequency (RF) sources and the low-frequency limit of optical emitters as well as by combining RF and optical techniques [20]. Photoconductive terahertz optoelectronics has been a catalyst in the development of terahertz emitters and detectors [20, 21].

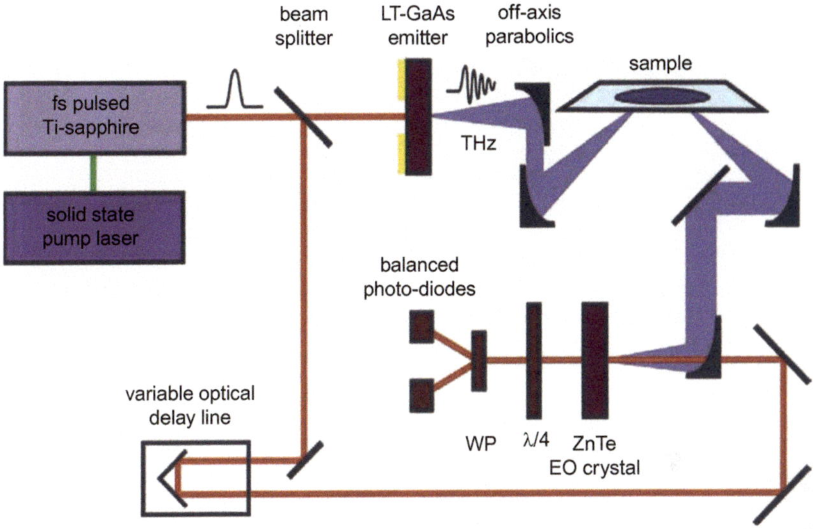

Fig. 9.2 Schematic of a THz-TDS system. In this example, the THz pulses are generated by a low-temperature grown (LT) GaAs photoconductive switch and subsequently detected, after reflection off the sample, by electro-optic sampling using a ZnTe electro-optic crystal [15].

A schematic diagram and operational concept of photoconductive terahertz emitters and detectors in shown in Fig. 9.3 [20].

9.2 Classification of terahertz detectors

Depending on the nature of the radiation-detector interaction, terahertz detectors are further classified into two types: photon detectors and thermal detectors. In photon detectors, the terahertz radiation is absorbed by the interaction of the incident photons with active carriers (free electrons, impurity atoms, or bound to the lattice) of the radiation absorber materials and an electric signal output is generated. In thermal detectors, the radiation is absorbed by a thin layer of the absorber material, where the radiation is converted into heat; this heat results in the change of an electric property of the thermosensing element that generates an output in the form of an electric signal. Fig. 9.4 shows the spectral detectivity characteristics of different types of detectors [22].

9.3 Terahertz microbolometer detectors

Historically, terahertz imaging systems were bulky and, like infrared imagers, required cooling. In recent years, with the significant development in the manufacture of uncooled microbolometers, they have led the way for the development of uncooled terahertz imaging technology. Some of these developments can be attributed to the exponential growth of applications of terahertz imaging and spectroscopy in various industries including biological sensing, food (chemical identification), material

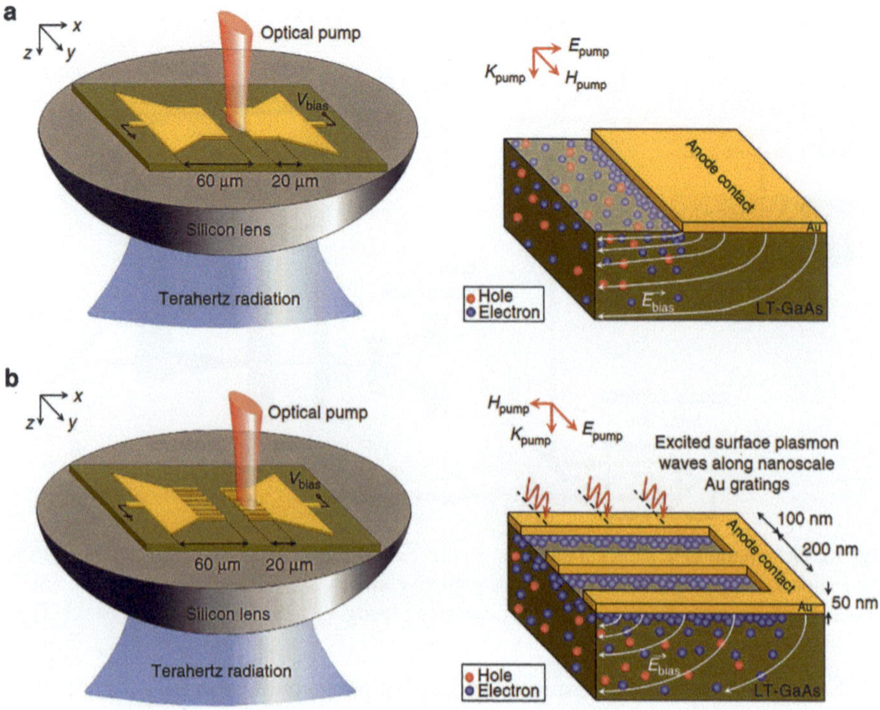

Fig. 9.3 Schematic diagram and operation concept of photoconductive terahertz emitters and detectors. *Left panels*: an ultrafast photoconductor connected to a bowtie terahertz antenna on a LT-GaAs substrate. The device is mounted on a silicon lens to collect the terahertz radiation from back side of the substrate. An optical pump is incident on the ultrafast photoconductor, generating electron-hole pairs inside the photoabsorbing substrate. *Right panels*: to operate as a terahertz emitter, a bias voltage is applied to the ultrafast photoconductor, which induces an electric field inside the substrate and drifts the photogenerated electrons and holes toward the anode and cathode contact electrodes, respectively. On reaching the contact electrodes, the photocurrent drives the terahertz antenna to produce radiation. To operate as a terahertz detector, a zero bias voltage is applied to the ultrafast photoconductor, and the bowtie terahertz antenna connected to the contact electrodes of the ultrafast photoconductor receives the incident terahertz radiation, inducing a terahertz electric field across the photoconductor contact electrodes. The photogenerated electrons and holes are drifted to the photoconductor contact electrodes under the influence of the induced terahertz field, generating an output photocurrent that is proportional to the magnitude of the incident terahertz field. (A) A conventional photoconductive terahertz emitter/detector. (B) A plasmonic photoconductive terahertz emitter/detector incorporating plasmonic contact electrodes to reduce photocarrier transport times to the contact electrodes and, thus, increase the photoconductor quantum efficiency [20].

characterization, medicine (medical imaging—burns, cancers, exposed skin, human bone, tooth decay, heart muscle fibers), and security (see through applications such as hidden bombs/explosives, drugs, guns, knives, etc.).

A schematic of an antenna coupled microbolometer is presented in Fig. 9.5. In this approach, the incident terahertz waves are converted into current by the antenna, which is subsequently dissipated in the form of heat through Joule heating in the load resistor [23].

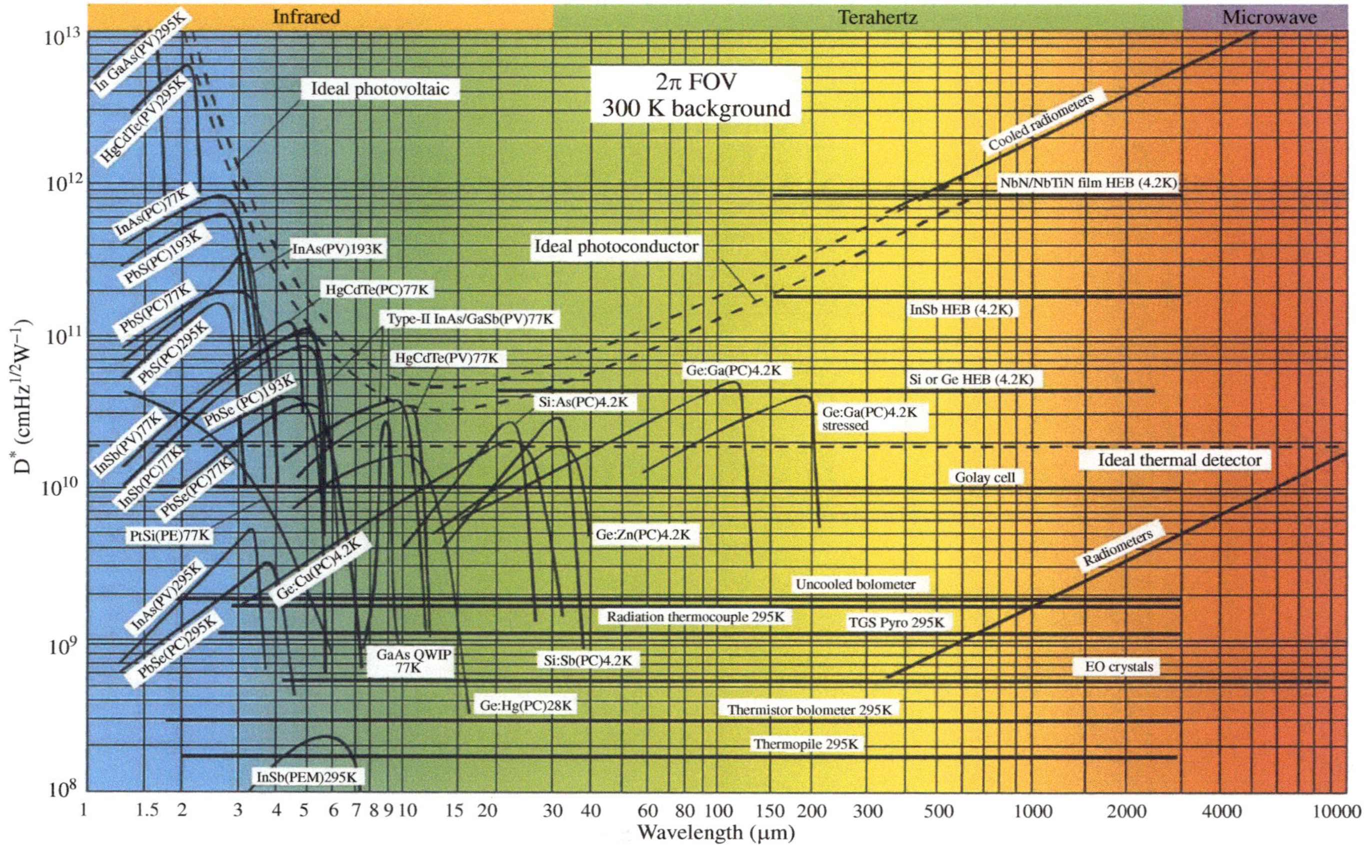

Fig. 9.4 Comparison of the D^* of various available detectors when operated at the indicated temperature. Chopping frequency is 1000 Hz for all detectors except the thermopile (10 Hz), thermocouple (10 Hz), thermistor bolometer (10 Hz), Golay cell (10 Hz), and pyroelectric detector (10 Hz). Each detector is assumed to view a hemispherical surrounding at a temperature of 300 K. Theoretical curves for the back-ground-limited D^* *(dashed lines)* for ideal photovoltaic and photoconductive detectors and thermal detectors are also shown. *PC*, photoconductive detector; *PV*, photovoltaic detector; *PEM*, photoelectromagnetic detector; *HEB*, hot electron microbolometer [22].

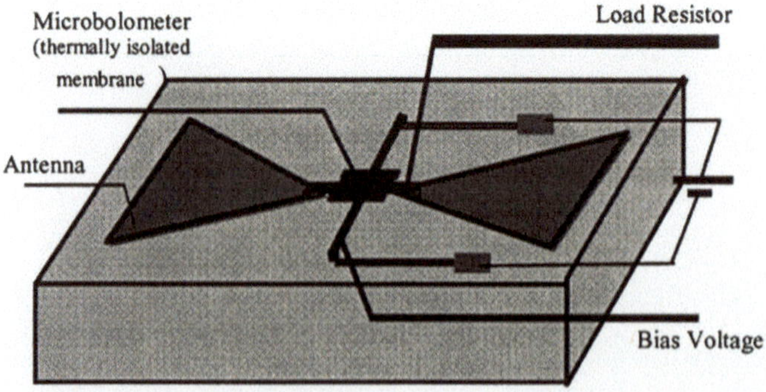

Fig. 9.5 Schematic to describe the principle of an antenna coupled microbolometer [23].

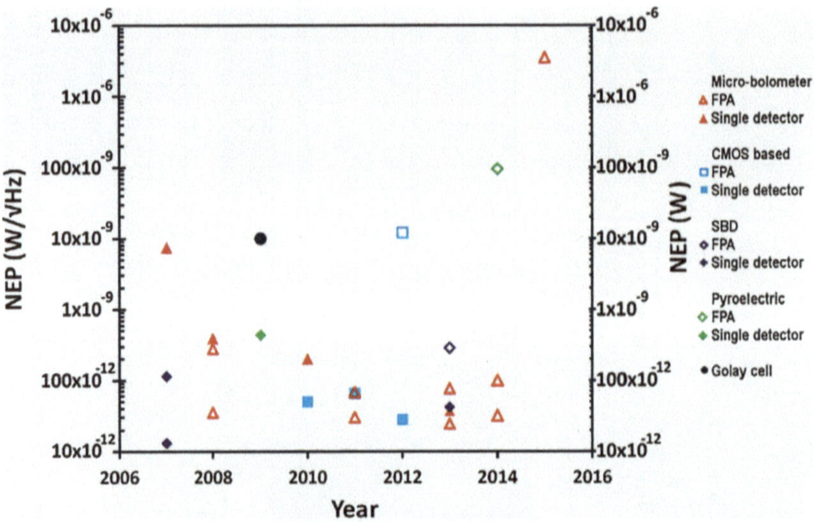

Fig. 9.6 Progress of terahertz detectors over recent years [24].

The 2017 THz science and technology roadmap [24] presents an excellent review of the growth and future of terahertz. Examples of these are presented in Fig. 9.6 and Table 9.1.

In Fig. 9.6, a schematic of the progress of terahertz detectors, over the years, is presented [24].

In Table 9.1, a summary of the various terahertz detector technologies is presented [24].

A summary of the specifications of four types of commercially available focal plane arrays, operating in the terahertz range of frequencies, is presented in Table 9.2 [25, 26].

Table 9.1 Summary of terahertz detectors (Ref. [24] and references therein).

Detector	Freq. (THz)	R_v ($V\ W^{-1}$)	NEP	Frame rate/response time	FPA size	Year	Commercial
Schottky barrier diode (SBD)	0.11–0.17	2000	13.2	42 ns	Single detector	2007	VDI
	0.9–1.4	100	113.7	25 ns			
	1.1–1.7	100	113.7	25 ns			
SBD	0.86	273	42	–1 μs	Single detector	2013	–
	0.28	336	290 pW		4 × 4		
Photo conductive	0.1–4.0	–	–	–	Single detector	2011 1985	EKSPLA
Folded dipole antenna	0.6–1.0	800 (1.027 THz)	66 (1.027 THz)	–	Single detector	2011	STM
FET FPA	0.7–1.1	115×10^3 (0.856 THz)	12 nW (0.856 THz)	25 Hz	32 × 32	2012	STM
VO_x micro bolometer	2.5	72×10^3 (2 μA)	37 (15 Hz mod.)	68 ms	Single detector	2013	–
	2.5	5620 (100 nA)	3.6 μW	194 ms	5 × 5	2015	–
Bolometer	4.25	–	24.7 pW	50 Hz	384 × 288	2013	INO
	2.54	–	76.4 pW				
Golay cells	0.2–20	10×10^3 (12.5 Hz modulation)	10×10^3	25 ms	Single detector	2009	Microtech
Microbolometer	1.0–7.0	–	<100 pW (4 THz)	30 Hz	320 × 240	2014	NEC
$LiTaO_3$ pyroelectric	0.1–300	–	96 nW (50 Hz mod.)	50 Hz	320 × 320	2014	Ophir Photonics
Pyroelectric	0.3, 1.0, 3.010	18.3×10^3 (10 Hz mod.)	440	10 Hz	Single detector	2009	QMC

Continued

Table 9.1 Continued

Detector	Freq. (THz)	R_v (V W^{-1})	NEP	Frame rate/ response time	FPA size	Year	Commercial
Hot electron bolometer	0.89	0.095	7.4×10^3	200 Hz	Single detector	2007	–
Si_xGe_y:H microbolometer	0.934	170	200	1 ms	Single detector	2010	–
α-Si microbolometer	2.4	14×10^6 5.9×10^6	30 pW 68 pW	–	320 × 240	2011	CEA-Leti
Nb_5N_6 microbolometer	0.1	100.5	398 (2 mA)	1 KHz	Single detector	2008	–
Vo_x microbolometer	2.8	200×10^3	35 pW	30 Hz	160 × 120	2008	Infrared systems
Antenna QW cavity	2.0–4.0	12.6 × 10 (2.5 THz)	32 pW (2.5 THz)	25 Hz	320 × 240	2014	CEA-Leti
VO_x microbolometer	3.1	–	280 pW	16 ms	640 × 480	2008	NEC
Folded dipole antenna	0.65	1.1×10^3 (0.15 V bias)	50	–	Single detector	2010	STM
FET	0.2–4.3	528 (1.4 THz)	28 (1.4 THz)	–	Single detector	2012	–

Table 9.2 A summary of the typical specifications of four types of commercially available focal plane arrays that can be used to visualize THz beams.

	CMOS	GaAs Heterostructure	Microbolometer	Pyroelectric
Number of pixels	32 × 32	64 × 64	320 × 240	320 × 320
Pixel size (μm)	80	1500	50	75
Operating range (THz)	0.3–1.3	0.05–0.7	0.6–4.0	0.1 and up
NEP (nW $Hz^{-1/2}$)	0.4	1	0.03	13
Responsivity (kV/W)	140	50	18 (at peak)	N/A
Frame rate (fps)	30	24	25	50

Experimental comparisons of some of these cameras have been discussed in [25, 26].

9.4 Case studies

Leading terahertz detector manufacturing companies such as NEC (Japan), INO (Canada), CEA-Leti (France), and TeraView (UK) have developed new generations of uncooled and cooled microbolometer configurations to be used in terahertz focal plane arrays and commercial cameras. Fig. 9.7 shows the schematic of the development of uncooled THz focal plane arrays [22].

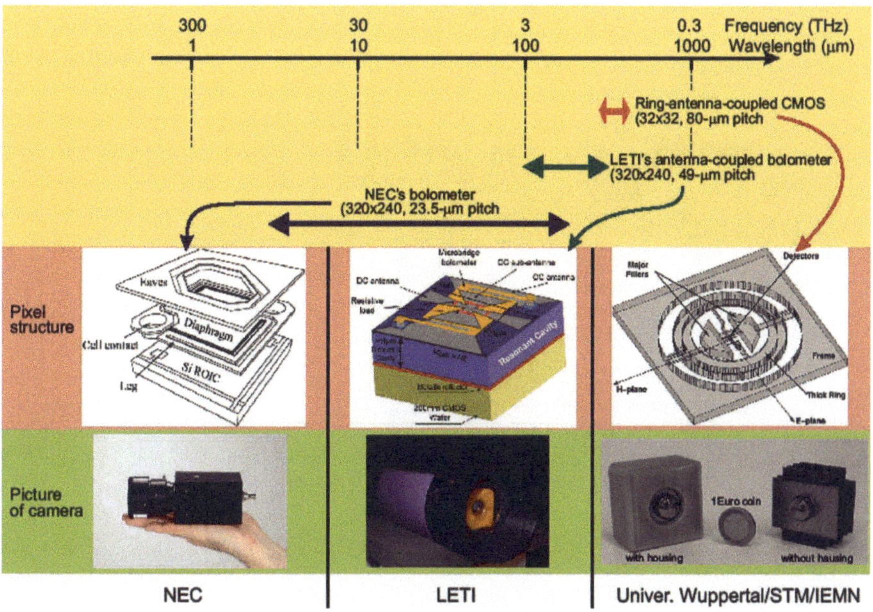

Fig. 9.7 Schematic of the development of uncooled THz focal plane arrays [22].

Terahertz microbolometer detectors are thermal detectors. They comprise of microbolometers that are suspended microbridge structures, consisting of a suspended multilayer membrane and two long supporting legs. The legs act as thermal isolators and form electric contact with the read-out circuit. The isolated and suspended membrane consists of a multilayer that includes a dielectric material layer, a thermosensing material layer, an electrode layer, and an infrared radiation absorption layer. The radiation absorber layer absorbs the incident THz radiation, converting the radiation into heat; the heat causes changes in the electrical properties of the thermosensing material [27, 28].

NEC has developed microbolometer terahertz focal plane arrays and cameras with improved sensitivity in the sub-terahertz region. The microbolometer pixel has a multilayer structure, suspended above a read-out integrated circuit (ROIC), that absorbs the terahertz electromagnetic radiation. The multilayer structure (diaphragm) contains a thermosensing layer (vanadium oxide) and a passivation layer (SiN_x) that is covered on the roof to increase the fill-factor. A thin metallic layer is formed in the roof, while a thick metallic layer (a perfect reflector) is formed on the ROIC surface [29]. The structure is composed of a thin metallic layer and a thick metallic layer acting as an optical cavity. The combination of the geometrical optical cavity length, 3–4 μm, with a certain range of sheet resistance for the thin metallic layer (50–100 Ω range), brings good minimal detectable power (MDP) values from ~1.5 to 100 THz. A thick SiN layer (~7 μm thick) and a layered electrical connection are inserted between the thick metallic layer and the air gap, so that the geometrical optical cavity length is made much longer by a factor of three than a previous version of the device. This modification can increase the sensitivity in the sub-THz region; this approach is utilized for the fabrication of both 320 × 240 and 640 × 480 THz-FPAs. Fig. 9.8 shows a pixel structure of a terahertz microbolometer focal plane array with

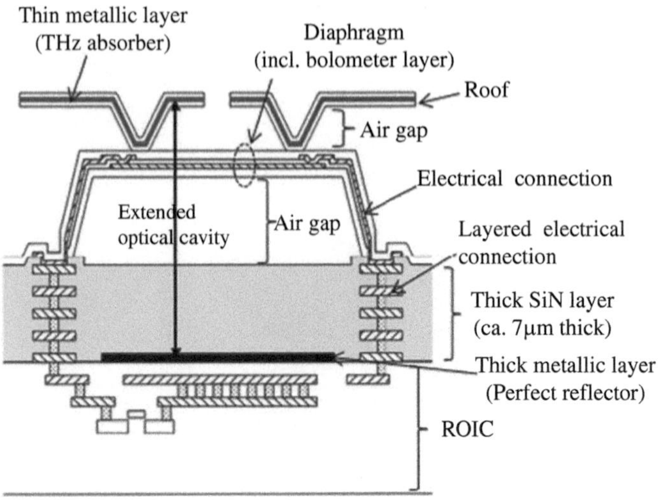

Fig. 9.8 Cross sectional view of a pixel structure of a terahertz microbolometer focal plane array with improved sensitivity in the sub-terahertz region [29].

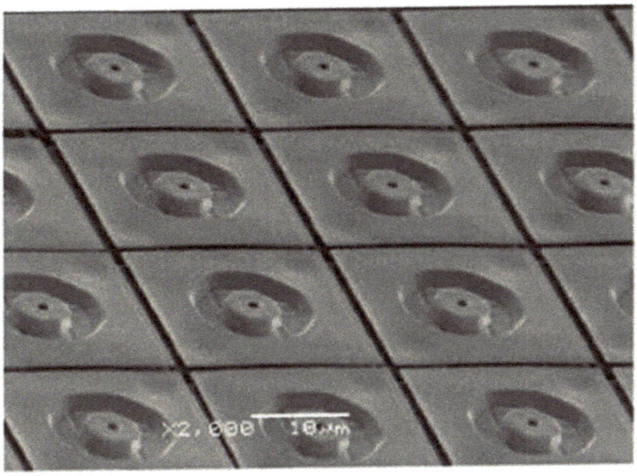

Fig. 9.9 SEM micrograph of a section of a 640 × 480 THz-FPA [29].

improved sensitivity in the sub-terahertz region [29]. A SEM (scanning electron microscope) micrograph of a section of 640 × 480 THz-FPA is shown in Fig. 9.9 [29].

CEA-Leti has developed a sensitive multispectral terahertz microbolometer that works below 1 THz; CEA-Leti has demonstrated the imager performance feasibility through preliminary simulation results. The detector consists of a suspended microbridge with an innovative antenna-coupled microbolometer. The sheet resistance is tuned to achieve impedance matching the terahertz wavelength [30] and a resonant cavity is added to enhance the terahertz absorption.

Fig. 9.10 shows a detailed structure of a 2-level antenna coupled bolometer. "The two-level antenna structure provides efficient coupling of the 2 cross-polarization components of the incoming radiation. One polarization excites the longer bowtie antenna (DC antenna). A lower stack level antenna (CC antenna) collects the other

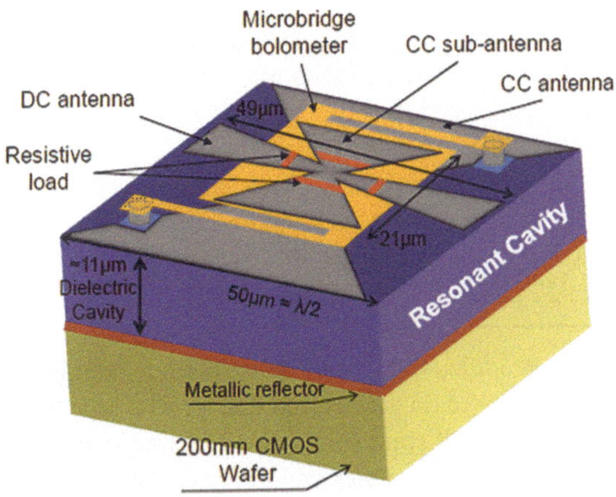

Fig. 9.10 Detailed structure of the 2-level antenna coupled bolometer [31].

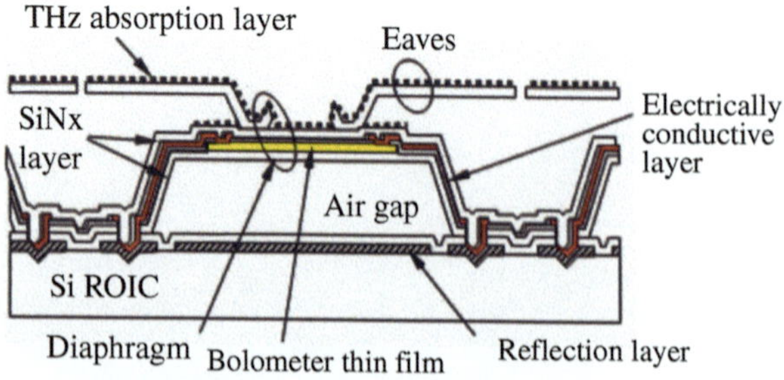

Fig. 9.11 Schematic diagram of pixel structure of uncooled bolometer-type THz-FPA [22].

polarization. A subantenna located on the microbridge capacitively couples this polarization-related current. Induced currents of both antennas heat up the suspended membrane through Joule dissipation in matched load resistances" [31]. The sizes and shapes of the antennas can be chosen to match a particular frequency.

A schematic of the cross sectional view of one CEA-Leti's pixel, with configuration similar to the microbolometer structure used by NEC, is shown in Fig. 9.11. In this structure, the pixel size is 50-μm pitch, the thermosensing material is amorphous, and the radiation absorber material is SiN_x. The membrane is suspended over the substrate by arms and pillars. The combination of an antenna gain and a quarter-wavelength resonant cavity enhances the radiation absorption [22].

INO has been actively working in the design and manufacture of uncooled microbolometer detectors for infrared imaging to produce terahertz (THz) imaging systems. INO has developed a leading-edge THz imaging system, the IRXCAM-THz-384 camera. The key to this technology is the development of custom microbolometer-based focal plane arrays with new absorber materials to enhance absorption in the THz bands coupled with terahertz imaging lenses. The results of the combination of these cameras with the appropriate terahertz sources show that the IRXCAM-THz-384 camera is able to image a wide variety of concealed objects of interest for applications such as nondestructive testing and detection of weapons [32].

INO uses vanadium oxide (VO_x) as the thermosensing material to manufacture MEMS-based microbolometer arrays for thermal infrared imaging. The uncooled microbolometer detectors fabricated at INO are obtained by MEMS surface micromachining techniques, using standard lithographic processes. An individual microbolometer consists of a suspended platform with VO_x thin film as thermosensing material. The vanadium oxide thin films are deposited by a DC sputtering technique. The detectors are fabricated on top of custom-built read-out integrated circuits (ROICs) to form arrays. Two THz array platforms have been developed at INO, based on 160 × 120 pixel (52 μm pixel pitch) and 384 × 288 pixel (35 μm pixel pitch) ROICs [32].

The IRXCAM main module, shown in Fig. 9.12(a), is composed of a set of electronic boards that handle raw data acquisition and processing, final data formatting, synchronization with an external source, and data transfer in accordance with various standards. These low noise camera cores can operate numerous FPAs at their maximum pixel rates. The first THz camera built at INO was the IRXCAM-THz-160,

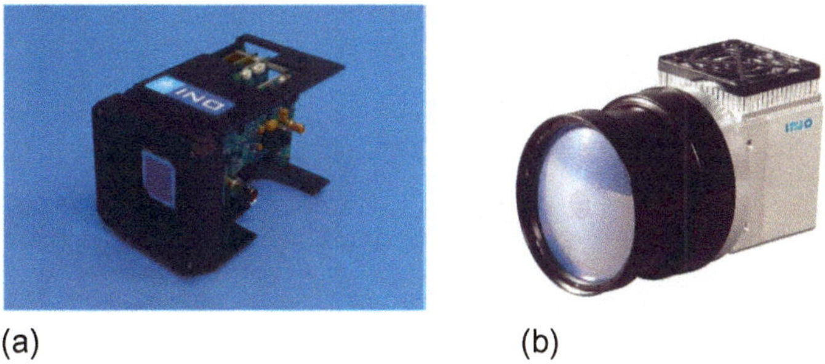

Fig. 9.12 Photograph of (A) IRXCAM camera core [32]; (B) MICROXCAM-384I-THZ TERAHERTZ camera [33].
Source: MICROXCAM-384I-THZ TERAHERTZ CAMERA, https://www.ino.ca/en/solutions/thz/microxcam-384i-thz/, Accessed 27 June 2020.

which supports the IRM160A 160 × 120 THz optimized uncooled microbolometer pixel array with a pixel pitch of 52 μm [32]. A recent addition to the THz family is the Microxcam-THz-384 that supports the IRM384A 384 × 288 THz optimized uncooled microbolometer FPA with pixel pitch of 35 μm. This has become the standard camera core used by INO in its development of THz detection and imaging applications. The technical specifications of this camera module are presented in Table 9.3 [32]. A photograph of IRXCAM-THz-384i is presented in Fig. 9.12(b) [33].

Table 9.3 IRXCAM-THz-384 technical specifications [32].

Parameter	Specification
Sensor	384 × 288 pixels, 35 μm pixel pitch Uncooled microbolometer THz optimized
Wavelength range	70–3198 μm (4.25–0.094 THz)
Video output	GigE RJ-45 16 bit raw data image
Frame rate	50 Hz
Control	GigE Link System operation control Loading of parameters
Power supply	9–14.4 V
Dimensions ($H \times W \times L$)	65 mm × 59 mm × 105 mm (excluding optics)
Weight (with sealed package)	230 g
Temperature	Operating: −10°C to 50°C
Mechanics	Integrated heatsink

Various approaches have been used by INO in order to enhance the absorption of terahertz radiation. These approaches include the following:

(a) use of metals, with different sheet resistances and thickness, as radiation absorbers;
(b) metallic antenna designs targeting the THz region; and
(c) the use of metamaterials.

The materials are chosen such that they are compatible with standard deposition and lithographic techniques. The material properties, device dimensions, and the optimization of the microbolometer structure contribute to the enhancement of terahertz radiation absorption and image resolution [32].

Teraview [34], a spin-off of Toshiba Corporation and Cambridge University, focuses on applications of terahertz in inspection of coatings, pharmaceuticals, semiconductor packaging, and R&D solutions. An example of a programmable terahertz chip-scale interface is described in Fig. 9.13 [35].

9.5 R&D, applications and market trends: Update

In a recent review, Salén and coworkers have reported a detailed review of the progress that has been made in THz technology, particularly with reference to the manipulation of matter with extreme terahertz light [36]. Fig. 9.14, from their study [36], presents an overview of the fundamental excitations in solids and molecular systems in the THz region and examples of applications of strong-field THz radiation. Depending on the nature of the chemical species and the interactions between photons and materials/device structures, studies such as these facilitate in utilizing terahertz spectroscopy and terahertz imaging for a variety of applications.

In Fig. 9.15, three main need-based application areas of terahertz spectroscopy have been summarized [37].

These are as follows:

- detection of objects and defects (size >1 mm) on large areas;
- thickness measurement over large areas;
- chemical and structural characterization of small objects and defects (~100 μm) over small areas or volumes.

A summary of the applications of THz imaging and spectroscopy, in various sectors, is presented in Fig. 9.16 [38].

A recent report of polarization-sensitive cross-nanowire THz detector (Fig. 9.17) [39] is anticipated to open up a wide range of new applications in various fields of engineering, science, and technology.

An overview of the global market revenue of THz market is presented in Fig. 9.18 [38]. The compound annual growth rate (CAGR) of global market of THz related products during 2012–2022 is ~26% [38].

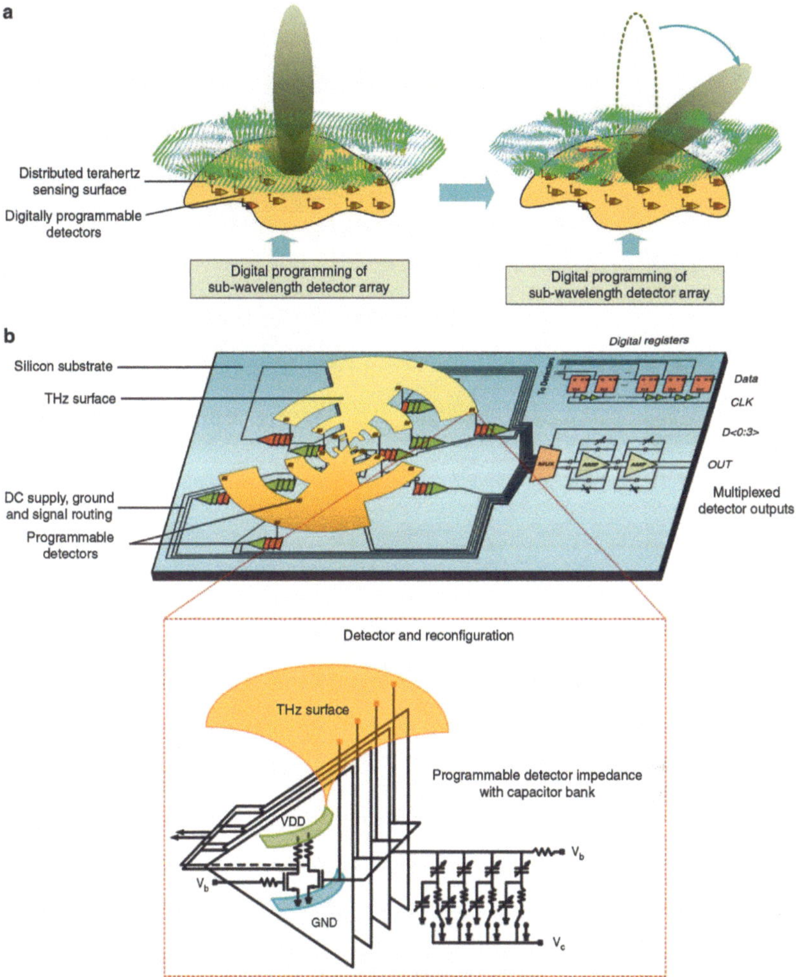

Fig. 9.13 Programmable terahertz chip-scale interface. (A) Concept of an universal terahertz sensing surface that maps a large discerte digital reconfiguration space to incident field properties including spectrum, angle of incidence and polarization. This is acheived by directly programming the surface currents induced on the interface at subwavelength scales. The incident power is absorbed in a distributed fashion at the sites of reconfiguration. (B) The system overview of the implemented terahertz sensor with a log-periodic tooth interface with 16 detectors distributed over the surface. Each site is reconfigurable with a coded capacitor bank that locally changes the boundary condition to redistribute the surface current distribution. This collectively programs the sensor response to allow for optimal reception against incident spectrum, polarization and angle of incidence [35].

Source: Wu, X., Lu, H. & Sengupta, K. Programmable terahertz chip-scale sensing interface with direct digital reconfiguration at sub-wavelength scales. Nat. Commun. 10 (2019) 2722. https://doi.org/10.1038/s41467-019-09868-6.

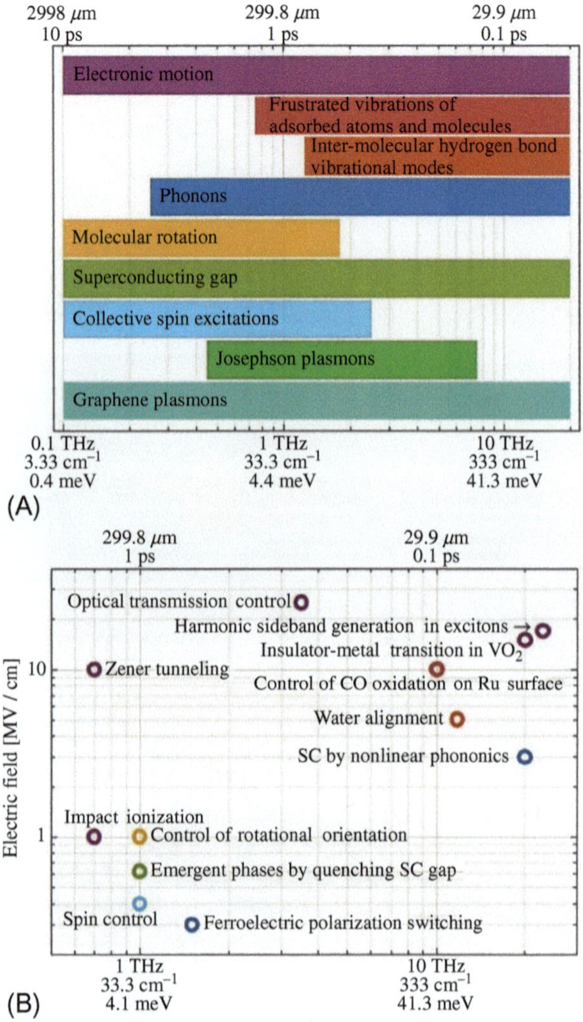

Fig. 9.14 (A) Fundamental excitations in solids and molecular systems in the THz region. (B) Examples of applications of strong-field THz light [36].

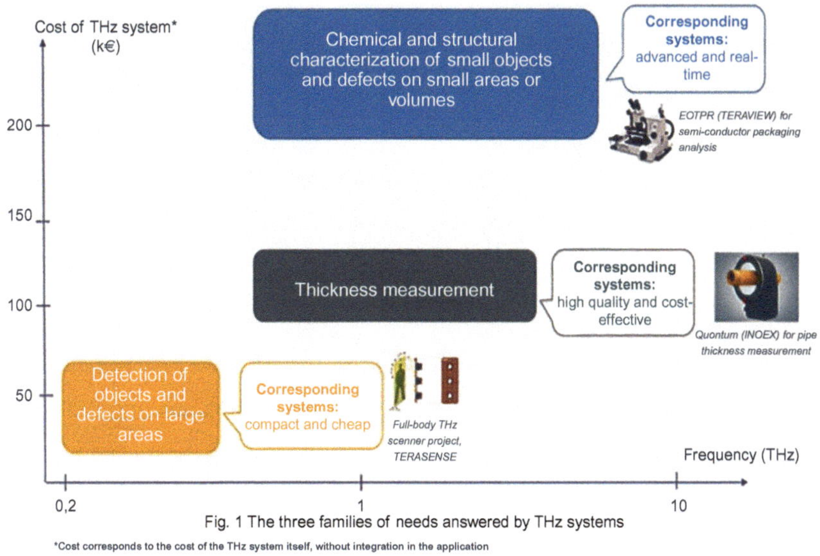

Fig. 9.15 The three families of needs answered by THz systems [37, 38].

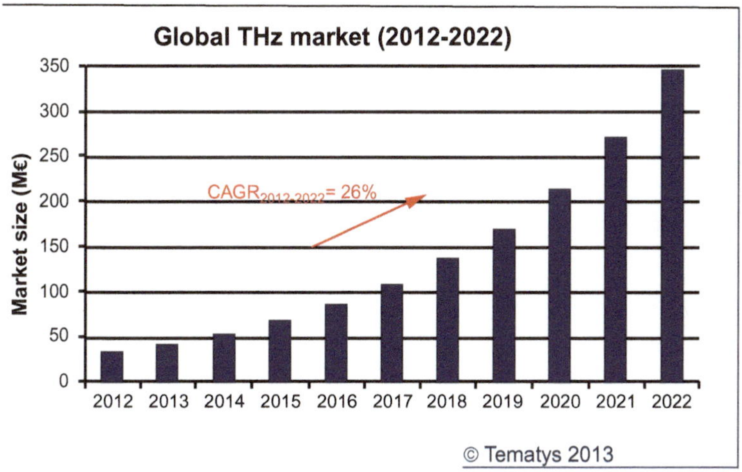

Fig. 9.16 Applications of THz imaging [38].

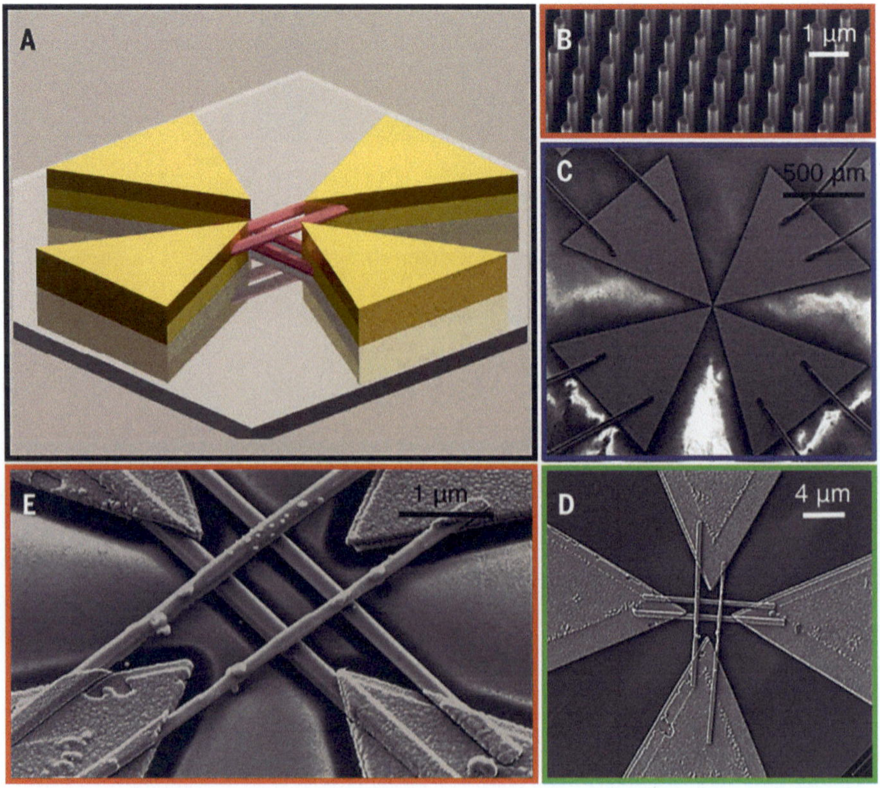

Fig. 9.17 Structure of the polarization-sensitive cross-nanowire THz detector. (A) Schematic illustration of device geometry. (B) Scanning electron micrograph (SEM) image of the as-grown InP nanowire array. (C to E) SEM images of the fabricated detector (*blue*: entire device; *green*: center of device; *orange*: close-up center of device under a tilted view of 25 degree) [39].

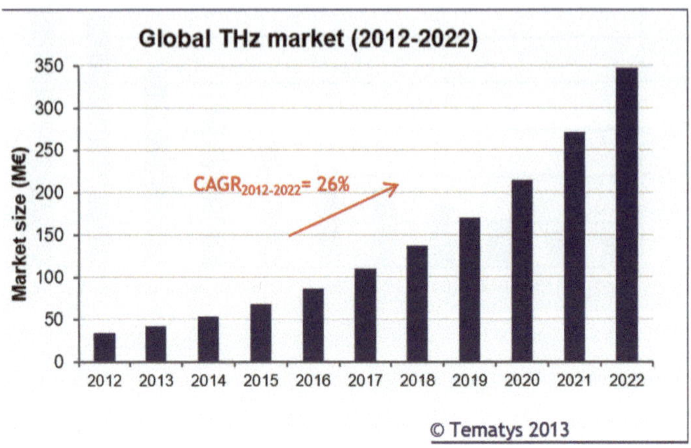

Fig. 9.18 Global THz market revenue (2015–2020) [38].

References

[1] G.P. Williams, Filling the THz gap—high power sources and applications, Rep. Prog. Phys. 69 (2) (2006) 301–326.

[2] R.H. Clothie, N. Bourne, Effects of THz exposure on human primary keratinocyte differentiation and viability, J. Biol. Phys. 29 (2–3) (2003) 179–185.

[3] A.G. Davies, E.H. Linfield, Bridging the Terahertz gap, Phys. World 14 (2004) 37–41.

[4] J.F. Federici, B. Schulkin, D.A. Zimdars, THz imaging and sensing for security applications—explosives, weapons, and drugs, Semicond. Sci. Technol. 20 (7) (2005) S266–S280.

[5] S.M. Kim, F. Hatami, J.S. Harris, Biomedical terahertz imaging with a quantum cascade laser, Appl. Phys. Lett. 88 (15) (2006) 153903.

[6] Z.D. Taylor, R.S. Singh, M.O. Culjat, J.Y. Suen, W.S. Grundfest, H. Lee, E.R. Brown, Reflective terahertz imaging of porcine skin burns, Opt. Lett. 33 (11) (2008) 1258–1260.

[7] M. Anderson, Graphene detectors bring terahertz astronomy to light, in: IEEE Spectrum, 2019.

[8] E.J. Martinez, Next Generation of Terahertz Sources and Detectors, *DARPA*, 2020. Accessed 29 June 2020.

[9] C.-S. Friedrich, et al., THz sources and detectors based on diode lasers, nonlinear frequency generation and conversion: materials, devices, and applications VIII, edited by Peter E. Powers, Proc. SPIE *7197* (2009) 71970B. pages 1–8 https://doi.org/10.1117/12.807853.

[10] H. Sarieddeen, et al., Next generation of terahertz communications: a rendevezous of sensing, imaging, and localization, IEEE Commun. Mag. 58 (5) (2020) 69–75.

[11] N. Saeed, et al., CubeSat communications: recent advances and future challenges, IEEE Commun. Surv. Tutorials (2020), https://doi.org/10.1109/COMST.2020.2990499.

[12] M. Tonouchi, Cutting-edge terahertz technology, Nat. Photonics 1 (2007) 97–105.

[13] A.Y. Pawar, et al., Terahertz technology and its applications, Drug Invent. Today 5 (2013) 157–163.

[14] A. Lee, Q. Qin, S. Kumar, B.S. Williams, Real-time terahertz imaging over a standoff distance (>25 meters), Appl. Phys. Lett. 89 (2006), 1411e1425.

[15] D. Giles, Terahertz spectroscopy of explosives and drugs, Mater. Today 11 (2008) 18e26.

[16] D. Saeedikia, Optoelectronic techniques for the generation and detection of terahertz waves, in: D. Saeedkia (Ed.), Handbook of Terahertz Technology for Imaging, Sensing and Communications, Series in Electronic and Optical Materials, Woodhead Publishing, 2013, https://doi.org/10.1533/9780857096494.1.3.

[17] S. Fathololoumi, E. Dupont, C.W.I. Chan, Terahertz quantum cascade lasers operating up to $\sim$ 200 K with optimized oscillator strength and improved injection tunneling, Opt. Express 20 (2012), 3866e3876.

[18] B.B. Hu, M.C. Nuss, Imaging with Terahertz waves, Opt. Lett. 20 (16) (1995) 1716–1718.

[19] P.R. Smith, D.H. Auston, M.C. Nuss, Subpicosecond photoconducting dipole antennas, *IEEE J. Quantum Electron.* 24 (2) (1988) 255–260.

[20] C.W. Berry, N. Wang, M.R. Hashemi, M. Unlu, M. Jarrahi, Significant performance enhancement in photoconductive terahertz optoelectronics by incorporating plasmonic contact electrodes, Nat. Commun. 4 (2013) 1622, https://doi.org/10.1038/ncomms2638.

[21] M. Jarrahi, Advanced photoconductive terahertz optoelectronics based on nano-antennas and nano-plasmonic light concentrators, IEEE Trans. Terahertz Sci. Technol. 5 (3) (2015) 391–397.

[22] A. Rogalski, Semiconductor detectors and focal plane arrays for far-infrared imaging, Optoelectron. Rev. 21 (4) (2013) 406–426.

[23] E. Peytavit, et al., Room temperature terahertz microbolometers, TB2–1, in: Sources, Detectors and Receivers Semiconductor, Superconductor and Other, Joint 30th International Conference on Infrared and Millimeter Waves & 13th International Conference on Terahertz Electronics, IEEE, 2005, pp. 257–258.

[24] S.S. Dhillon, et al., The 2017 terahertz science and technology roadmap, J. Phys. D Appl. Phys. 50 (2017), 043001.

[25] D.M. Mittleman, Twenty years of terahertz imaging (invited), Opt. Express 26 (8) (2018) 9417.

[26] E. Hack, et al., Comparison of thermal detector arrays for off-axis THz holography and real-time THz imaging, *Sensors* (Basel) 16 (12) (2016) 221.

[27] J. Wang, W. Li, J. Gou, Z. Wu, Y. Jiang, Fabrication and parameters calculation of room temperature terahertz detector with micro-bridge structure, J. Infrared Millim. Terahertz Waves 36 (2015) 49–59.

[28] X. Zheng, Z. Wu, J. Gou, Z. Liu, J. Wang, J. Zheng, Z. Luo, W. Chen, L. Que, Y. Jiang, Enhancement of real-time THz imaging system based on 320×240 uncooled microbolometer detector, J. Infrared Millim. Terahertz Waves 37 (2016) 965–976.

[29] K. Doi, T. Ishi, T. Sudo, T. Morimoto, H. Goto, T. Sasaki, N. Oda, S. Kurashina, M. Miyoshi, Microbolometer terahertz focal plane array and camera with improved sensitivity in the sub-terahertz region, J. Infrared Millim. Terahertz Waves 36 (2015) 947–960.

[30] N. Oda, Development of terahertz focal plane arrays and handy camera, Proc. SPIE (2011) 8012.

[31] J. Meilhana, et al., Terahertz frequency agility of uncooled antenna-coupled microbolometer arrays, in: 37th International Conference of Infrared, Millimeter, and Terahertz Waves, 2012.

[32] D. Dufour, et al., Review of terahertz technology development at INO, J. Infrared Millim. Terahertz Waves 36 (2015) 922–946.

[33] MICROXCAM-384I-THZ TERAHERTZ CAMERA, https://www.ino.ca/en/solutions/thz/microxcam-384i-thz/, Accessed 27 June 2020.

[34] https://teraview.com/about/, Accessed 1 July 2020.

[35] X. Wu, H. Lu, K. Sengupta, Programmable terahertz chip-scale sensing interface with direct digital reconfiguration at sub-wavelength scales. Nat. Commun. 10 (2019) 2722. https://doi.org/10.1038/s41467-019-09868-6.

[36] P. Salén, et al., Matter manipulation with extreme terahertz light: progress in the enabling technology, Phys. Rep. 836–837 (2019) 1–74.

[37] THz System Sales to Grow at 16% Through 2020, https://optics.org/news/7/7/27, dated July 19th 2016, Accessed 27 June 2020.

[38] Terahertz Components and Systems: Technology and Market Trends, https://www.tematys.fr/reports/en/accueil/30-terahertz-components-systems-technology-and-market-trends.html, Accessed 1 July 2020.

[39] K. Peng, et al., Three-dimensional cross-nanowire networks recover full terahertz state, Science 368 (2020) 510–513.

Antennas

10

10.1 Introduction

Generally, terahertz radiation refers to electromagnetic waves with frequencies in the range of 0.1 to 10 THz or 30 to 3000 μm or 0.4 to 40 meV [1]. Due to the associated wavelengths being in the range of 0.03 to 3 mm, terahertz radiation is also known as submillimeter waves, particularly in areas related to astronomy [2]. A schematic of the electromagnetic spectrum is presented in Fig. 10.1 [3]. As can be seen in the figure, the terahertz range of frequencies lie in between those that are of interest to electronics and photonics [3,4]. In addition to several characteristics, terahertz has two unique features that facilitates in the following: (a) penetration of materials that are conventionally deemed to be opaque; and (b) high chemical selectivity [4].

Microbolometers operate on the basis of the fundamental principle of change in resistance with temperature. Thus, the temperature coefficient of resistance (TCR) of the resistive element of the bolometer plays a critical role in determining its sensitivity. During its operation, a bias current is passed through the bolometer element/detector and the output voltage is monitored as a function of the incident infrared or terahertz radiation. The change in the output voltage is the indicator of the change in resistance of the bolometer detector. The responsivity of the bolometer detector to the incident infrared or terahertz radiation can be improved by reducing its size. However, this also reduces the total incident energy on the bolometer element. The interaction between the incident radiation and the bolometer element, and the collection area of the bolometer element, can be improved by coupling an antenna or arrays of antennas to it. This results in enhancement and sensitivity of the bolometer detector that can operate at room temperature [5–7].

Reducing the size of the bolometer detector and increasing the number of detectors in the imager helps to improve the collection area and increase the resolution. The resolution of the imager array determines its ability to generate a high-quality image [8]. However, increasing the number of detectors in the imager leads to increase in noise as well as increase in its total resistance [7]. The principal source of noise in a microbolometer is Johnson noise; it is the associated thermal noise and is given by [7,9]

$$V_j = \left\{ \sqrt{(4kTR\Delta f)} \right\}, \tag{10.1}$$

where k is the Boltzmann constant, T is the temperature in Kelvin, R is the resistance of the microbolometer detector, and Δf is the bandwidth of the measurement. The total Johnson noise is directly proportional to the number of detectors in the array [7].

Microbolometers. https://doi.org/10.1016/B978-0-08-102812-4.00016-4

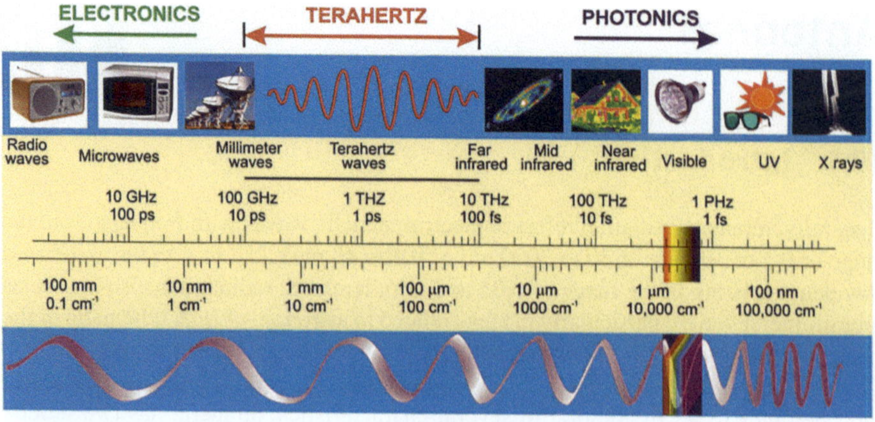

Fig. 10.1 Terahertz region in the electromagnetic spectrum [3].

The thermal time constant, τ, of a bolometer detector is defined by the following equation [10]:

$$\tau = \{C/G\}, \tag{10.2}$$

where C is the heat capacity and G is the thermal conductance of the device.

The responsivity, R_v, is given by.

$$R_v = \{V_s/P_i A_d\}, \tag{10.3}$$

where V_s is the output voltage, P_i is the intensity of input power, and A_d is the active area of the detector. The detector detectivity, D^*, is

$$D^* = \left\{\left(\sqrt{A_d \Delta f}\right)/\mathrm{NEP}\right\} \tag{10.4}$$

where NEP is the noise-equivalent power and is given by

$$\mathrm{NEP} = \{V_i/R_v\}. \tag{10.5}$$

10.2 Antenna fundamentals, characteristics, and applications

In addition to their use in microbolometers for infrared and terahertz imaging, due to their vast range of coverage of frequencies, antennas are utilized in a variety of applications. For completeness, the general definitions of characteristics of antennas are described here.

By its basic function, an antenna is a device that:

- converts RF (Radio Frequency) power that is applied to its feed point into electromagnetic radiation;
- intercepts energy from a passing electromagnetic radiation, which is then subsequently converted into an RF voltage across the feed point [11].

Antennas are also defined as devices that provide a transition between guided electromagnetic waves in wires and electromagnetic waves in free space [12]. Antennas follow the reciprocity theorem. The implication of the reciprocity theorem is that the properties of antenna remain the same in transmission or reception mode. The key characteristics that form the basis of design of antennas and their definitions are as follows:

- Bandwidth—Bandwidth is defined as range of frequencies over which the characteristics of the antenna are maintained.
- Beamwidth—Beamwidth of an antenna is the angular separation between the half-power points of the radiated pattern. It is determined from its two dimensional radiation pattern.
- Directivity or Gain—Directivity or Gain is defined as maximum power intensity radiated in a certain direction divided by the power intensity radiated by an isotropic antenna in that direction; gain of the antenna is the difference in directivity (in dB) of the antenna and the losses in the antenna (in dB); for an isotropic antenna or an omnidirectional antenna,

$$P_{\text{avg}} = \left\{P_{\text{rad}}/4\prod R^2\right\}, \tag{10.6}$$

where P_{avg} is the average power of the antenna, R is its radius, P_{rad} is its radiated power.

- Effective Aperture—Effective aperture, associated with a receiving antenna, is the measure of the ability of the antenna to extract energy from the electromagnetic wave.
- Effective Length—Effective Length is a parameter of wire antennas that characterizes the efficiency of the antenna in transmitting and receiving electromagnetic waves. The effective length for a receiving antenna is defined as the ratio of the electromotive force at the receiver input to the intensity of the electric field incident on the antenna; for a transmitting antenna, it is defined as the length in free space of a conductor with a uniform and in-phase current distribution along its entire length.
- Effective Area—In general, antennas capture power from the passing electromagnetic waves; some of this power is delivered to the terminals. It is given by

$$P_d = S A_{eff}, \tag{10.7}$$

where P_d is the power delivered to the terminals, S is the radiated power density, and A_{eff} is the effective area of the antenna.

- Input Impedance—Input impedance of an antenna is a measure of its resistance to an electrical signal in the antenna. Several factors influence the ability of the antenna to transmit a signal, including the environment that the antenna is in and the design and composition of the antenna. Understanding the input impedance of an antenna is critical in designing components that connect an antenna to a receiver or transmitter. It is simply a ratio of voltage across the two terminals of the antenna to the current flowing in the antenna. Mathematically, the input impedance of an antenna is

$$Z_{in} = R_{in} + jX_{in}. \tag{10.8}$$

Z_{in} is the impedance of the antenna at its terminals, R_{in} is the resistance of the antenna at its terminals, and X_{in} is the reactance of the antenna at its terminals. The power stored in the near field of the antenna is represented by the imaginary part X_{in}.

- Polarization—Polarization of an antenna is defined as the orientation of the transmitted or received electric field. Linear and circular polarization are the most common types of polarization. For electromagnetic waves, the polarization represents the plane in which the electric wave vibrates.
- Radiation Efficiency and Power Gain—Radiation efficiency, η, is the ratio of the radiated power P_{rad} to the input power, P_{in}, i.e.,

$$\eta = \{P_{\text{rad}}/P_{\text{in}}\} \tag{10.9}$$

Power gain of an antenna is defined as the ratio of the power radiated in a direction to the total input power.

- Radiation Intensity—Radiation intensity of an antenna is the power per unit solid angle. It is independent of the distance from the antenna.
- Radiation Pattern—Radiation pattern represents the energy radiated by an antenna. They are diagrammatical representations of the distribution of radiated energy into space, as a function of direction.
- Return Loss—Return loss is the total amount of power that is lost to the load and does not return as a reflection. It is a measure of the matching between the transmitter and the antenna.

The fundamentals of antennas, their types, properties and applications have been described in several presentations, theses, and papers [11–20].

The types of antennas include the following [11]:

- Hertz antenna or dipole
- Marconi antenna or vertical monopole
- Large loop antenna
- Yagi-Uda array

Antenna arrays represent group of antennas or antenna elements. Their role is to provide the desired directional characteristics as well as shape a beam [13]. An example of an antenna is a metal structure whose dimensions are comparable to the wavelength of the electromagnetic radiation and can convert the electromagnetic radiation into an electrical current. There have been studies on infrared antenna metrology in which infrared antennas have been developed for a variety of applications [18].

In general, antenna technologies include the following [21]:

- Flat panel/short-range antenna;
- fractal antenna;

- high-gain reflector antenna;
- medium-gain antenna;
- patch, miniature, embedded, and other low-power antenna;
- smart antenna.

In the literature, reconfigurable liquid metal antennas, using liquid eutectic of gallium and indium, have been demonstrated to operate over a large bandwidth [22] by simply inducing a change in physical length of the alloy.

While historically, antennas were made using metals, recent developments in low-dimensional materials have led to the use of 1D and 2D materials [21,23,24] and metamaterials [25] in the fabrication of antennas. Various materials including 2D materials such as graphene [26] and transition metal dichalcogenides (TMDCs) [27] as well as polymers [28] are being utilized for the manufacture of flexible antennas in wearable devices. The exponential growth in geosynchronous navigations systems, telecommunications, and wireless communications, in particular, Internet of Things (IoT), satellite communications, and wireless technology, coupled with wearable technology and its applications to healthcare, drones, and autonomous vehicles, has led to the use of several forms of antennas including the following,

- cavity slot monopoles [29];
- dipole antennas [30];
- electromagnetic band gap (EBG)-based antennas [31];
- integrated inverted-F antennas (IIFA) [32];
- magnetoelectric dipole antennas [33];
- microstrip patch antennas [34];
- planar inverted-F antennas (PIFA) [35–37];
- planar monopole antennas [38];
- substrate-integrated waveguide (SIW) antennas [39,40]

in the lower frequency bands. At the mm-wave, Yagi-Uda antennas have been proposed as an appropriate candidate for on-body communications due to a good compromise in terms of size and gain performance [41].

The applications of antennas in aerospace and defense, consumer electronics, healthcare, and telecommunications is growing rapidly. It is anticipated to reach a value of $31.79 Billion by 2025, representing a compound annual growth rate (CAGR) of 6.96% over the 2020–2025 period [42].

Products and applications relating to antennas are constantly evolving with the result that the industry has been keeping pace with the antenna design requirements as well as its applications. These design tools are based on electromagnetic simulation. Some of these software industries include the following:

- IE3D—Scalable EM Simulation Solution from Mentor Graphics, formerly Zeland Software
- Ansys High Frequency Structure Simulator (HFSS)—3D Electromagnetic Field Simulator for RF and Wireless Design
- Remcom (USA) (Mfg.)—State College, PA, 3D Electromagnetic Simulation Software—XFdtd
- Sonnet Software, Inc. (Mfg.)—Syracuse, NY—Electronic Design Automation Software—Sonnet Suites, Level 2 Gold

- EM Software & Systems (Mfg., Svc.)—Hampton, VA Antenna Magus
- CST—Computer Simulation Technology (Mfg., Svc.)—Framingham, MA Antenna Magus
- Integrated Engineering Software (Mfg.)—Winnipeg, Manitoba CHRONOS
- COMSOL, Inc. (Mfg.)—Burlington, MA, COMSOL Multiphysics—RF Module

10.3 Antenna-coupled microbolometers

An example of a bow-tie antenna-coupled microbolometer is presented in Fig. 10.2 [43].

In utilizing planar antennas, the orientation of the electromagnetic field determines the degree of coupling between the radiation and the antenna [43]. Once the antenna is aligned appropriately with the electromagnetic radiation, a current is induced. An illustration of the bow-tie antenna, the orientation of the electromagnetic field and the resulting current is shown in Fig. 10.3 [43].

The integration of antennas with infrared sensors provides significant opportunities for utilizing a variety of infrared sensors. These infrared sensors include the following: conventional bolometers, metal-insulator-metal (MIM) or metal-oxide-metal (MOM) or metal-barrier-metal (MBM) diodes, metal-semiconductor contacts/Schottky barrier diodes, p-n junction diodes, pyroelectric bolometers [44], quantum dot detectors, thermocouples, thermopile, etc. [45]. From the perspective of infrared detection and imaging, historically, there exist two atmospheric windows: 3 to 5 μm and 8–14 μm. The choice of the detectors or imagers is determined by their ability to operate in these range of wavelengths. Examples of single metal antenna coupled nanothermocouple are illustrated in Fig. 10.4 [45–47].

In Table 10.1, a summary of the various designs of antennas in the literature and their highlights is presented (Ref. [45] and references therein).

The key performance parameters of antennas include the following: directivity, gain, impedance, and resonance.

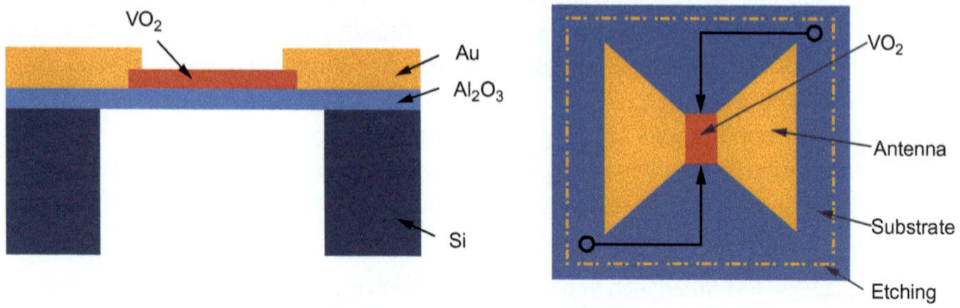

Fig. 10.2 Bow-tie antenna-coupled microbolometer. In this quasioptical receiver, the radiation is focused on the bow-tie antenna with a hemispherical lens. The coupled radiation is dissipated in the bolometer element, causing a temperature rise. The resulting change in resistance is sensed by the biasing network [43].
From 2019 IEEE International Symposium on Antennas and Propagation and USNC-URSI Radio Science Meeting.

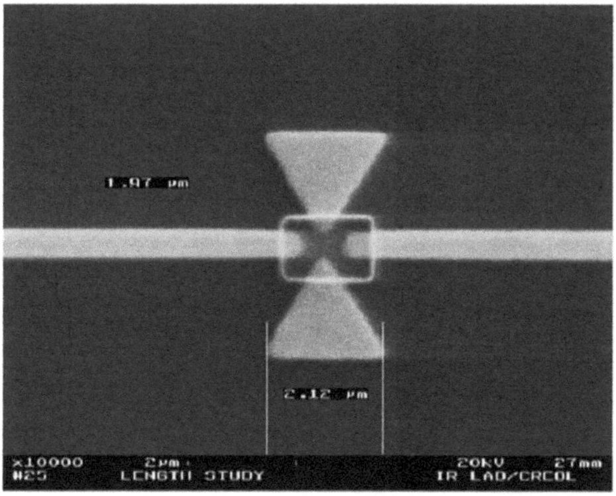

Fig. 10.3 Illustration of a bow-tie antenna along with the direction of the electromagnetic field and the resulting current. Electromagnetic radiation induces current in the bow-tie antenna. It must be noted that the orientation of the electromagnetic field is important for an optimal functioning of the antenna [43].
From S. Chen, M. Lust, N. Ghalichechian, A vanadium dioxide microbolometer in the transition region for millimeter wave imaging, 2019 IEEE International Symposium on Antennas and Propagation and USNC-URSI Radio Science Meeting, 2019, pp. 1641–1642, https://doi.org/10.1109/APUSNCURSINRSM.2019.8888891.

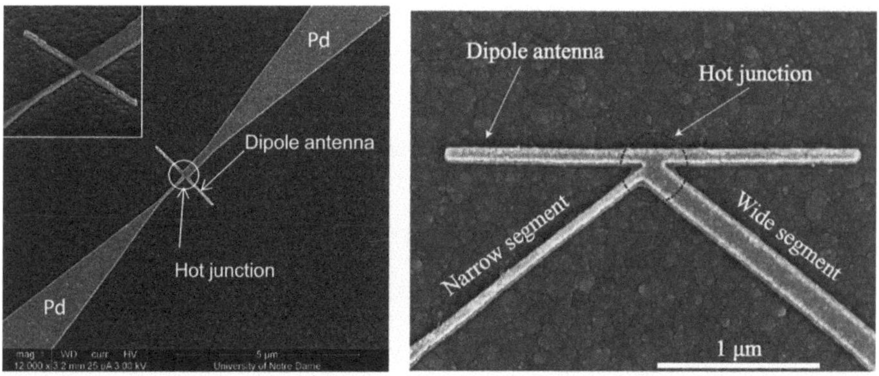

(a) (b)

Fig. 10.4 Examples of single metal antenna coupled nanothermocouple [45–47].

Example of an antenna coupled to a nickel microbolometer is shown in Fig. 10.5 [7]. Fig. 10.6 shows a two-dimensional array of antenna-coupled microbolometers [7]. A Fresnel zone plate that has been utilized to focus infrared energy on a single element antenna-coupled microbolometer is shown in Fig. 10.7 [7]. Recently, Liu et al. [53] have demonstrated polarization-independent infrared microlens array based on all-silicon metasurfaces. These microlens arrays have been designed to operate at

Table 10.1 Summary of various designs of antennas in the literature and their highlights (Ref. [45] and references therein).

Contribution/study	Structure and λ (μm)	Measurement
Fabricate lithographic antenna, 1991	Spiral and Nb bolometer at 9.5 μm	$NEP = 7 \times 10^{-11}$ W
Develop the matching layer, 1994	Dipole and Ni MOM at 10.8 μm	transmission increase 31%
Fabricate on air membrane, 1997	Log-periodic and poly-Si at 10.2 μm	Directivity = 9 dB
Polarization tunable device, 1998	Spiral and MOM at 10.6 μm	Polarization response
Substrate-side illumination, 2000	Dipole and Nb bolometer at 9.2–10.8 μm	$\mathfrak{R}_v$ increase
Wavelength tunable device, 2001	Microstrip and bolometer at 10.6 μm	Spectral response
Fabricate on silica gel (isolation), 2003	Bowtie and Nb bolometer at 10.6 μm	$\mathfrak{R}_v$ increase 20%
Develop MMW/LWIR dual band device 2004	Dipole and MOM at 10.6 μm	$D^* = 1 \times 10^6$ cm Hz$^{1/2}$ W^{-1}
Fabricate on Si_3N_4 bridge (isolation), 2006	Square spiral and Ni, bolometer and Ni bolometer at 9–11 μm	$D^* = 2.89 \times 10^7$ cm Hz$^{1/2}$ W^{-1}
Fabricate on hemispherical lens (surface waves), 2006 [102]	Dipole and V bolometer at 10.6 μm	$\mathfrak{R}_v$ increase 11.7 ×
Fabrication using a single EBL (process optimization), 2008	Dipole and MOM at 10.6 μm	$D^* = 2.15 \times 10^6$ cm Hz$^{1/2}$ W^{-1}
Develop traveling wave diode (impedance matching), 2010	Traveling wave MIM TWMIM at 3 μm	$D^* = 4 \times 10^6$ cm Hz$^{1/2}$ W^{-1}
Study substrate configurations, 2010	Dipole and MOM at 10.6 μm	Radiation pattern
Develop beam steerable device, 2010	Phased array and MOM at 10.6 μm	Radiation pattern
Optimize MOM barrier formation (process optimization), 2011	Dipole and Al/Pt MOM at 10.6 μm	$D^* = 9.65 \times 10^6$ cm Hz$^{1/2}$ W^{-1}
Develop CPS coupled device (impedance matching), 2012	Bowtie and thermocouple with CPS at 10.6 μm	D^* increase 2.4 ×
Simulate Seebeck antenna, 2014 [103]	Square spiral and thermocouple at 10.6 μm	Efficiency simulation
Develop a rectenna using graphene geometric diode (response time), 2014	Dipole and geometric diode at 10.6 μm	$D^* = 2.6 \times 10^6$ cm Hz$^{1/2}$ W^{-1}

Table 10.1 Continued

Contribution/study	Structure and λ (μm)	Measurement
Develop a single metal/EBL device, 2015	Dipole and thermocouple at 10.6 μm	Spectral response
Fabricate parabolic reflector antenna (surface waves), 2018	Parabolic reflector and Nb bolometer at 10.6 μm	Fabrication process recipe

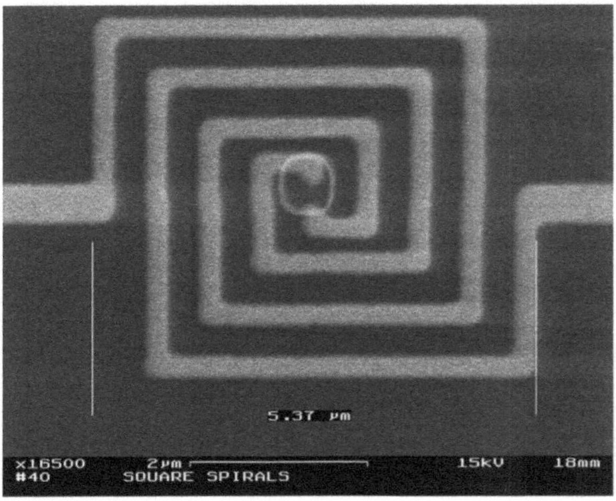

Fig. 10.5 Single-element square-spiral antenna coupled to a nickel microbolometer [7].

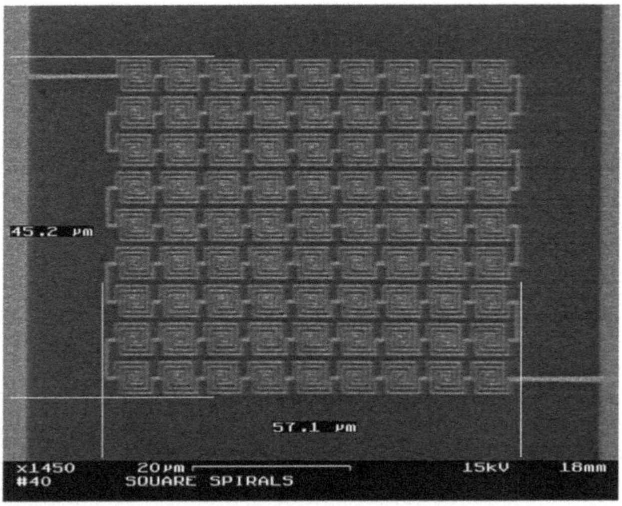

Fig. 10.6 Two-dimensional array of antenna-coupled microbolometers [7].

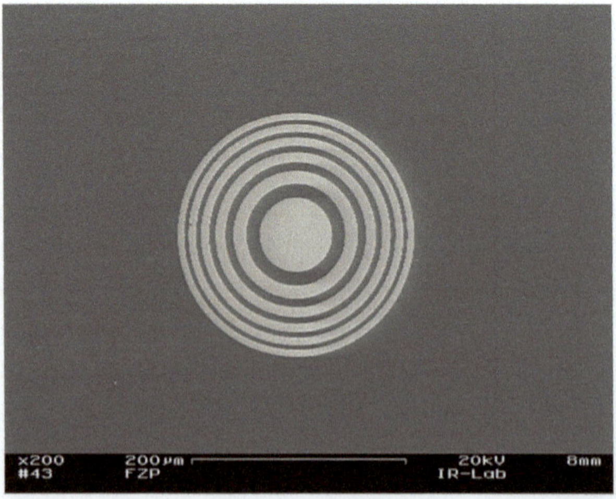

Fig. 10.7 Fresnel zone plate lens used to focus infrared energy on a single element antenna-coupled microbolometer [7].

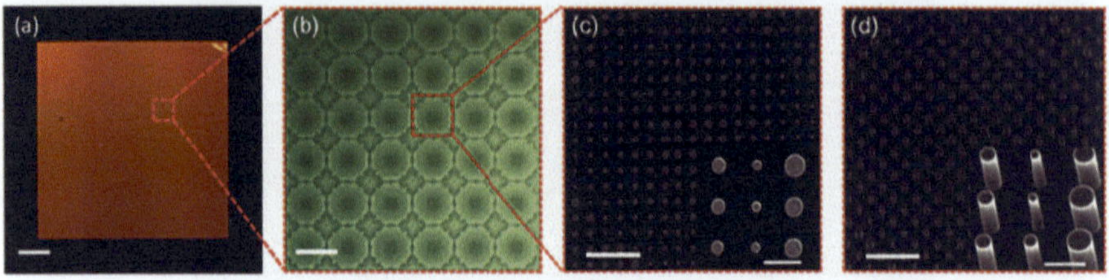

Fig. 10.8 (A) Top-view optical photograph of the fabricated 60 × 60 LWIR microlens array. Scale bar: 1 mm. (B) Top-view optical micrograph of the microlens array. Scale bar: 100 μm. (C) Top-view scanning electron micrograph of a single micrometalens. Scale bar: 20 μm. The inset is zoomed-in SEM image. Scale bar: 5 μm. (D). Side-view scanning electron micrograph of a single micrometalens. Scale bar: 20 μm. The inset is the zoomed-in SEM image. Scale bar: 5 μm [53].

10.6 μm. In Fig. 10.8, the details of the microlens array are presented [53]. In each of these approaches, the goal is to maximize the collection area of the incident electromagnetic radiation for the imager, enhancing thereby, the incident photon intensity.

Traditionally, based on the technology that is being utilized, terahertz antennas have been classified into two broad areas [54]. These include the following:

- Wavelength scale beam forming or feed elements that enable the coupling of energy from free space into or out of subwavelength generators or receivers such as bolometers, diodes, photoconductive elements, transistors, etc.
- Signal collection schemes such as much larger apertures that are utilized to shape or focus the beams for numerous applications such as high resolution imaging or scanning to gathering and integrating power required by large radio telescopes [54].

In recent years, efforts have been concentrated on the development of high sensitivity terahertz (THz) uncooled and cooled [48] detection microbolometers [49,55–57].

Terahertz detectors offer noninvasive and nondestructive ways of evaluation and testing of materials and are well-suited for applications in the field of security screening (security and military), diagnostic medicine, astronomy, materials science, and gas and explosive detection.

Antenna-coupled infrared radiation sensing detectors represent a potential solution in the development of infrared detector focal plane arrays (FPA). In comparison to the micrometer wavelength photons in the near and mid-infrared radiation, the longer wavelength in the millimeter range of the terahertz photons represent comparatively weak radiation. The absorption of terahertz radiation has been achieved by antenna-coupled thermal detectors, specifically via microbolometers. Terahertz detectors using microbolometers as detection structures are reported as one of the most sensitive uncooled terahertz detectors that are suitable for imaging applications. These light-weight and low-cost infrared detectors are widely used in the field of infrared and terahertz (THz) wave detection and imaging. Typically, these detectors can be operated uncooled throughout the infrared frequency range.

In order to improve the responsivity of the microbolometer, the traditional design of a suspended infrared sensing bridge with a cavity as an optical resonator, in combination with a backside reflective material (gold, aluminum), is enhanced by the addition of a frequency tuned antenna geometry. During the last decade, uncooled terahertz sensing structures based on antenna-coupled suspended bridges have been developed and have achieved a high performance. In Fig. 10.9, a schematic of an antenna-coupled Ti-microbolometer with a THz antenna, a resonating cavity and a mirror is presented [58].

Based on its thermal infrared bolometer technology, CEA-Leti has been working on terahertz real-time imaging [59]. An example of CEA-Leti's terahertz real-time imaging uncooled array, based on antenna- and cavity-coupled bolometer, is illustrated in Fig. 10.10 [52].

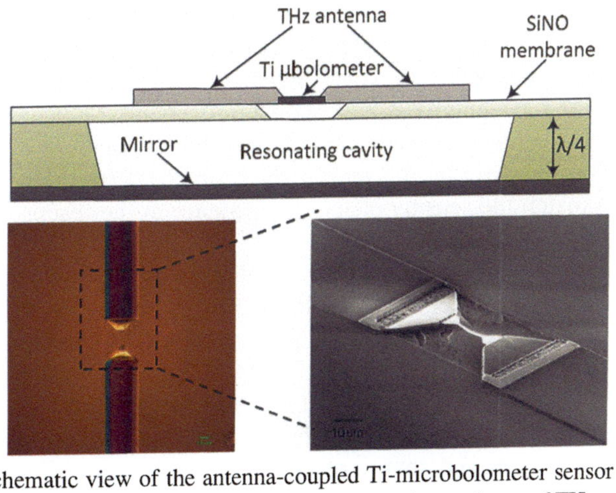

Fig. 10.9 A schematic view of the antenna-coupled Ti-microbolometer sensor *(top)*; microscope image of the fabricated Ti-microbolometer and central part of THz antenna: flat-top view *(bottom left)* and side-3D view *(bottom right)*. Note a tiny air-bridged Ti wire positioned in the center of the antenna [58].

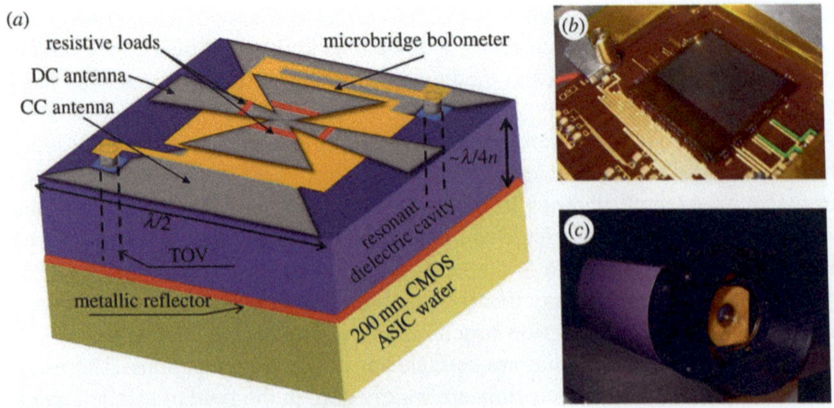

Fig. 10.10 (A) Uncooled THz antenna- and cavity-coupled microbolometer pixel structure. (B) Prototyped 320 × 240 pixel array chip. (C) Camera prototype integrating the THz bolometer array in its vacuum packaging [52].

10.4 Types of antennas

The terahertz region has a very low radiation energy and various terahertz detector configurations have been developed to achieve the maximum sensitivity required [58,60]. At the moment, the uncooled antenna-coupled infrared microbolometers are the most promising terahertz detectors [60,61].

In a related study, the microbolometer suspended bridge is coupled to a receiving terahertz antenna, which has its own impedance that changes with the absorbed radiation frequency. The impedance of the microbolometer should match with the antenna to achieve the interaction with the maximum radiation [62]. Studies have shown that the terahertz detector response can be enhanced by using additional transmission line for matching [63] and matching load and antenna impedances [64]. The size of the wavelengths of the terahertz frequencies is longer than the physical dimensions of the microbolometer bridge, and the microbolometer sensing element impedance is resistive; for this reason, the coupled antenna impedance should be optimized to match the sensing element impedance at the resonance frequency required.

Several antenna design variants have been developed and studied, concerning the antenna shape (dipole, bow-tie, or log-periodic, depending on the desired detector bandwidth) and the presence and position of an optional terahertz reflector situated on the wafer backside (Fresnel lens) or at a position that is just below the antenna on the front side. Different types of antenna configurations can be used to achieve broad-band and narrow-band terahertz detectors with high gain and low noise. The sizes and shapes of antennas can be chosen to match a particular frequency. Fig. 10.11 shows three of the most used configurations of terahertz antennas [65].

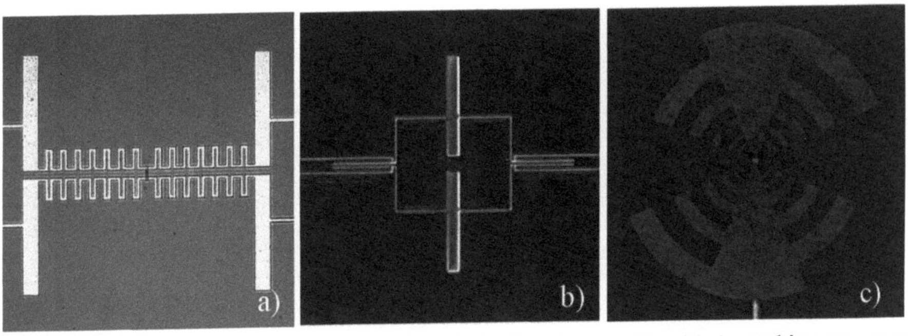

Fig. 10.11 Different types of antennas: (A) dual dipole antenna with added matching structure for 300 GHz, (B) dipole antenna for 600 GHz, and (C) log-periodic antenna [65].

10.5 Terahertz antenna-coupled microbolometer detectors

Terahertz microbolometer detectors are usually fabricated using various modifications of MEMS technology. The maturity of the microbolometer suspended bridge technology fabrication and its full compatibility with silicon CMOS processes allow the integration of antennas to enhance the absorption of terahertz radiation.

Various antenna-coupled microbolometer approaches have been developed in the industry in order to enhance the absorption of terahertz radiation. The use of different antenna configurations, with varying terahertz metallic absorber materials with different sheet resistances and thickness as radiation absorbers, has been explored and used.

Microbolometers for terahertz region have a multilayer structure that is suspended above a read-out integrated circuit (ROIC). This suspended structure absorbs the terahertz electromagnetic radiation. The multilayer structure (diaphragm) contains a thermosensing layer (VO_x, a-Si, Ti, Bi) and a radiation absorber material such as SiN_x or/and SO_2. A thin metallic layer deposited on the substrate acts as a reflective layer, which in combination with a resonant cavity between the suspended structure and the substrate enhances the absorption of the incident radiation. At terahertz frequencies, millimeter and submillimeter wavelengths, the coupling of antennas to the traditional suspended microbolometers improves their absorptance and sensitivity. The terahertz radiation that is not absorbed by the antennas is absorbed by the absorber material layer and by the resonant cavity.

Fig. 10.12 shows a shows a scanning electron microscope (SEM) image of a thin-film spiral-antenna-coupled VO_x microbolometer on Si_3N_4/SiO_2 (300 nm)/SiO_2 (300 nm) membrane. The microbolometer consists of a thermosensing element of VO_x thin film on $Si_3N_4/SiO_2/Si$ substrate fabricated by metal-organic decomposition (MOD) [66].

The antenna has a configuration of a 2-arm Archimedean spiral antenna [66,67]. The antenna has been designed so that it covers a frequency range of 140–220 GHz. The arm width of the antenna is 42 μm and is equal to the space between 2 arms because of its self-complementary structure. The area of the membrane is $15 \times 75\ \mu m^2$, and a

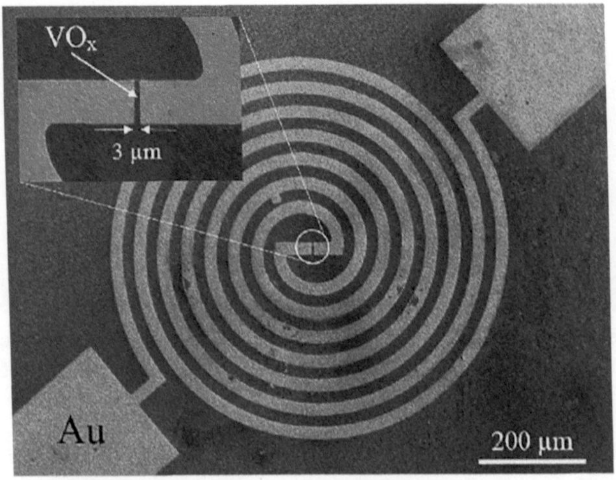

Fig. 10.12 SEM micrograph of the fabricated spiral-antenna-coupled VOx microbolometer on Si_3N_4/SiO_2 membrane [66].

microbolometer with the size of $3 \times 42\ \mu m^2$ is formed on the membrane. The outer and inner diameters are 1.42 mm and 0.2 mm, respectively, and the angle at termination is selected to be 1300 degree (3-turns +220 degree). VO_x as a thermosensing element with the size of $3 \times 42\ \mu m^2$ is located at the center of the antenna. The antenna is fabricated on the membrane plus 20 μm-thick Si. Because the thickness of the membrane plus 20 μm-thick Si (nondoped) is much thinner than a wavelength in operation, the antenna can be expected to operate as an antenna in the air [66].

This thin-film spiral-antenna-coupled VO_x microbolometer is fabricated on Si_3N_4/SiO_2 membrane by metal-organic decomposition (MOD). The membrane structure is fabricated by the wet etching technique using KOH solution. The sensitivity of the microbolometer is reported to be $3120\ W^{-1}$. The responsivity of the bolometer is 370 V/W at 190 GHz with a current bias of 50 mA. A high performance of the VOx microbolometer, integrated with the spiral antenna, is obtained by introducing the Si_3N_4/SiO_2 membrane structure in the terahertz frequency region [66].

A design concept of a terahertz microsensor that is widely used in the industry for biomedical applications is shown in Figs. 10.13 and 10.14 [51]. This concept is used for a broadband low-cost microbolometer detector for terahertz radiation. Fig. 10.13 shows the concept of the detector, which is based on an antenna-coupled bolometer design. The antenna (Au) collects the terahertz radiation and delivers the power to a niobium matched load; the choice of this metal is not due to its superconductivity properties, since the bolometer is designed to work at ambient temperature, but because of its resistivity, which allows to obtain the correct resistance value within the required antenna size. The heat from the load is transferred through a thin silicon nitride membrane to a mesoscopic thermosensing material wire of polysilicon (26 μm long and 0.6–1 μm wide), acting as a bolometer. The ensemble is suspended over the substrate in order to increase the thermal isolation and therefore to enhance the responsivity [51].

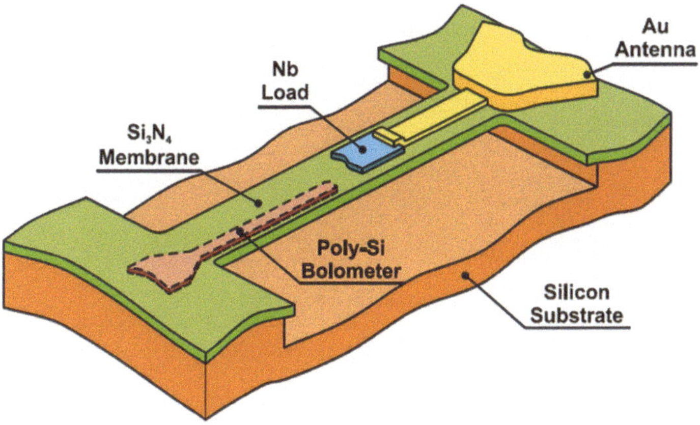

Fig. 10.13 Concept of the THz antenna-coupled bolometer. For clarity reasons, only half of polysilicon bridge, gold antenna, and niobium load are shown [51].

Fig. 10.14 Microphotograph of one of the fabricated devices [51].

Fig. 10.14 shows a microphotograph of one of the fabricated device. An important feature is represented by a dummy bolometer without the antenna, close to the active one. This additional device has the purpose of compensating any drift in the temperature of the substrate and attenuating the detection of other wavelengths that are not related to the antenna (e.g., photoconductivity due to visible light, infrared) [51].

Figs. 10.15 and 10.16 show the concept of the detector which is based on an antenna-coupled bolometer design in which titanium is used as the thermosensing material. Fig. 10.15 shows a cross-section of the microbolometer device with a narrow-band titanium microbolometer terahertz detector developed to operate at room-temperature [68].

In the fabrication of a microbolometer, the thermosensing material that is selected for its functionality must have a high-temperature coefficient of resistance and low

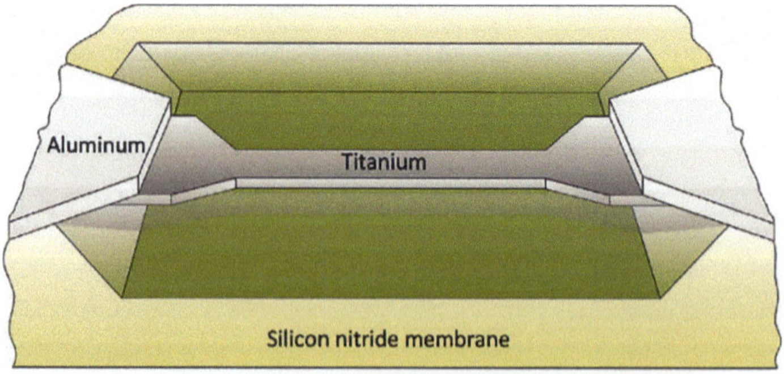

Fig. 10.15 Cross-section of the microbolometer device [68].

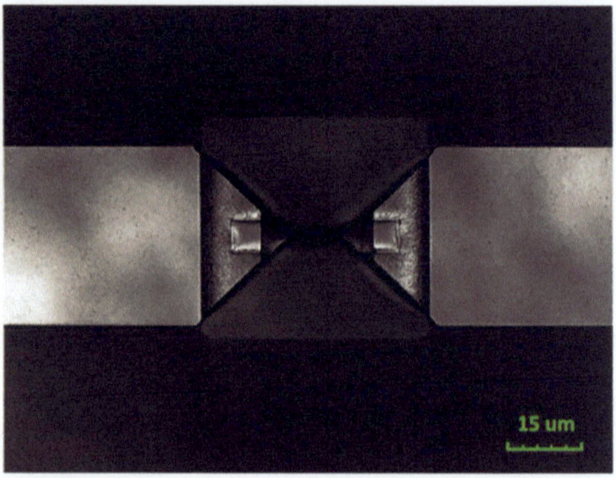

Fig. 10.16 Fabricated Titanium microbolometer with the antenna contact and noticeable cavity in the membrane [68].

noise (metals have only the Johnson noise); it must be stable with no deterioration, and ideally, it should be compatible with the CMOS fabrication technology. The thermal conductivity, electrical resistivity, and temperature coefficient of resistance, TCR, are the most important properties of the thermosensing materials. Vanadium oxide (VO_x) has high values of TCR; however, it is not compatible with CMOS process. Platinum seems to be the best solution due to its high TCR of 0.39%/K [68].

This detector configuration design is suitable for several applications, such as standoff THz imaging, dual-color terahertz imaging, and spectroscopic terahertz imaging. The detector displays a high sensitivity up to 1000 V/W, and low NEP down to 5 $pW/\sqrt{Hz}$ [68].

Examples of dipole antenna coupled infrared detector are illustrated in Fig. 10.17.

The design and simulation [10,13,71–74] and fabrication [74–76] of antenna-coupled microbolometers have been discussed in several publications. While microelectromechanical structures (MEMS) technology has been the driver for the

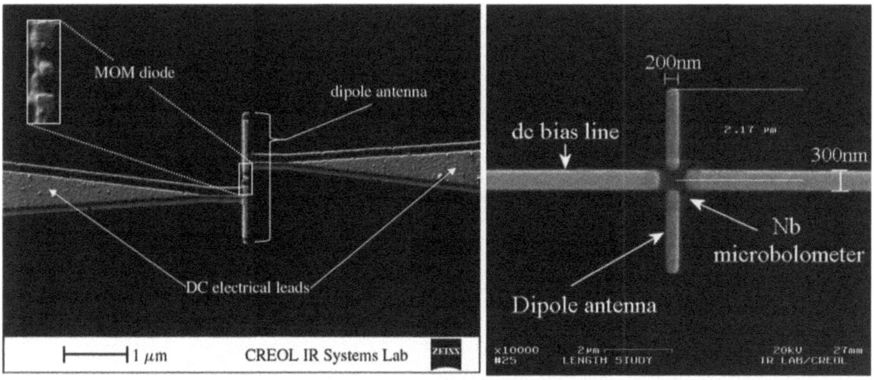

(a) (b)

Fig. 10.17 Scanning electron micrograph of a dipole antenna-coupled infrared detector example: (A) antenna-coupled MOM diode [13,69] and (B) antenna-coupled bolometer [13,70].

manufacture of antennas, in recent years, antennas have been fabricated utilizing 3D printing technologies [77–82]. An example of an integrated process of micromachined antenna-coupled uncooled microbolometer for terahertz imaging arrays is presented in Fig. 10.18 [75]. These arrays have been demonstrated to exhibit an improvement over the current substrate-supported bolometers by a factor of two [75].

Details of a MEMS-based wavelength-selective bolometer are presented in Fig. 10.19 [76]. In this approach, wavelength-selective absorbers have been utilized as efficient spectroscopic IR light-to-heat transducers; these transducers facilitate the amorphous silicon bolometers to respond at the desired resonance wavelengths [76].

Shi, Yang, Xu, and Li have reported their studies on a passive temperature-sensing antenna based on a bimetal strip coil [83]. An example of the prototype of a bimetal strip coil is presented in Fig. 10.20 [83]. This coil, consisting of Mn72Cu18Ni10 on one side and Ni36Fe64 on the other side, works on the fundamental principle of differences in thermal expansion coefficients due to the different alloys, that results in the bending of the bimetal strips due to change in temperature. By keeping one end of the bimetal coil fixed, the change in temperature results in a rotation of the strip coil. In this study, the authors report a rotation of 100 degree when the temperature changes from 20°C to 50°C with no friction [83]. Antennas such as these belong to the family of reconfigurable sensing antennas (RSAs). Like traditional antennas, RSAs can transmit and receive electromagnetic radiation. Additionally, they can also measure the surrounding parameters of the environment such as gas concentration, humidity, and temperature [83].

In recent years, there has been significant interest in the up-conversion of terahertz to near-infrared range of frequencies for applications in photon-type terahertz imaging [50,84]. An example of such a device that facilitates in the up-conversion from low energy to high energy photons, at approximately liquid helium temperatures, by

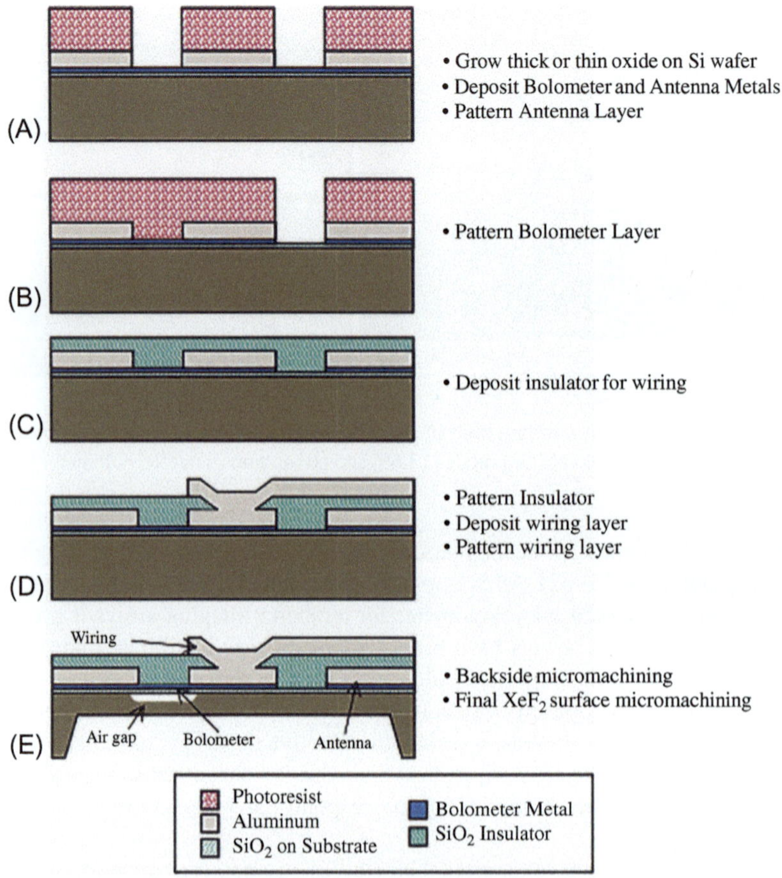

Fig. 10.18 Diagram of the complete device fabrication process including the antenna and bolometer definition and final substrate removal steps [75].

utilizing a gallium arsenide homojunction interfacial workfunction internal photoemission (HIWIP)-detector-LED up-converter and silicon CCD, is illustrated in Fig. 10.21. In providing a different imaging scheme in the terahertz band, the authors have successfully demonstrated the up-conversion imaging of a 1000 K blackbody pinhole.

The continued evolution in antenna technology represents work in progress. 3D printing, autonomous vehicles, drones, IoT, 5G mobile antennas, and personal health care are just a few of the recent technologies and applications that are yet to dominate the market place.

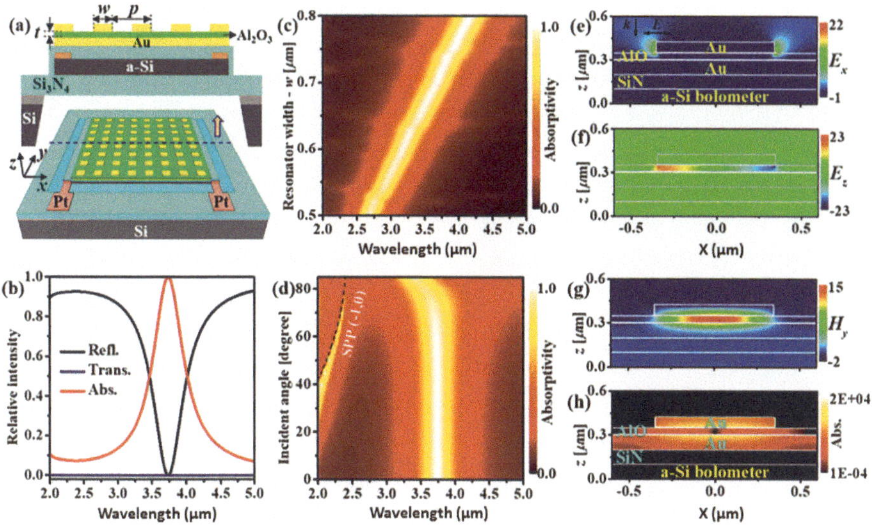

Fig. 10.19 (A) Schematic illustrations, tilted view *(bottom)* and cross-sectional view *(top)*, of the proposed MEMS-based wavelength-selective bolometer. (B) Simulated reflectance, transmittance and absorptivity of a device having geometrical parameters of $p = 1.2$ µm, $t = 0.045$ µm, and $w = 0.72$ µm exhibits a resonance at 3.73 µm with a unity absorptivity. (C) Simulated absorptivity map of the chosen absorber for the proposed device show wide resonance tunability just simply by changing the size of the resonator while keeping other parameters unchanged ($p = 1.2$ µm, $t = 0.045$ µm). (D) Simulated angle-dependent absorptivity of the absorber reveals a wide-range working angle up to 70 degree of the proposed devices. The peak appearing in the shorter wavelength region indicates SPP in periodic Au square array. (E–H) Simulated electric field (E_x, E_z), magnetic field—H_y and absorption maps of a device excited at the resonance. In the simulations, the incident electric field propagated along the z-axis and oscillated along the x-axis. The thicknesses of the bottom Au layer and a-Si film were fixed at 0.1 µm and 0.2 µm for all simulations [76].

Fig. 10.20 A prototype of a bimetal strip coil [83].

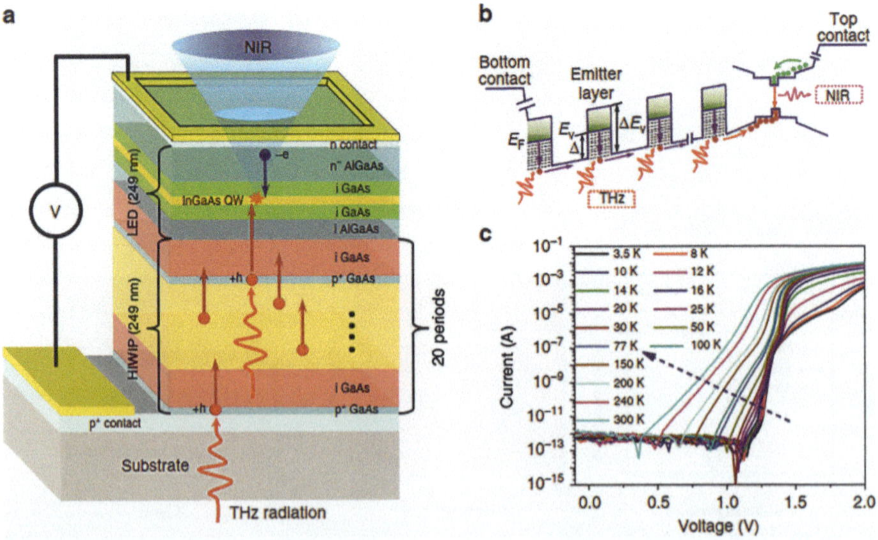

Fig. 10.21 Device structure and detection mechanism of the integrated homojunction interfacial workfunction internal photoemission (HIWIP) detector and light emitting diode (LED). (A) Structure of the HIWIP-LED up-converter device. (B) Band diagram of HIWIP-LED up-converter device. (C) The I-V characteristic (with 300 K background radiation) of the up-converter at different temperatures [84].

References

[1] https://nelson.mit.edu/blog/terahertz-polaritonics, Accessed 17 July 2020.

[2] C. Zhang, Optical Simulation of Terahertz Antenna Using Finite Difference Time Domain Method (Thesis, Master of Science), Rochester Institute of Technology, December 2014.

[3] A. Rogalski, F. Sizov, Terahertz detectors and focal plane arrays, Opto-Electron. Rev. 19 (3) (2011) 346–404.

[4] http://www.toptica.com/products/terahertz_generation/terahertz_technologysources_andthz_generation_methods/properties_of_terahertz_radiation.html, Accessed 17 July 2020.

[5] T.-L. Hwang, S.E. Schwarz, D.B. Rutledge, Microbolometers for infrared detection, Appl. Phys. Lett. 34 (11) (1979) 773–776.

[6] C. Fumeaux, M.A. Gritz, I. Codreanu, W.L. Schaich, F.J. González, G.D. Boreman, Measurement of the resonant lengths of infrared dipole antennas, Infrared Phys. Technol. 41 (2000) 271–281.

[7] F.J. González, B. Ilic, J. Alda, G.D. Boreman, Antenna-coupled infrared detectors for imaging applications, IEEE J. Sel. Top. Quantum Electron. 11 (1) (2005).

[8] Opgal, Intro to IR (Part 3): Sensitivity, Resolution and Frame Rate, 2020. https://www.opgal.com/blog/thermal-cameras/intro-to-ir-part-3-sensitivity-resolution-and-frame-rate/. (Accessed 13 July 2020).

[9] M. Alhussein, S.I. Haider, Simulation and analysis of uncooled microbolometer for serial readout architecture, J. Sens. 2016 (2016) 13 (Article ID 9751056).

[10] A. Colin, E. Pérez-Tijerina, F. Solis-Pomar, Design and simulation of an antenna-coupled microbolometer at 30THz, Int. J. Antennas Propag. 2017 (2017) 6 (Article ID 9783541).

[11] M.C. Tarplee, Introduction to Antennas, 2020. https://slideplayer.com/slide/9225107/. (Accessed 24 July 2020).
[12] USNA, EE302 Lesson 13: Antenna Fundamentals, 2020. www.usna.edu. (Accessed 25 July 2020).
[13] USNA, EE302 Lesson 14: Antenna Fundamentals, 2020. www.usna.edu. (Accessed 25 July 2020).
[14] J.-C. Bolomey, Qualitative versus quantitative approaches to microwave-based imaging techniques for medical applications, in: Workshop on Qualitative and Quantitative Approaches to Inverse Scattering Problems, Institute for Mathematical Sciences, National University of Singapore, Singapore, 24–28 September, 2018.
[15] Antenna design overview, https://ardupilot.org/copter/docs/common-antenna-design.html, Accessed 24 July 2020.
[16] M. Fallahpour, Synthetic Aperture Radar-Based Techniques and Reconfigurable Design for Microwave Imaging of Layered Structures (PhD Dissertation), Department of Electrical and Computer Engineering, Missouri University of Science & Technology, 2013.
[17] N. Singh, M. Sharma, Antenna and its application, Int. J. Comput. Sci. Technol. 6 (1) (2015) 95–97.
[18] J.M. López-Alonso, et al., Infrared antenna metrology, electro-optical and infrared systems: technology and applications II, in: R.G. Driggers, D.A. Huckridge (Eds.), Proc. of SPIE, vol. 5987, 2005, p. 59870L, https://doi.org/10.1117/12.631217.
[19] Antenna: Including Patch, Embedded, Fractal, Smart and Reflectors, https://www.bccresearch.com/market-research/information-technology/antennas-systems-devices-markets report.html#:~:text=The%20global%20antenna%20market%20will,the%20period%20of%202016%2D2021.
[20] P. Ramachandran, T.S. Keshav, L. Minz, V. Parupalli, S. Chakravarty, Antenna Design, Simulation and Fabrication (Project Report), Department of Electronics and Computer Science Engineering, Visvesvaraya National Institute of Technology, Nagpur, India, 2006-2007.
[21] C. Tripon-Canseliet, J. Chazelas, Low-Dimensional Materials for Disruptive Microwave Antennas Design, Intechopen, 2018, https://doi.org/10.5772/intechopen.79514.
[22] M. Wang, et al., A reconfigurable liquid metal antenna driven by electrochemically controlled capillarity, J. Appl. Phys. 117 (2015) 194901-1–194901-5, https://doi.org/10.1063/1.4919605.
[23] Novel 2D antenna can be sprayed onto flexible materials, Materials Today, 10th October (2018).
[24] A. Sarycheva, et al., 2D titanium carbide (MXene) for wireless communication, Sci. Adv. 4 (9) (2018), https://doi.org/10.1126/sciadv.aau0920.
[25] A. Erentok, R.W. Ziolkowski, Metamaterial-inspired efficient electrically small antenna, IEEE Trans. Antennas Propag. 56 (3) (2008) 691–707.
[26] R. Bala, R. Singh, A. Marwaha, S. Marwaha, Wearable graphene based curved patch antenna for medical telemetry applications, ACES J. 31 (5) (2016) 543–550.
[27] P. Bharadwaj, et al., Antenna-coupled optoelectronics with two-dimensional materials, in: Proceedings Conference on Lasers and Electro-Optics, 5-10, June (2016), San Jose, CA CLEO; Optical Society of America, 2016.
[28] B. Mohamadzade, et al., Recent advances in fabrication methods for flexible antennas in wearable devices: state of the art, Sensors 19 (10) (2019) 2312, https://doi.org/10.3390/s19102312.
[29] S.P. Pinapati, D.C. Ranasinghe, C. Fumeaux, Textile multi-layer cavity slot-monopole for UHF applications, IEEE Antennas Wirel. Propag. Lett. 16 (2017) 1, https://doi.org/10.1109/LAWP.2017.2731978.

[30] M.R. Islam, M. Ali, A 900 MHz beam steering parasitic antenna array for wearable wireless applications, IEEE Trans. Antennas Propag. 61 (2013) 4520–4527, https://doi.org/10.1109/TAP.2013.2269764.
[31] A.Y. Ashyap, Z. Abidin, S.H. Dahlan, H.A. Majid, S. Shah, M.R. Kamarudin, A. Alomainy, Compact and low-profile textile EBG-based antenna for wearable medical applications, IEEE Antennas Wirel. Propag. Lett. 16 (2017) 1.
[32] W. El Hajj, C. Person, J. Wiart, A novel investigation of a broadband integrated inverted-F antenna design; application for wearable antenna, IEEE Trans. Antennas Propag. 62 (2014) 3843–3846.
[33] S. Yan, P.J. Soh, G.A.E. VandenBosch, Wearable dual-band magneto-electric dipole antenna for WBAN/WLAN applications, IEEE Trans. Antennas Propag. 63 (2015) 1.
[34] A. Owadally, C.G. Parini, Y. Nechayev, C.C. Constantinou, P.S. Hall, A. Alomainy, Y. Hao, Statistical analysis and performance evaluation for on-body radio propagation with microstrip patch antennas, IEEE Trans. Antennas Propag. 55 (2007) 245–248.
[35] S. Yan, V. Volskiy, G.A.E. Vandenbosch, Compact dual-band textile PIFA for 433-MHz/2.4-GHz ISM Bands, IEEE Antennas Wirel. Propag. Lett. 16 (2017) 2436–2439.
[36] G. Gao, B. Hu, S. Wang, C. Yang, Wearable planar inverted-F antenna with stable characteristic and low specific absorption rate, Microw. Opt. Technol. Lett. 60 (2018) 876–882.
[37] A. Michel, R. Colella, G.A. Casula, P. Nepa, L. Catarinucci, G. Montisci, G. Mazzarella, G. Manara, Design considerations on the placement of a wearable UHF-RFID PIFA on a compact ground plane, IEEE Trans. Antennas Propag. 66 (2018) 3142–3147.
[38] M.A.B. Abbasi, S.S. Nikolaou, M.A. Antoniades, M.N. Stevanovi'c, P. Vryonides, Compact EBG-backed planar monopole for BAN wearable applications, IEEE Trans. Antennas Propag. 65 (2017) 453–463.
[39] R. Moro, S. Agneessens, H. Rogier, A. Dierck, M. Bozzi, Textile microwave components in substrate integrated waveguide technology, IEEE Trans. Microwave Theory Tech. 63 (2015) 422–432.
[40] R. Moro, S. Agneessens, H. Rogier, M. Bozzi, Circularly-polarized cavity-backed wearable antenna in SIW technology, IET Microwaves Antennas Propag. 12 (2018) 127–131.
[41] N. Chahat, M. Zhadobov, L. Le Coq, R. Sauleau, Wearable Endfire textile antenna for on-body communications at 60 GHz, IEEE Antennas Wirel. Propag. Lett. 11 (2012) 799–802.
[42] Antenna Market, Growth, Trends and Forecast (2020–2025), 2020. www.mordorintelligence.com/.../antenna-market. (Accessed 28 July 2020).
[43] S. Chen, M. Lust, N. Ghalichechian, A vanadium dioxide microbolometer in the transition region for millimeter wave imaging, 2019 IEEE International Symposium on Antennas and Propagation and USNC-URSI Radio Science Meeting, 2019, pp. 1641–1642, https://doi.org/10.1109/APUSNCURSINRSM.2019.8888891.
[44] U. Sassi, et al., Graphene-based mid-infrared room-temperature pyroelectric bolometers with high temperature coefficient of resistance, Nat. Commun. 8 (14311) (2017) 1–10.
[45] M.H. Mubarak, et al., Nano-antenna coupled infrared detector design, Sensors 18 (3714) (2018) 1–25.
[46] G.P. Szakmany, A.O. Orlov, G.H. Bernstein, W. Porod, Polarization-dependent response of single- and bi-metal antenna-coupled thermopiles for infrared detection, IEEE Trans. Terahertz Sci. Technol. 5 (2015) 884–891.
[47] J.A. Russer, C. Jirauschek, G.P. Szakmany, M. Schmidt, A.O. Orlov, G.H. Bernstein, W. Porod, P. Lugli, P. Russer, High-speed antenna-coupled terahertz thermocouple detectors and mixers, IEEE Trans. Microwave Theory Tech. 63 (2015) 4236–4246.
[48] A. Luukanen, J.P. Pekola, A superconducting antenna-coupled hot-spot microbolometer, Appl. Phys. Lett. 82 (22) (2003) 3970–3972.

[49] D.P. Neikirk, Planar antennas and far infrared detectors, Appl. Phys. Lett. 41 (5) (1998) 400–402.
[50] K. Kawase, K. Murate, Highly sensitive THz parametric detection, in: Proceedings Volume 11348, Terahertz Photonics, 2020, p. 113480G, https://doi.org/10.1117/12.2553039 (SPIE Photonics Europe (2020)).
[51] M. Perenzoni, D. Perenzoni, D. Stoppa, V. Mulloni, F. Solazzi, G. Resta, B. Margesin, Terahertz microsensor for biomedical applications, in: 2011 International Workshop on Biophotonics, 2011, https://doi.org/10.1109/IWBP.2011.5954813.
[52] F. Simoens, J. Meilhan, Terahertz real-time imaging uncooled array based on antenna- and cavity- coupled bolometers, Phil. Trans. R. Soc. A 372 (2013) 12, https://doi.org/10.1098/rsta.2013.0111 (20130111).
[53] M. Liu, Q. Fan, L. Yu, T. Xu, Polarization-independent infrared micro-lens array based on all-silicon metasurfaces, Opt. Express 27 (8) (2019) 7 (10738).
[54] P.H. Siegel, P. de Maagt, A.I. Zaghloul, Antennas for terahertz applications, in: Proceedings—IEEE Antennas and Propagation Society International Symposium, July 9–14, 2006, Albuquerque, New Mexico, 2006, https://doi.org/10.1109/APS.2006.1711074.
[55] A.J. Miller, D.J. Burdette, J. Alverbro, Z. Zhang, P. Fay, Y. Ni, P. Potet, K. Sertel, G. Trichopoulos, K. Topalli, J. Volakis, And Mosbacker H L., development of an 80 x 64 pixel, broadband, real-time THz imager, Proc. SPIE 5411 (2004) 18–24.
[56] J.A. Cox, R. Higashi, F. Nusseibeh, C. Zins, Passive uncooled subMMW focal plane array imager for person-borne explosives detection, in: Proceedings of SPIE, 7311, 2009.
[57] E. Peytavit, l.P. Agnese, J.-L.O. Buffet, A. Beguin, Room temperature terahertz microbolometers, in: 30th International Conference on Infrared and Millimeter Waves & 13th International Conference on Terahertz Electronics, 2005, pp. 257–258.
[58] I. Kašalynas, et al., Spectroscopic terahertz imaging at room temperature employing microbolometer terahertz sensors and its application to the study of Carcinoma tissues, Sensors 16 (2016) 432. 15 pages.
[59] Array detectors in the THz waveband based on antenna-coupled microbolometers, http://www.leti-cea.com/cea-tech/leti/english/Pages/Industrial-Innovation/Demos/THz-imaging.aspx, Accessed 21 July 2020.
[60] A. Lisauskas, M. Bauer, S. Boppel, M. Mundt, B. Khamaisi, E. Socher, R. Venckevičius, L. Minkevičius, I. Kašalynas, D. Seliuta, G. Valušis, V. Krozer, H.G. Roskos, Exploration of terahertz imaging with silicon MOSFETs, J. Infrared Millimeter Terahertz Waves 35 (2014).
[61] F. Schuster, D. Coquillat, H. Videlier, M. Sakowicz, F. Teppe, L. Dussopt, B. Giffard, T. Skotnicki, W. Knap, Broadband terahertz imaging with highly sensitive silicon CMOS detectors, Opt. Express (2011) 7827–7832.
[62] M. Bauer, R. Venckevicius, I. Kasalynas, S. Boppel, M. Mundt, L. Minkevicius, A. Lisauskas, G. Valusis, V. Krozer, H.G. Roskos, Antenna-coupled field-effect transistors for multi-spectral terahertz imaging up to 4.25 THz, Opt. Express 22 (2014).
[63] J. Simón, E.S. Kolosovas-Machuca, G. Vera-Reveles, E. Briones, J. Flores-Troncoso, F.J. González, Enhancement of antenna-coupled microbolometers response by impedance matching, J. Appl. Res. Technol. 13 (2015) 523–525.
[64] P.M. Krenz, B. Tiwari, G.P. Szakmany, A.O. Orlov, F.J. Gonzalez, G.D. Boreman, Response increase of IR antenna-coupled thermocouple using impedance matching, IEEE J. Quantum Electron. 48 (2012) 659–664.
[65] Y. Xu, B.A. Lail, Absorption efficiency in impedance matched infrared dipole antenna-coupled microbolometers, in: 2015 IEEE International Symposium on Antennas and Propagation USNC/URSI National Radio Science Meeting, 2015, pp. 1650–1651.

[66] T. Uchida, A. Matsushita, T. Tachiki, Detection properties of VO_x microbolometer on Si_3N_4/SiO_2 membrane fabricated by MOD at 200 GHz Band, in: 41st International Conference on Infrared, Millimeter, and Terahertz Waves (IRMMW-THz), 2016, https://doi.org/10.1109/IRMMW-THz.2016.7758929.
[67] J.A. Kaiser, The Archimedean two-wire spiral antenna, IRE Trans. Anntenas Propag. AP-8 (1960) 312–323.
[68] A. Seseka, I. Kasalynasb, A. Zemvac, J. Trontelj, Antenna-coupled Ti-microbolometers for high-sensitivity terahertz imaging, Sensors Actuators A 268 (2017) 133–140.
[69] J.A. Bean, A. Weeks, G.D. Boreman, Performance optimization of antenna-coupled Al/AlO_x/Pt tunnel diode infrared detectors, IEEE J. Quantum Electron. 47 (2011) 126–135.
[70] F.J. González, G.D. Boreman, Comparison of dipole, bowtie, spiral and log-periodic IR antennas, Infrared Phys. Technol. 46 (2005) 418–428.
[71] D.-T. Nguyen, et al., Broadband THz uncooled antenna-coupled microbolometer array – electromagnetic design, simulations and measurements, IEEE Trans. Terahertz Sci. Technol. 2 (3) (2012) 299–305.
[72] I. Malhotra, Analysis, Design and Characterization of Small-Gap Photoconductive Dipole Antenna for Terahertz Imaging Applications (PhD Thesis), Electronics & Communication Engineering, Jaypee University of Information Technology, 2017.
[73] A. Abohmra, et al., Flexible and wearable graphene-based terahertz antenna for body-centric applications, in: 2019 Second International Workshop on Mobile Terahertz Systems (IWMTS), Bad Neuenahr, Germany, 2019, pp. 1–4, https://doi.org/10.1109/IWMTS.2019.8823687.
[74] M. Gadalla, M. Abdel-Rahman, A. Shamim, Design, Optimization and Fabrication of a 28.3 THz Nano-Rectenna for Infrared Detection and Rectification, Sci. Rep. 4 (2015) 4270, https://doi.org/10.1038/srep04270.
[75] A.J. Miller, A. Luukanen, E.N. Grossman, Micromachined antenna-coupled uncooled microbolometers for terahertz imaging arrays, in: Proceedings of SPIE, 5411, Terahertz for Military and Security Applications II, 2004, https://doi.org/10.1117/12.543236.
[76] T.D. Dao, et al., MEMS-based wavelength-selective bolometers, Micromachines 10 (2019) 416, https://doi.org/10.3390/mi10060416.
[77] B. Zhang, Y. Guo, H. Zirath, Y.P. Zhang, Investigation on 3-D-printing technologies for millimeter-wave and terahertz applications, Proc. IEEE 105 (4) (2017) 723–736, https://doi.org/10.1109/JPROC.2016.2639520.
[78] P. Nayeri, et al., 3D printed dielectric reflectarrays: low-cost high-gain antennas at sub-millimeter waves, IEEE Trans. Antennas Propag. 62 (4) (2014) 2000–2008.
[79] X. Huang, T. Leng, M. Zhu, et al., Highly flexible and conductive printed graphene for wireless wearable communications applications, Sci. Rep. 5 (2016) 18298, https://doi.org/10.1038/srep18298.
[80] M. Rizwan, et al., Flexible and stretchable brush-painted wearable antenna on a three-dimensional(3-D) printed substrate, IEEE Antennas Wirel. Propag. Lett. 16 (2017) 3108–3112.
[81] S. Kim, et al., Inkjet-printed antennas, sensors and circuits on paper substrate, IET Microwaves Antennas Propag. 7 (10) (2013) 858–868.
[82] J. Kimionis, et al., 3D-printed origami packaging with inkjet-printed antennas for RF harvesting sensors, IEEE Trans. Microwave Theory Tech. 63 (12) (2015) 4521–4532.
[83] X. Shi, F. Yang, S. Xu, M. Li, A passive temperature-sensing antenna based on a bimetal strip coil, Sensors 17 (2017) 665, https://doi.org/10.3390/s17040665.
[84] P. Bai, et al., Broadband THz to NIR up-converter for photon-type THz imaging, Nat. Commun. 10 (2019) 3513. 9 Pages https://doi.org/10.1038/s41467-019-11465-6.

Infrared focal plane arrays

11

11.1 Introduction

Focal plane array or staring array is an array of light detectors, which are placed in the focal plane of an imaging system. A focal plane array is a rectangular two-dimensional array containing many thousands or even a several million detectors. The detectors may be either photon or thermal detectors. FPAs also consist of readout and processing electronics. They provide an economical method for high-density integration and packaging of detectors. Additionally, the signal processing takes place in the focal plane itself. Choice of spectral range is very important in the selection of FPAs.

Example of an (a) analog read out integrated circuit (ROIC) architecture with (b) simplified unit-cell or pixel circuit diagram and (c) unit-cell layout from the literature is shown in Fig. 11.1 [1]. In Fig. 11.1(b), I_{det} is the photocurrent that is generated by the photodiode. It is subsequently integrated onto the capacitance C_{int} through the injection transistor M_i, which also provides the photodiode bias V_{bias}. The signal voltage across C_{int} can be switched onto the multiplexor bus for readout via control signal ϕ_{sel} on switch M_{sel} [2]; the signal voltage across C_{int} can be reset by control signal ϕ_{rst} using switch M_{rst}. In this case study, a maximum of 2.2-V process (set by V_{dd}) and C_{int} equal to 1850 femtofarads results in a maximum stored photocharge equal to 25 million electrons. It may be noted that the circuit capacitance dominates the pixel footprint as shown in Fig. 11.1(c) [1].

An example of a (a) digital read out integrated circuit (DROIC) architecture with (b) simplified unit-cell circuit diagram and (c) unit-cell layout is shown in Fig. 11.2 [1]. In this example, (b), an injection transistor is considered to be feeding a small integrating capacitor (~1 femtofarad). This corresponds to an electron well of ~6000 electrons at a maximum of 1 V across the integrating capacitor. Once the well fills, the integrating capacitor charge is reset, and a pulse is generated. This results in a pulse train with frequency that is proportional to the input photocurrent. Each pulse increments or decrements a counter depending on the initial configuration. Simple digital signal processing (DSP) operations are applied during signal integration. Once the shutter is closed, the counter represents the total number of times the electron well is filled. With the preknowledge of the well size, the digital representation of the number of integrated photoelectrons is contained in the counter. The contents of the 16-bit counter/register is then shifted to any of its four nearest neighbors. High-speed digital readout circuits are utilized to read the contents. It may be noted from Fig. 11.2 that the (c) 16-bit counter is dominated by the pixel footprint [1].

The test-related issues of individual detectors, FPAs, and related electronics are summarized in Table 11.1 [3].

From the perspective of the fundamentals of discrete IR detector structures, their energy band diagrams are presented in Fig. 11.3 [3].

Microbolometers. https://doi.org/10.1016/B978-0-08-102812-4.00009-7

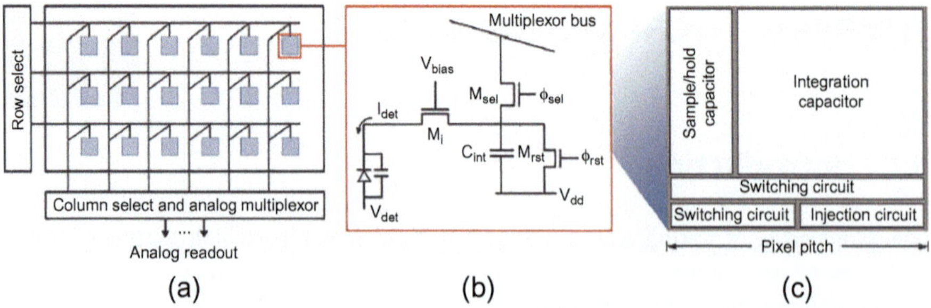

Fig. 11.1 (a) Analog ROIC architecture with (b) simplified unit cell or pixel circuit diagram and (c) unit cell layout [1].

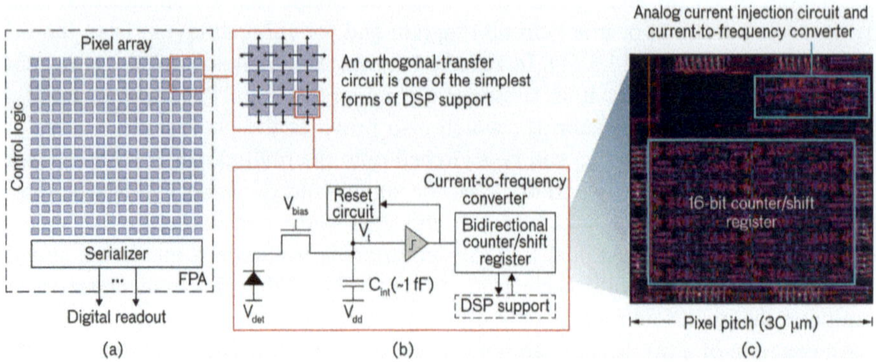

Fig. 11.2 (a) Digital ROIC architecture with (b) simplified unit cell diagram and (c) unit cell layout [1].

Table 11.1 Test related issues of individual detectors, FPAs and related electronics [3].

Detectors	Input circuits	Readouts
Quantum efficiency	Charge storage	Charge handling
Spectral response	Uniformity	Transfer efficiency
Noise spectrum	Linearity	Data rate
Uniformity	Noise spectrum	Cross talk
	Injection efficiency	
	FPA operability	
Performance efficiency	Nonuniformity correctibility	Reliability
Dead pixel count	Dynamic range	Hardness
Temporal noise	Intrascene	
Saturation level	Intrascene	

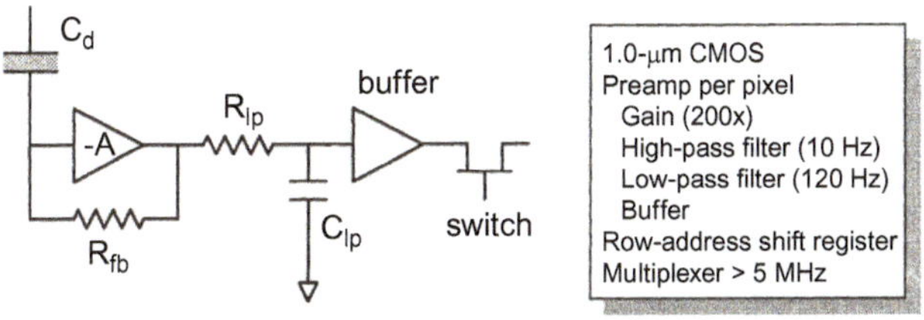

Fig. 11.3 Four common IR detector structures and their related energy band diagrams: (a) photoconductive, (b) photovoltaic, (c) metal-insulator-semiconductor, and (d) Schottky barrier [3].

The optical excitation processes in various device structures in illustrated in Fig. 11.4 [4].

Fig. 11.5 presents a performance overview of the quantum efficiency of infrared detectors over a range of wavelengths spanning the ultraviolet, visible, and infrared [5–7].

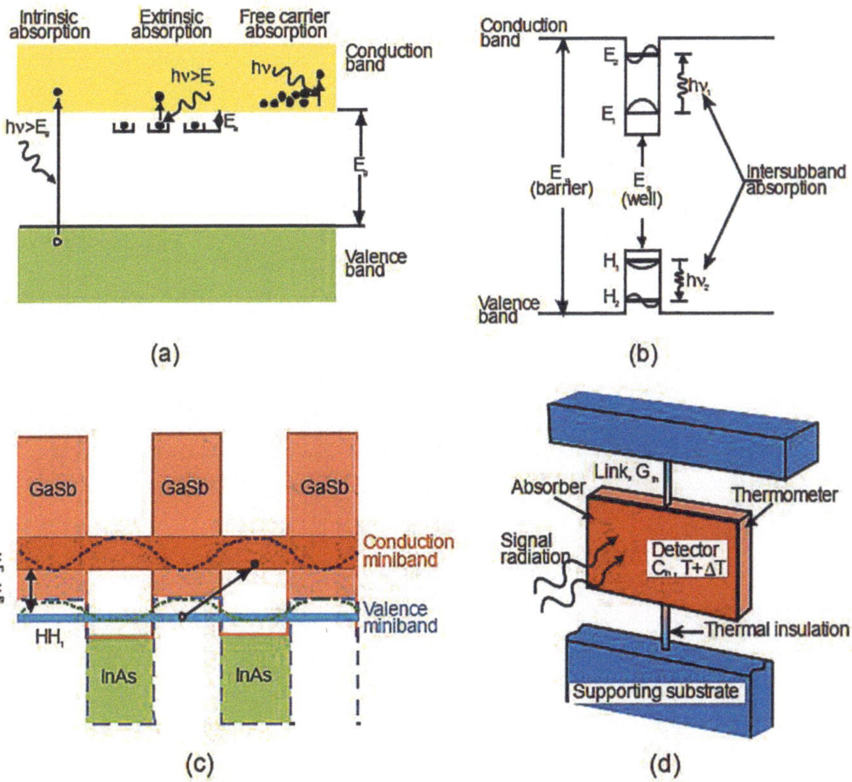

Fig. 11.4 Optical excitation processes in (a) bulk semiconductors, (b) quantum wells, (c) type-II InAs/GaSb superlattices, and (d) thermal detectors [4].

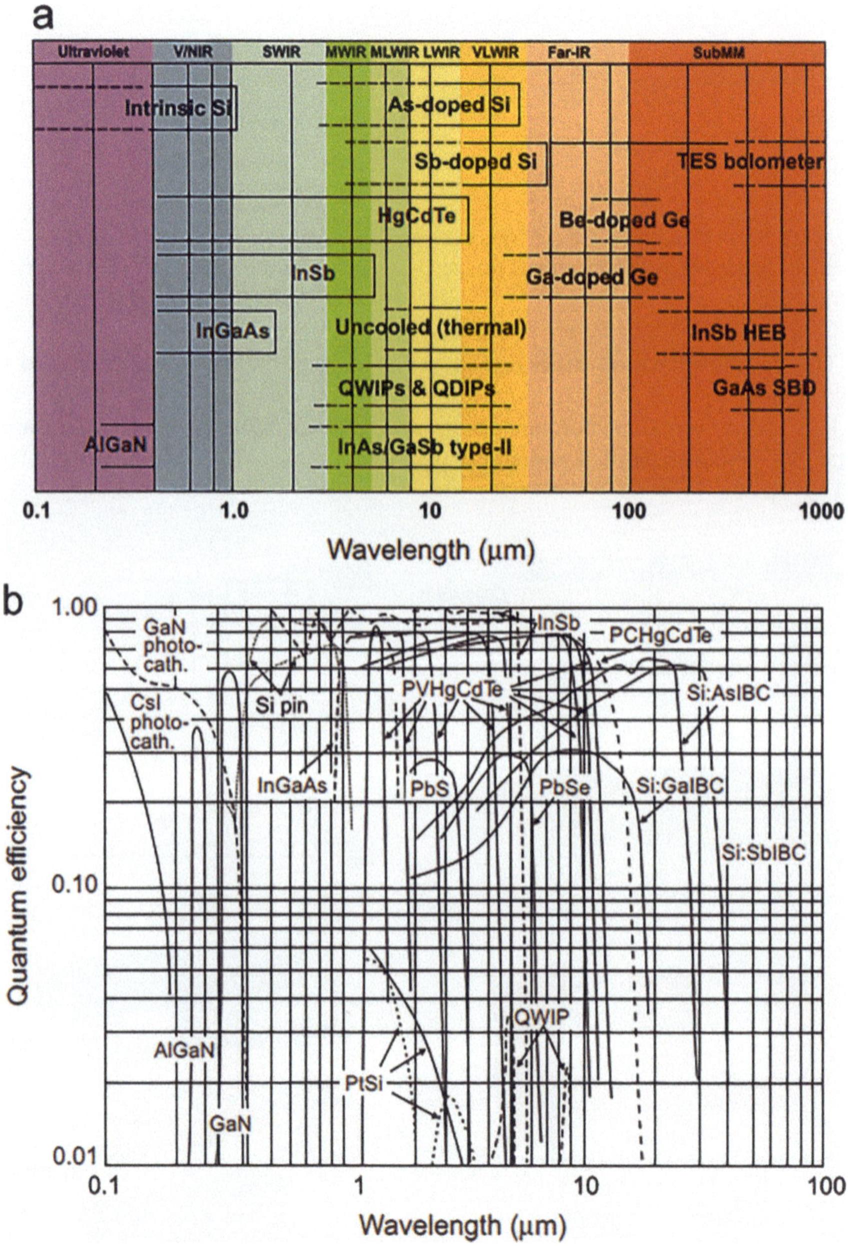

Fig. 11.5 Overview of UV, visible, and infrared detectors: (a) various material systems [5, 6] and (b) quantum efficiency of detectors [5, 7].

In general, infrared systems are characterized by the detector size d and the F-number ($F/\#$) [4, 8]. The detection/identification range and the noise equivalent difference temperature ($NEDT$) are influenced by d and ($F/\#$) due to their dependence on $F\lambda/d$ [4, 9]. The equations for range and $NEDT$ are given by the following [4]:

$$\text{Range} = [D\Delta x/M\lambda]\,[F\lambda/d], \tag{11.1}$$

$$\text{NEDT} \approx \left[2/C\lambda\left(\eta\varphi_B{}^{2}\prod\tau_{\text{int}}\right)\right][F\lambda/d], \tag{11.2}$$

where C is the scene contrast, D is the aperture, M is the number of pixels that are required to identify a target Δx, $\varphi_B^{2\prod}$ is the background flux into a $2\prod$ field of view (FOV), λ is the wavelength, η is the detector collection efficiency, and τ_{int} is the integration time. Thus, an optimal design of an IR system is defined by $F\lambda$ and d [4].

Pixel pitch is plotted as function of $F\lambda$ in Fig. 11.6 [4, 10]. For most defense applications, detector size is in the range of 15 to 50 μm [4].

A schematic of an imaging system illustrating the various critical components of the system is shown in Fig. 11.7 [5, 6].

Fig. 11.8 presents a summary of the operating temperature and the wavelength range of various IR detector technologies [5].

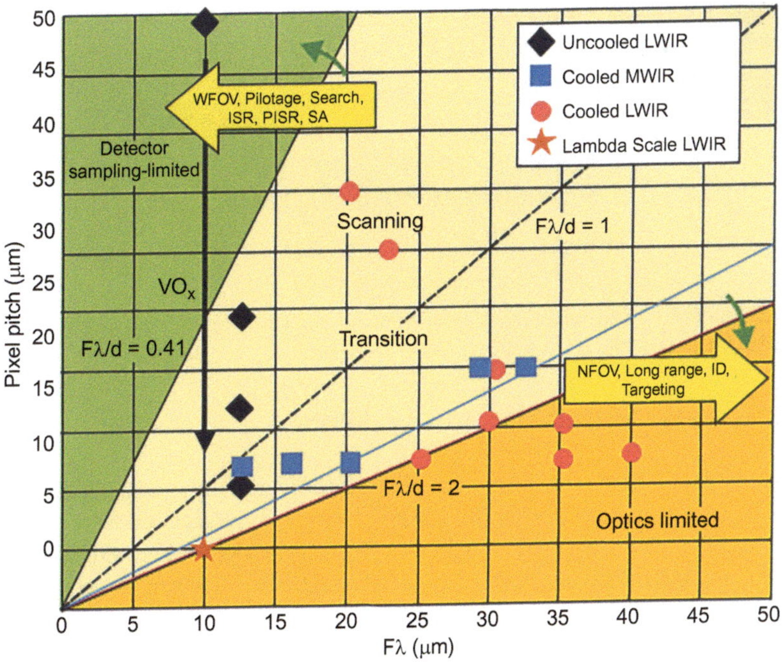

Fig. 11.6 *Fλ/d* space for infrared system design. Straight lines represent constant *NEDT*. There are an infinite number of combinations that provide the same range [4, 8].

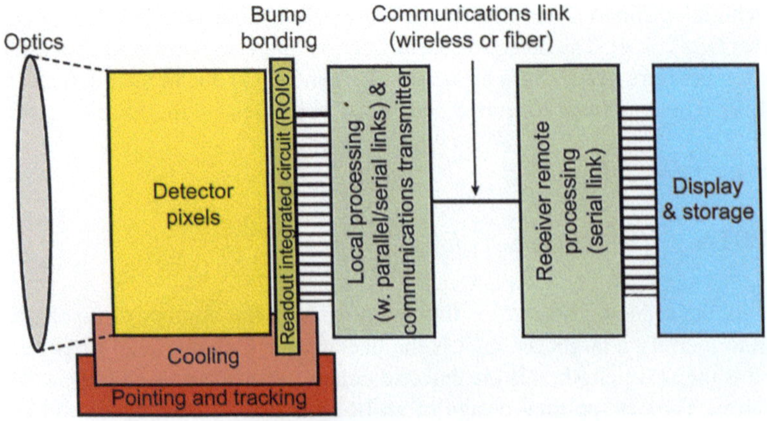

Fig. 11.7 Schematic of an imaging system illustrating the various critical components [5, 6].

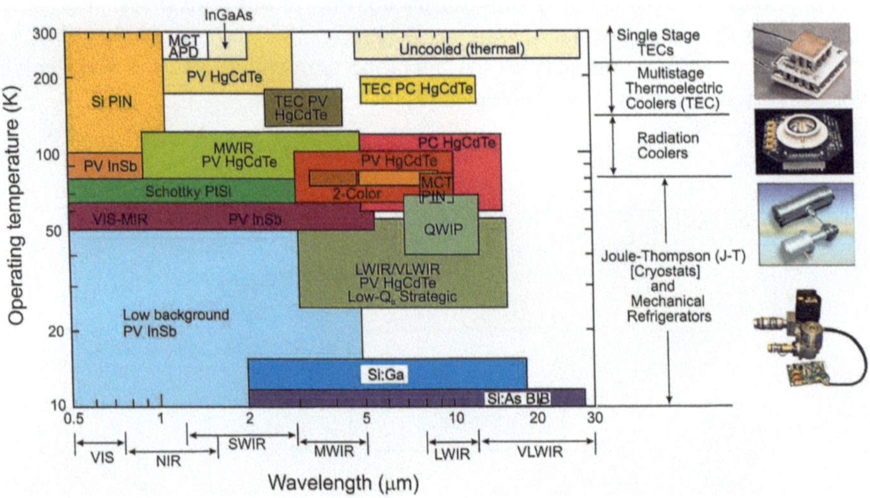

Fig. 11.8 Summary of the operating temperature and the wavelength range of various IR detector technologies [5].

11.2 Types, materials, and systems

IRFPAs, as photon detectors, with ultrahigh sensitivity, usually operate in the 3–5 μm and 8–12 μm wavelength transmission bands. They require semiconductor detectors with bandgap in the range of 0.1–0.25 eV. Thus, these IRFPAs are fabricated with narrow band gap semiconductors. An infrared image is normally of low contrast with high background noise. Because of small bandgap and the necessity to obtain low electronic noise level, cryogenic cooling is required in photon assisted IRFPAs. High-density detector configurations lead to higher image resolution as well as greater sensitivity.

11.2.1 Dual-band IRFPAs

Numerous applications require MWIR and LWIR dual-band IRFPAs. Such dual-band IRFPA-based cameras provide accurate temperature of targets with unknown emissivity. Dual-band IRFPAs also play a major role in earth and planetary remote sensing and astronomy. Monolithically integrated pixels, colocated simultaneously with readable dual-band FPAs, eliminate the need for beam splitters, filters, moving filter wheels, and rigorous optical alignment conditions that are required by dual-band systems based on two separate single-band FPAs or a broad-band FPA system with filters. Dual-band FPAs also offer advantages in the form of reduced mass, volume, and power requirements [11, 12].

The antimonides-based type-II superlattices have a significant potential in realizing high performance large format highly uniform IR FPAs at lower cost due to the availability of large area epiready GaSb substrates and relatively easy III–V materials growth and processing technology compared to their II–VI counterparts.

An example of a dual-band III–V compound semiconductor strained-layer superlattice (SLS)-based IRFPA, on GaSb substrates, is presented in Fig. 11.9 [12]. In this research work, the authors report their studies on 3 in. and 4 in. SLS wafers that were processed into 1280 × 720 12 μm pitch detector arrays. By implementing improvements in lithography and dry etching processes, the size of the trenches separating the pixels were reduced from 3 μm to <1.5 μm. This resulted in increased physical fill factor from 56% to 76% [12]. This is shown in Fig. 11.10 [12].

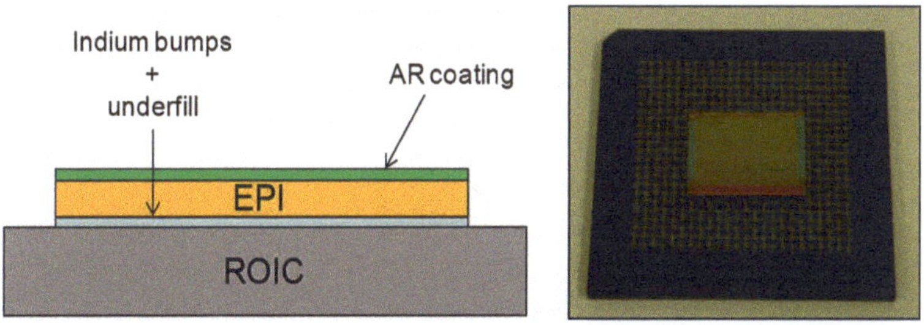

Fig. 11.9 (a) Schematic of a dual band FPA after complete substrate removal and antireflection coating (ARC); (b) photograph of a dual band FPA after complete substrate removal and ARC [12].

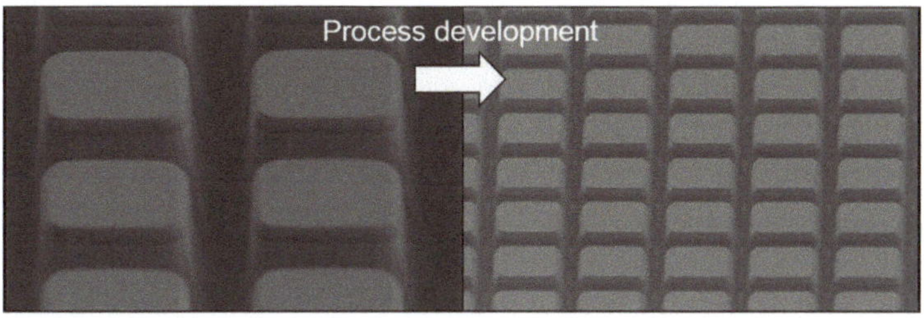

Fig. 11.10 Improvements in lithography and dry etching processes reduced the size of the trenches separating pixels from 3 μm to <1.5 μm. As a result, the physical fill factor was increased from 56% to >76% [12].

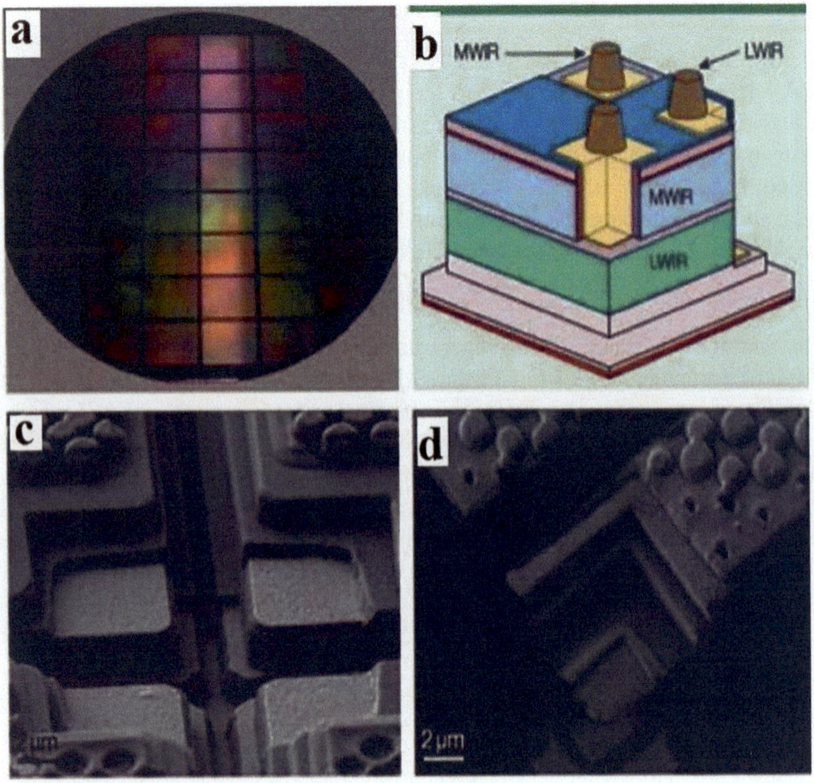

Fig. 11.11 Two-color 320 × 256 MWIR/LWIR QWIP FPA: (a) individual dies as FPAs fabricated on a 4-in. GaAs wafer, (b) isometric three-dimensional outlook of the pixel cross-section, (c) SEM picture showing access and electrical connections for the common contact, and (d) SEM picture showing that pixel connections are routed to the top of each pixel using the gold [13, 14].

A wide variety of architectures can be used for the development of IRFPAs [3]. Each design considers several functionalities of FPAs such as photon detection, charge storage, and readout.

Another example of a two-color quantum well infrared photodetector (QWIP) is shown in Fig. 11.11. In this figure, a two-color 320 × 256 MWIR/LWIR QWIP FPA is presented. Fig. 11.11(a) shows individual dies as FPAs fabricated on a 4-in. GaAs wafer; an isometric three-dimensional outlook of the pixel cross-section is shown in Fig. 11.11(b). In Fig. 11.11(c), an SEM picture shows access and electrical connections for the common contact, and in Fig. 11.11(d), an SEM micrograph shows pixel connections that are routed to the top of each pixel using the gold [13, 14].

An example of a multicolor detector pixel is shown in Fig. 11.12 [4]. A three-color detector pixel is illustrated in Fig. 11.12(a). In Fig. 11.12(b), a dual-band InAs/GaSb/AlSb is shown [4, 15].

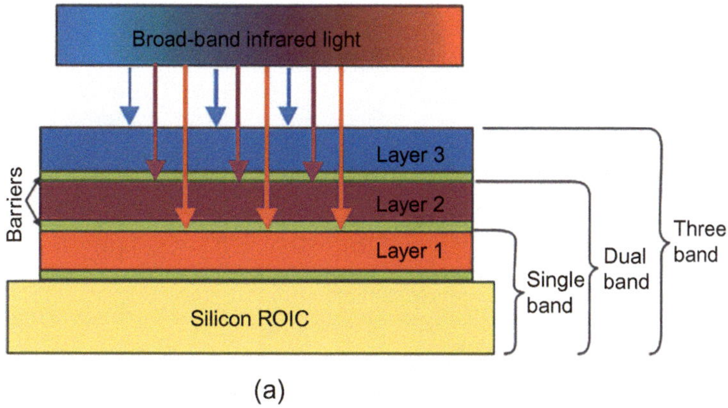

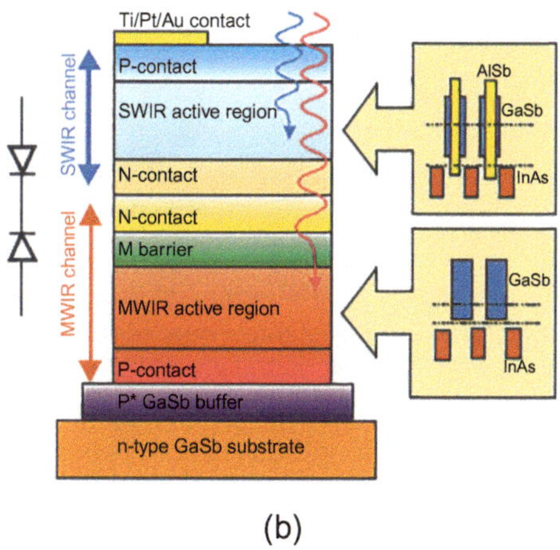

Fig. 11.12 Multi-color detector pixels: (a) structure of a three-color detector pixel. IR flux from the first band is absorbed in layer 3, while longer wavelength flux is transmitted through to the next layers. Thin barriers separate the absorbing bands. (b) Dual-band SW/MW InAs/GaSb/AlSb T2SL back-to-back p-i-n-n-i-p photodiode structure and schematic band alignment of superlattices in the two absorption layers (*dotted lines* represent the effective band gaps of superlattices) [4, 15].

11.2.2 FPA Architectures

Various types of FPA architectures are shown in Fig. 11.13. In monolithic devices, these functions are performed on the same chip. This allows fabrication using silicon like processing techniques that leads to a significant reduction in costs of manufacturing. With the advent of vanadium oxide and amorphous silicon technology, MEMS (microelectromechanical structures) related processing techniques have taken over much of the manufacturing of uncooled microbolometers in the IR imaging industry. The hybrid architecture of the FPAs consists of two approaches. In flip-chip technology, the bonding of the detector array to a silicon multiplexer is performed by aligning

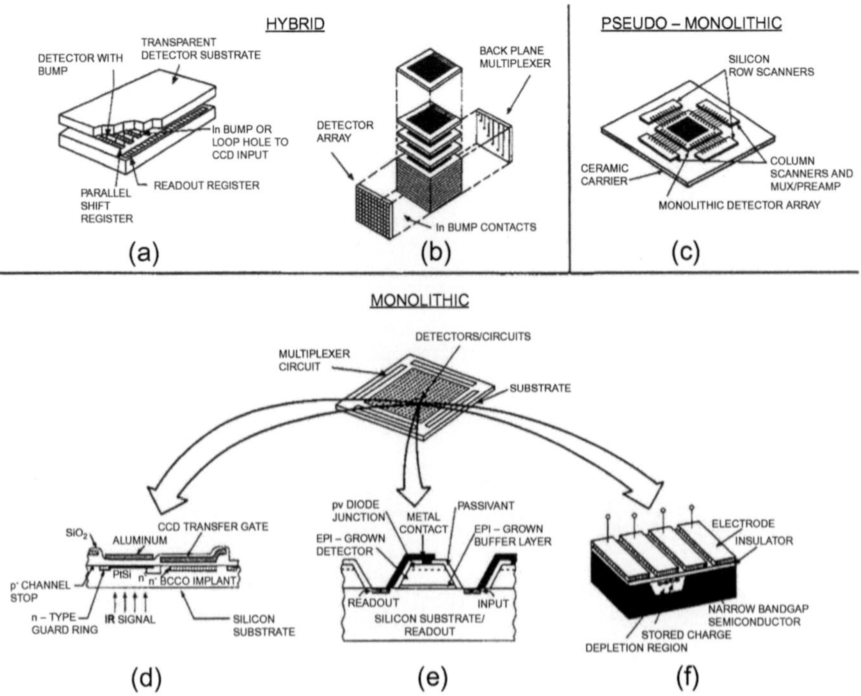

Fig. 11.13 IRFPA architectures. For hybrid arrays: (a) flip-chip; (b) Z-technology. For pseudomonolithic arrays: (c) xy-addressable. For monolithic arrays: (d) all-silicon; (e) heteroepitaxy-on silicon; (f) nonsilicon-based (e.g., nonsilicon CCD) [3].

the front-side of the detector array with the multiplexer; indium bumps are utilized as contacts between each detector and the silicon multiplexer. In this approach, the detector array can be illuminated from the front or the back, although illumination from the detector side is preferred due to the areas of metallization and other opaque regions in the multiplexer [3]. In the Z-plane approach, the integrated circuits are stacked on top of each other—this leads to a third dimension of real estate for the electronics. The detector array is bonded to the edges of the stacked IC chips; the IC chips have one channel per detector. All the signal processing functions of the FPA, such as analog-to-digital conversion, band-pass filtering, gain and offset correction, preamplification, and some image processing functions are performed on the focal plane [3].

Recent developments in microbolometer IRFPAs have allowed increase in sensitivity with increase in pixel count combined with smaller pixel size. Reduction in pixel size allows higher resolution and longer range in telescope designs as well. Smaller pixel size also reduces the cost of fabrication of FPAs. The trend in decreasing pixel pitch during the last two decades for HgCdTe detector arrays is illustrated in Fig. 11.14(a) [4, 16]. A similar decreasing trend in the pixel pitch, as function of the production year, for the last 25 years, is also observed for uncooled amorphous silicon microbolometers in Fig. 11.14(b) [4, 16].

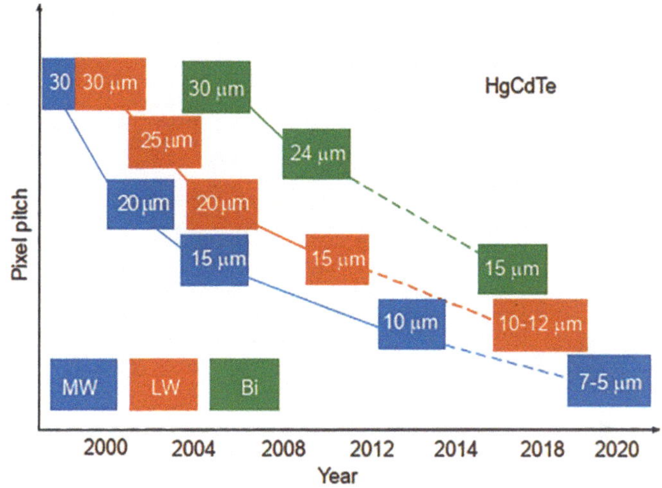

(a)

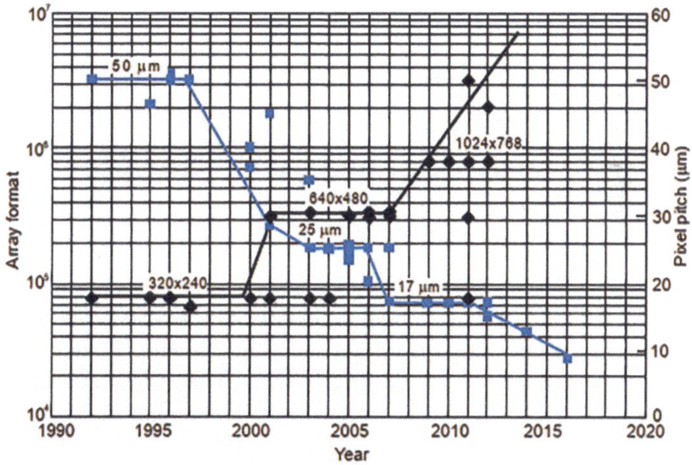

(b)

Fig. 11.14 Pixel pitch for (a) HgCdTe photodiodes and for (b) amorphous silicon microbolometes have continued to decrease due to technological advancements [4, 16].

A focal plane array consists of several individual microbolometer pixels. In a monolithic IRFPA, each microbolometer pixel consists of a thin infrared absorbing membrane suspended above a CMOS ROIC (read out integrated circuit). The bridge is connected by two long and narrow legs. The ROIC sends a current pulse through the bridge to sense the change in resistance of the bridge material resulting by its temperature change. The microbridge structures can be fabricated with very low thermal conductance. Lower thermal conductance implies higher detectivity and lower noise.

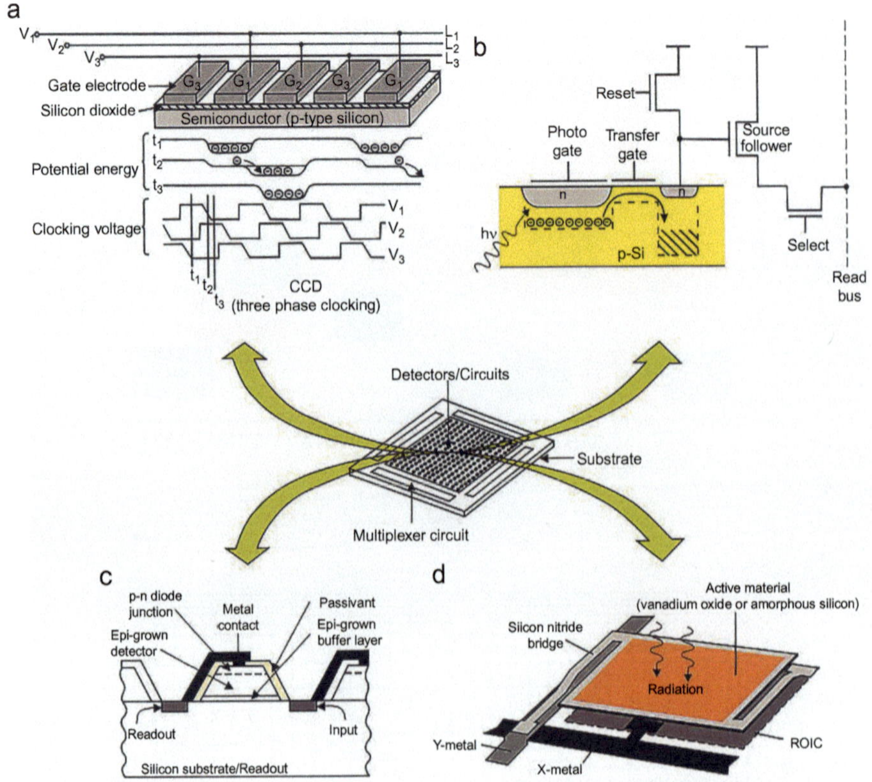

Fig. 11.15 Monolithic focal plane arrays: (a) CCD, (b) CMOS, (c) heteroepitaxy-on-silicon, and (d) microbolometer [5].

Both of these are required to achieve high sensitivity. The microbridge structure also minimizes thermal crosstalk between adjacent sensing elements. In monolithic FPAs, both detection and signal read out are performed in the detector material rather than in an external ROIC. This type of integration reduces the number of fabrication steps, increases yield and reduces cost as well. Materials such as Si, PbS, PbTe, and PtSi are used in developing monolithic structures. Monolithic FPA architecture is described in Fig. 11.15.

In the case of hybrid technology, the detector and multiplexer are optimized independently. Other advantages of hybrid-packaged FPAs are near-100% fill factors and increased signal to processing area on the multiplexer chip. InGaAs, InSb, HgCdTe, GaAs/AlGaAs, QWIP, and extrinsic silicon are used in hybrid structure configurations. Various hybrid FPA techniques are shown in Fig. 11.16.

11.2.3 InGaAs

$In_xGa_{1-x}As$ photodetectors are mainly used in systems operating in the spectral region of 1–1.7 μm. InGaAs arrays have uses in application of low light level night vision and new applications such as remote sensing, eye-safe range finding and process control.

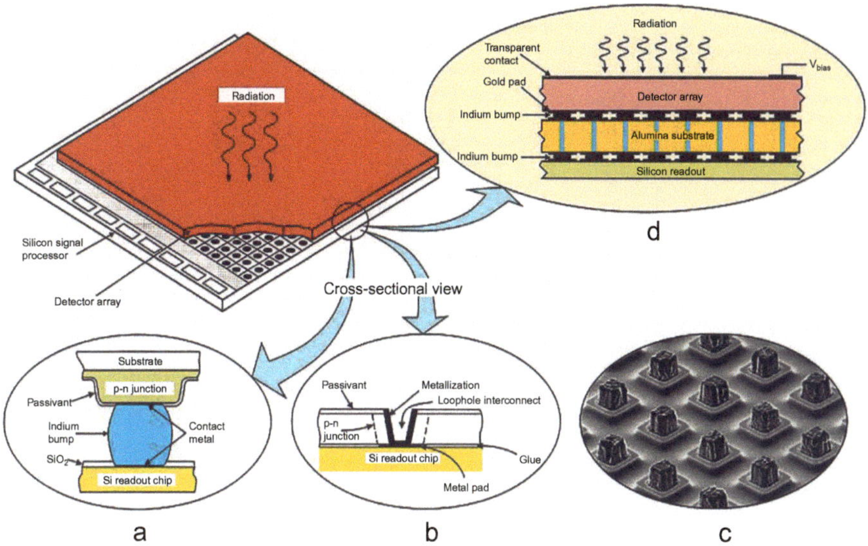

Fig. 11.16 Hybrid IR FPA interconnect techniques between a detector array and silicon multiplexer: (a) indium bump technique, (b) loophole technique, (c) SEM photo shows mesa photodiode array with indium bumps, and (d) layered-hybrid design suitable for large format far IR and sub-mm arrays [5].

Another advantage in InGaAs is the processing technology being similar to that of silicon technology. The active material of the InGaAs detector is deposited onto a substrate using chloride Vapor Phase Epitaxy (VPE) or Metal-Organic Chemical Vapor Deposition (MOCVD) techniques and is adjusted for thickness and background doping. InGaAs FPAs are fabricated by several manufacturers including Goodrich Corporation (previously Sensors Unlimited), Indigo Systems (merged with FLIR Systems), Teledyne Judson Technologies, Aerious Photonics, and Xen ICs [17–21]. Fig. 11.17 shows the (a) schematic representation of the cross section of a metal-semiconductor-metal (MSM) InGaAs/InP infrared photodetector and (b) its energy diagram [22].

InSb arrays are used in the 1–5 μm spectral region. In InSb photodiode fabrication, epitaxy is not used; instead, the standard manufacturing technique begins with bulk n-type single crystal wafers with donor concentration of about 10^{15} cm^{-3}. In-bumps are deposited on both detector array and silicon ROIC dies. These dies are hybridized together using a flip-chip bonder. The gap between bumps is backfilled with epoxy. An example of a four two-side buttable 2052 × 2052 InSb arrays in a mosaic configuration 4104 × 4104 array pixels is shown in Fig. 11.18 [20].

11.2.4 HgCdTe arrays

From a historical point of view, $Hg_xCd_{1-x}Te$ represents an excellent example of bandgap engineering wherein the composition x is chosen to match the desired wavelength of operation of the detector/imager. This is illustrated in Fig. 11.19. In this figure, an

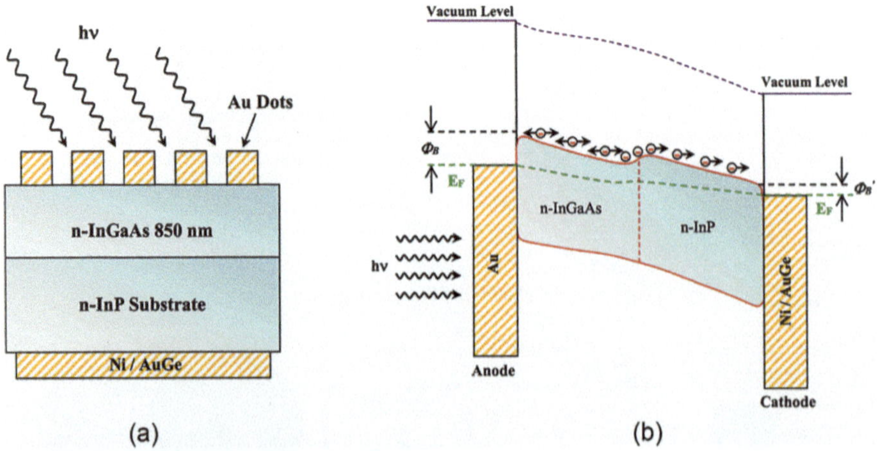

Fig. 11.17 (a) Schematic representation of the cross section and (b) energy band diagram of an InGaAs/InP MSM device. ϕ_B denotes Schottky barrier height of the device and ϕ_B' have to be almost zero for good ohmic contact behavior [22].

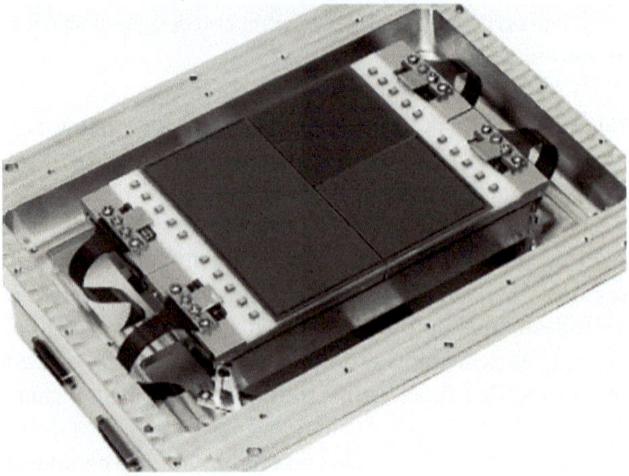

Fig. 11.18 Four two-side buttable 2052 × 2052 InSb arrays in a mosaic configuration 4104 × 4104 array pixels [20].

illustration of the change in energy gap and the corresponding cut-off wavelength of $Hg_xCd_{1-x}Te$ is shown as a function of the CdTe mole fraction in $Hg_xCd_{1-x}Te$ [23].

HgCdTe photodiodes operate in the spectral range of 1–20 μm. Epitaxial techniques are used in the fabrication of HgCdTe IRFPAs. Liquid Phase Epitaxy (LPE) method is more suitable for HgCdTe FPAs. n-on-p junctions are fabricated in two different manner using Hg vacancy doping and extrinsic doping. An example of a 180 element common module linear array of HgCdTe photoconductive detectors is illustrated in Fig. 11.20 [24].

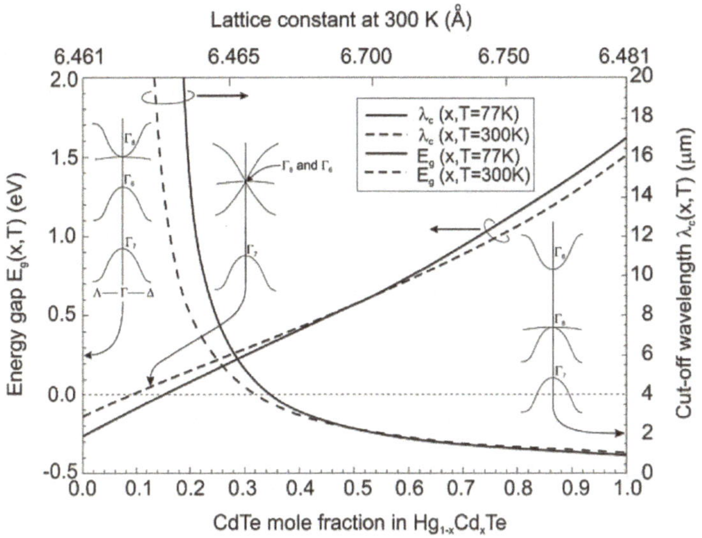

Fig. 11.19 The band gap structure of $Hg_{1-x}Cd_xTe$ near the Γ-point for three different values of the forbidden energy gap. The energy band gap is defined at the difference between the Γ_6 and Γ_8 band extrema at $\Gamma = 0$ [23].

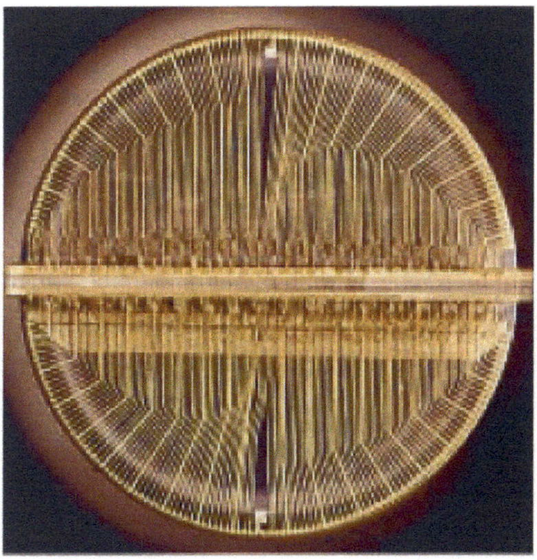

Fig. 11.20 180 element common module linear array of HgCdTe photoconductive detectors. The array is mounted on the end of a glass cold finger and wire bonded to metallized leads which interface to warm preamplifiers and bias circuits through a vacuum interface. Over 200 wire bonds are required on the surface shown [24].

In double-layer heterojunction (DLHJ) structure, about 10-μm thick absorber layer is doped with indium at 10^{15} cm^{-3} or less. The formation of planar p-on-n photodiodes is achieved by selective area arsenic ion-implantation through the cap layer into the narrow gap base layer. The implantation is done by a two-step thermal anneal under Hg overpressure: one at high temperature activates the dopant by substituting As

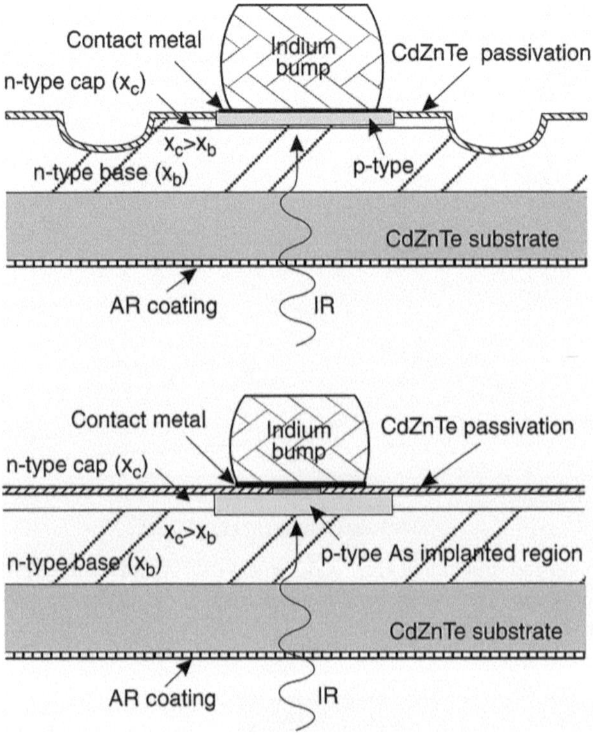

Fig. 11.21 Cross sections of p-on-n DLHJ HgCdTe photodiodes: (a) mesa structure and (b) planar structure [20].

atoms on the Te sublattice, while the second at lower temperature annihilates the Hg vacancies formed in the HgCdTe lattice during the growth and high temperature annealing step. Examples of such DLHJ HgCdTe photodiodes are illustrated in Fig. 11.21 [20].

Fig. 11.22 presents a detailed cross section views of the unit cells for various back-illuminated dual-band HgCdTe detector approaches in the infrared imaging industry [5, 25–30].

There are several approaches to enhance the performance of IRFPAs in IR imaging. One of the approaches involves the use of arrays of microlenses. These passive optical devices facilitate in increasing the fill factor of the individual detectors and, therefore, the FPAs. This is illustrated in Fig. 11.23 [5].

Another example of improving the performance of the detector involves the use of a metalens. This is shown in Fig. 11.24 [31]. In this study, the authors report the use of a polarization-independent dielectric meta-lens to monolithically integrate with an HgCdTe photodetector. This monolithic integration results in performance enhancement of the photodetector due to the ability to concentrate power flux into a reduced photosensitive area by the use of the metalens [31].

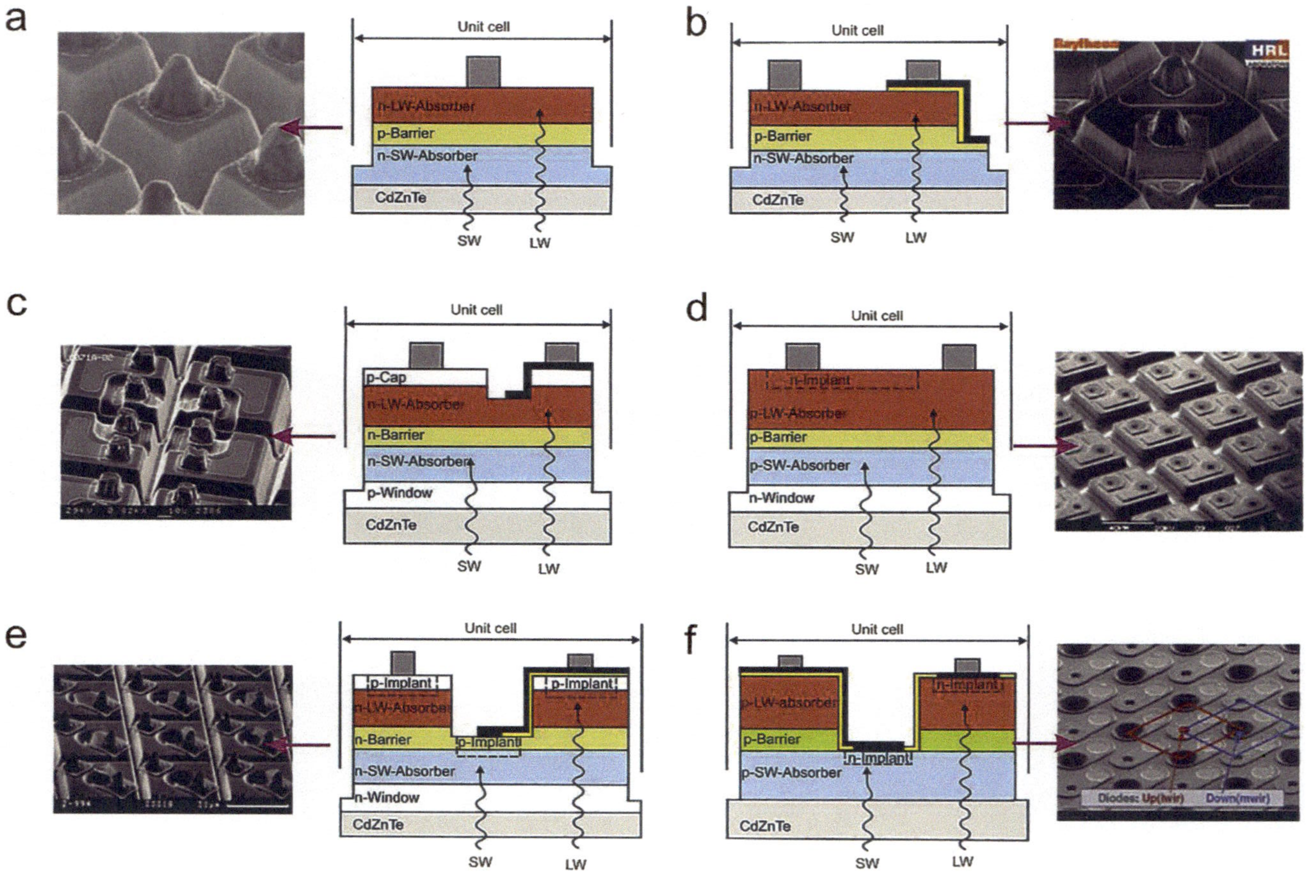

Fig. 11.22 Cross section views of unit cells for various back-illuminated dual-band HgCdTe detector approaches: (a) bias-selectable n-p-n structure reported by Raytheon [5, 25], (b) simultaneous n-p-n design reported by Raytheon [5, 26], (c) simultaneous p-n-n-p reported by BAE Systems [5, 27], (d) simultaneous n-p-p-p-n design reported by LETI [5, 28], (e) simultaneous structure based on p-on-n junctions reported by Rockwell [5, 29], and simultaneous structure based on n-on-p junctions reported by LETI [5, 30].

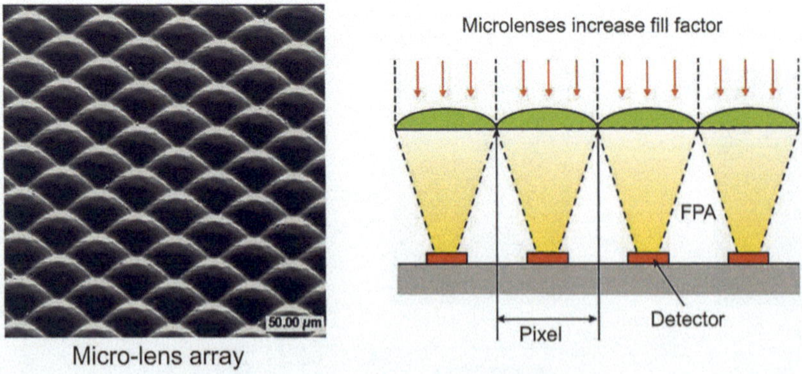

Fig. 11.23 Micrograph and cross-sectional drawing of microlensed FPA [5].

Fig. 11.24 Schematic diagram of the meta-lens integrated HgCdTe infrared detector. The meta-lens forms at the top by etching into the CdZnTe substrate. The infrared incident light is focused through the CdZnTe substrate and onto the HgCdTe active region, which consists of P, N, N+ type of HgCdTe. The CdZnTe substrate has a height of 73 μm. The contact, passivation and indium bumps are also presented [31].

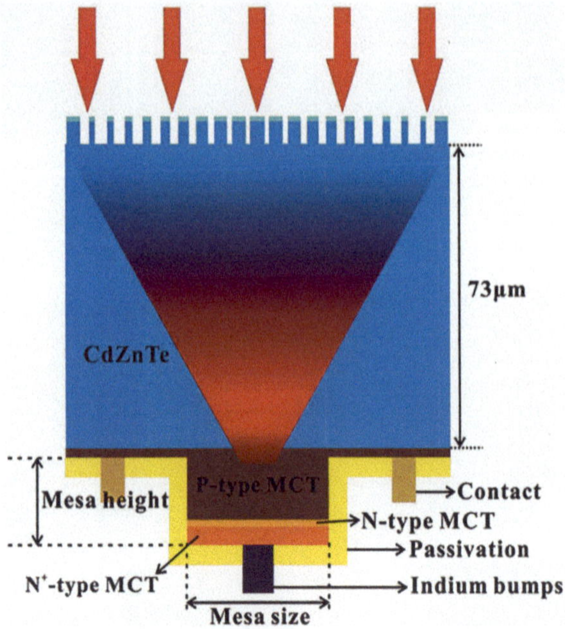

11.2.5 GaAs/AlGaAs quantum well infrared photodetector (QWIP)

The detector structures are molecular beam epitaxy (MBE) grown; the MBE system is equipped with valved cracking sources for the group V Sb_2 and As_2 fluxes, as well as dual In sources for independently varying the growth rates of GaInSb and InAs. Growth is performed on 75 mm diameter Te-doped n-type GaSb (100) substrates. The device fabrication recipe consists of a 0.5 μm Be-doped GaSb buffer layer, followed by a p-i-n superlattice (18 Å, 22 Å)-InAs/GaSb with the first 80 periods

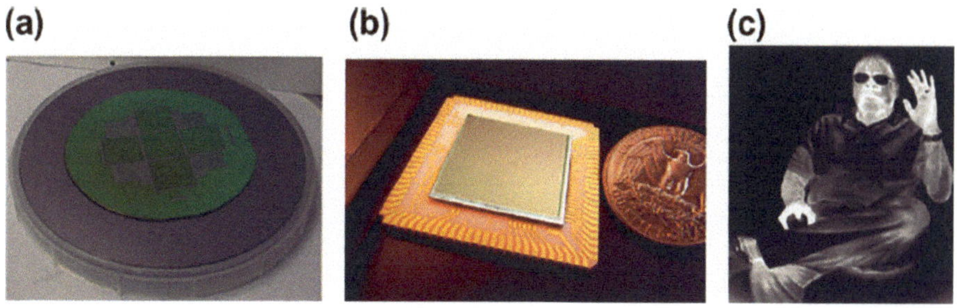

Fig. 11.25 (a) This picture shows five 1 K × 1 K superlattice detector arrays fabricated on 100 mm GaSb wafer, (b) 1 K × 1 K superlattice FPA on 100-pin lead-less-chip carrier, and (c) an image taken with the first LWIR superlattice FPA [21].

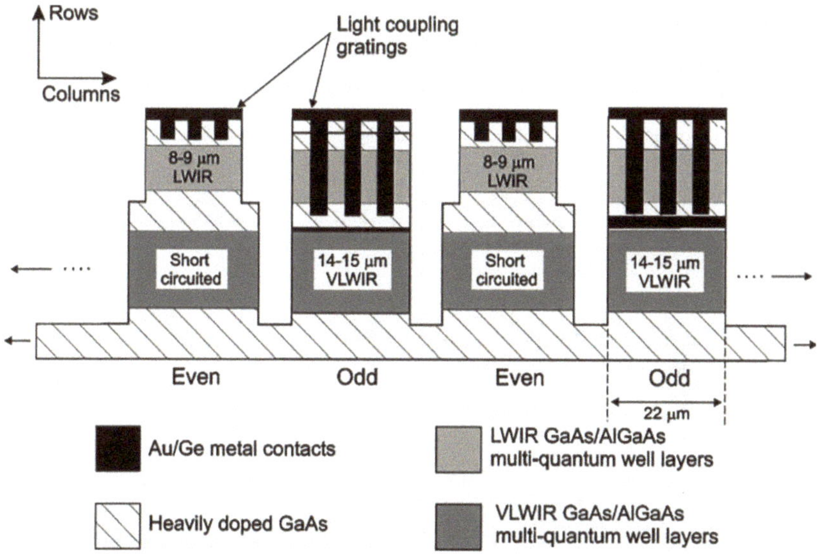

Fig. 11.26 Structure cross-section of the interlace dual-band FPA [32, 33].

Be doped in the GaSb layers, 200 undoped periods, and the final 80 periods doped with Te in the InAs layers [21]. Images of superlattice arrays fabricated on 100 mm GaSb wafer are presented in Fig. 11.25 [21].

An example of a dual-band GaAs/AlGaAs IRFPA is shown in Fig. 11.26 [32, 33].

A summary of hybrid IRFPAs from some major manufacturers is presented in Table 11.2.

11.2.6 Uncooled focal plane arrays

Amorphous silicon and vanadium oxide continue to dominate the manufacturing sector of uncooled infrared focal plane arrays. The major players in this area include the following companies: BAE, DRS, FLIR Systems, IRay Technology, L-3, Mitsubishi,

Table 11.2 Summary of hybrid IRFPAs from some major manufacturers [4].

Manufacturer/Web site	Size/architecture	Pixel size (μm)	Detector material	Spectral range (μm)	Oper. Temp. (K)	$D'(\lambda_p)$ (cm Hz$^{1/2}$/W)/ *NETD* (mK)
Sensors Unlimited	320 × 256	12.5 × 12.5	InGaAs	0.7–1.7	300	12.9×10^{13}
www.sensorsine.com	320 × 256	25 × 25	InGaAs	0.4–1.7	300	$<5 \times 10^{12}$
	640 × 512	25 × 25	InGaAs	0.7–1.7	300	4.2×10^{13}
	1280 × 1024	12.5 × 12.5	InGaAs	0.4–1.7	300	
Raytheon Vision	1024 × 1024	30 × 30	InSb	0.6–5.0	50	
Systems	2048 × 2048 (Orion II)	25 × 25	HgCdTe	0.6–5.0	32	
www.raytheon.com/	2048 × 2048 (Virgo-2 k)	20 × 20	HgCdTe	0.8–2.5	4–10	
businesses/ncs/rvs/	2048 × 2048	15 × 15	HgCdTe/Si	3.0–5.0	78	23
index.html	1024 × 1024	25 × 25	Si:As	5–28	6.7	
	2048 × 1024	25 × 25	Si:As	5–2S		
Teledyne Imaging	4096 × 4096 (H4RG)	10 × 10 or 15 × 15	HgCdTe	1.0–1.7	120	
Sensors	4096 × 4096 (H4RG)	10 × 10 or 15 × 15	HgCdTe	1.0–2.5	77	
http://teledynesi.com/	4096 × 4096 (H4RG)	10 × 10 or 15 × 15	HgCdTe	1.0–5.4	37	
imaging/	2048 × 2048 (H2RG)	18 × 18	HgCdTe	1.0–1.7	120	
	2048 × 2048 (H2RG)	18 × 18	HgCdTe	1.0–2.5	77	
	2048 × 2048 (H2RG)	18 × 18	HgCdTe	1.0–5.4	37	
Sofradir/	640 × 512	15 × 15	InGaAs	09–1.7	300	
www.sofradir.com/	640 × 512	15 × 15	InSb	3.7–4.8	80	<18
	1280 × 1024 (Jupiter)	15 × 15	HgCdTe	3.7–4.8	77–110	18
	1280 × 720 (Daphnis)	10 × 10	HgCdTe	3.4–4.9	110	<20
	640 × 512 (Scorpio)	15 × 15	HgCdTe	1.5–5.1	<90	≤16
	640 × 512 (Leo)	15 × 15	HgCdTe	3.7–4.8	110	20
	640 × 512	20 × 20	QWIP	8.0–9.0	73	31
	640 × 512	24 × 24	HgCdTe	MW/L(dual)	77–80	15–20
	640 × 512/	24 × 24	HgCdTe	MW/LW(dual)	77–80	20–25

Selex	320 × 256 (Saphira)	24 × 24	HgCdTe APD	0.8–2.5		
www.leonardocompany.	640 × 512 (Hawk)	16 × 16	HgCdTe	3–5	up to 170	17
com	1280 × 720 (Horizon)	12 × 12	HgCdTe	3.7–5		
	640 × 512 (Hawk)	16 × 16	HgCdTe	8–10	up to 90	32
	640 × 512 (Condor II)	24 × 24	HgCdTe	MW/LW(dual)	80	24/26
IAM	640 × 512	15 × 15	HgCdTe	1–5	95–120	17
www.aim-ir.com	640 × 512	15 × 15	HgCdTe	8–9	67–80	30
	640 × 512	20 × 20	HgCdTe	MW/LW	80	18/25
	384 × 288	40 × 40	Type II SL	MW(dual)	80	20/25
SCD	640 × 512	15 × 15	InSb	3–5		20
www.scd.co.il	1280 × 1024	15 × 15	InSb	3–5	77	20
	1920 × 1536	10 × 10	InSb	1–5.4		<25
	1280 × 1024	15 × 15	InAsSb nBn	3.6–4.2	150	20
	640 × 512	15 × 15	InAs GaSb T2SL	$\lambda_c = 9.5$	80	15
FLIR Systems	640 × 512	15 × 15	InGaAs	0.9–1.7	300	10^{10}pH/
http://www.flir.com						cm^2s(*NEI*)
	640 × 512	15 × 15	InSb	3.4–5.2	80	<25
DRS Technologies	1280 × 720	12 × 12	HgCdTe	3–5		20
	640 × 480	12 × 12	HgCdTe	3–5		25
	2048 × 2048	18 × 18	Si:As	5–28	7.8	
	1024 × 1024	25 × 25	Si:As	5–28	7.8	
	2048 × 2048	18 × 18	Si:Sb	5–40	7.8	

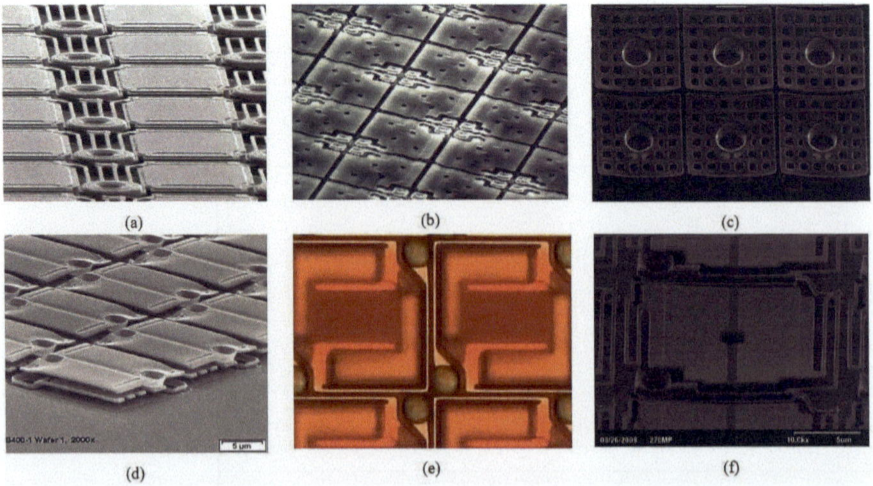

Fig. 11.27 Commercial bolometer design: (a) VO_x bolometer from BAE, (b) a-Si bolometer from Ulis, (c) VO_x umbrella design bolometer from DRS, (d) VO_x bolometer from Raytheon, (e) VO_x bolometer from SCD, and (f) a-Si/a-SiGe bolometer from L-3 Communications [4].

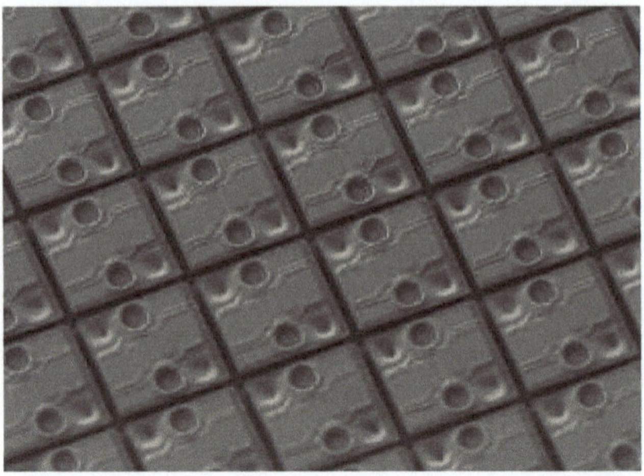

Fig. 11.28 SEM image of the fabricated 640 × 480 VO_x-based microbolometer FPA at ASELSAN, Inc. [34].

NEC, North Guangwei, Raytheon, SCD, Toshiba, ULIS and Zhejiang Dali. Examples of these commercial bolometer designs are presented in Fig. 11.27 [4].

ASELSAN Inc. has recently reported the development of VO_x-based 640 × 480 uncooled IRFPAs with 17 μm pixel pitch [34]. Fig. 11.28 presents an SEM image of the fabricated microbolometer FPA [34]. These detectors have been demonstrated to exhibit NETD values of <30 mK at *f*/1 and time constant of <12 mS.

The responsivity, R_i, of a detector is given by [4]

$$R_i = [FF\varepsilon TCRV_{\text{bias}}]/(G_{\text{th}}R), \tag{11.3}$$

where FF is the fill factor, ε is the absorption factor, TCR is the thermal coefficient of resistance, V_{bias} is the applied bias voltage, G_{th} is the thermal conductance, and R *is* the electrical resistance of the thermistor. FLIR's Seek thermal and microthermal camera (MTC) represent excellent examples of convergence of reduction in cost, power, size, and weight. These technological innovations are the result of extensive collaborations between FLIR and Raytheon that has resulted in 12 μm sensor chip along with the various components of the camera [4].

Table 11.3 presents a summary of commercially available uncooled microbolometer arrays from major manufacturers [4].

An illustration of the MEMS process for the manufacture of the microbolometer is presented in Fig. 11.29 [35]. The details of the process flow of this surface micromachining technology is discussed in Ref. [35].

Table 11.3 Examples of commercial uncooled infrared microbolometer arrays [4]

Company	Bolometer type	Array format	Pixel pitch (μm)	Detector *NEDT* (mK) (*f*/1, 20–60 Hz)	Time constant (ms)
L-3 (USA) www.l-3com.com	VO_x bolometer	320 × 240	37.5	50	
	a-Si bolometer	160 × 120–640 × 480	30	50	
	a-Si/a-SiGe bolometer	320 × 240–1024 × 768	17	30–50	
BAE (USA) http://www.baesystems.com	VO_x bolometer	320 × 240	28	<35	<10–15
	VO_x bolometer (standard design)	640 × 480	12	<50	<10–15
	VO_x bolometer (standard design)	1024 × 768	17		
DRS (USA) www.drsinfrared.com	VO_x bolometer (standard design)	320 × 240	25	≤40	≤18
	VO_x bolometer (umbrella design)	320 × 240	17	≤40	≤14
	VO_x bolometer (umbrella design)	640 × 480, 1024 × 768	17	<40	≤14

Continued

Table 11.3 Continued

Company	Bolometer type	Array format	Pixel pitch (μm)	Detector *NEDT* (mK) (*f*/1, 20–60 Hz)	Time constant (ms)
Raytheon (USA) http://www.raytheon.com/capabilities/products/uncooled	VO_x bolometer	320 × 240, 640 × 480	25	30–40	
	VO_x bolometer (umbrella design)	320 × 240, 640 × 480	17	50	
	VO_x bolometer (umbrella design)	1024 × 480, 2048 × 1536	17		
ULIS (France) www.ulis-ir.com	a-Si bolometer	80 × 80	34	<100	
	a-Si bolometer	160 × 120	25	<60	<10
		320 × 240	12	<60	<10
		384 × 240	17	<55	<10
		640 × 480, 1024 × 768	17	<50	<10
SCD (Israel) www.scd.co.il	VO_x bolometer	384 × 288	25	<20	22
	VO_x bolometer	384 × 288	25	<35	16
	VO_x bolometer	640 × 480	17	<35	16
	VO_x bolometer	1024 × 768	17	<35	14
FLIR Systems http://www.flir.com	VO_x bolometer	640 × 512	17	<60	<12
	VO_x bolometer	336 × 256	17	<50	<15
NEC (Japan) http://www.nec.com	VO_x bolometer	320 × 240	23.5	<75	
	VO_x bolometer	640 × 480	23.5	<75	
	VO_x bolometer	640 × 480	12	60	
	VO_x bolometer	320 × 240	23.5	NEP < 100 pW*	

* at 4 THz

A brief summary of the details of IRFPAs manufactured by a few companies is presented in Table 11.4.

From a historical point of view, the efforts on the development of uncooled focal plane array (UFPA), based on the field-enhanced pyroelectric effect in ferroelectrics, at Texas Instruments, is illustrated in Fig. 11.30 [37].

In Fig. 11.31, an example of a UFPA readout IC unit cell is described [37].

While research on pyroelectric infrared imagers continues on a small scale, its limited dynamic range and the need for a chopper have not helped the growth of pyroelectric-based thermal detectors.

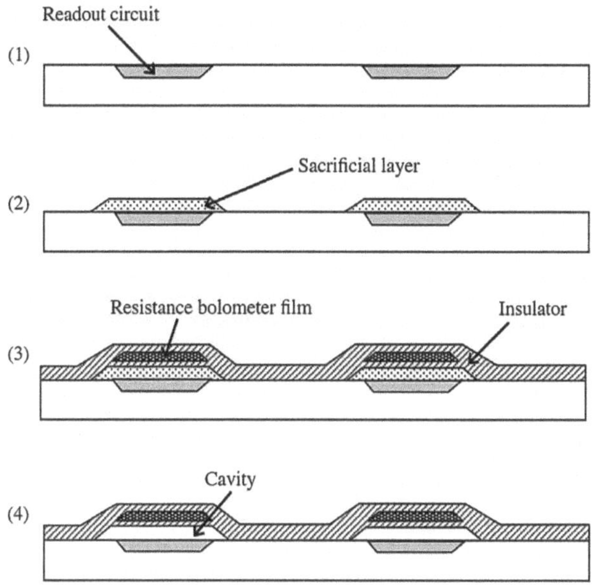

Fig. 11.29 MEMS process for manufacture of microbolometer [35].

Table 11.4 Examples of details of IRFPAs [36].

Reference	Material	Array size	Pixel size	ROIC type	Frame rate[a]	NETD
Mitsubishi [20]	Diode	320 × 240	17 μm	CTIA	60 Hz	50 mK
MikroSens [45]	CMOS-compatible resistance	120 × 160	35 μm	CTIA	11 Hz	117 mK
Toshiba [37]	Diode	320 × 240	22 μm	GMI	25 ms	200 mK
Raytheon [60]	VO_x	640 × 512	20 μm		30 Hz	<50 mK
DRS [61]	VO_x	1024 × 768	17 μm	DI	30 Hz	<50 mK
FLIR [99]	VO_x	640 × 512	12 μm		60 Hz	<40 mK
L-3 Communications [85]	a-Si/a-SiGe	1024 × 768	17 μm	DI	10 ms	35 mK
UL1S [65]	a-Si	1024 × 768	17 μm	CTIA	30 Hz	46 mK

ROIC type: *CTIA*,Capacitive trans impedance amplifier; *GMI*, Gate modulation input; *DI*, Direct injection.
[a] Or time constant in case the frame rate is not mentioned.

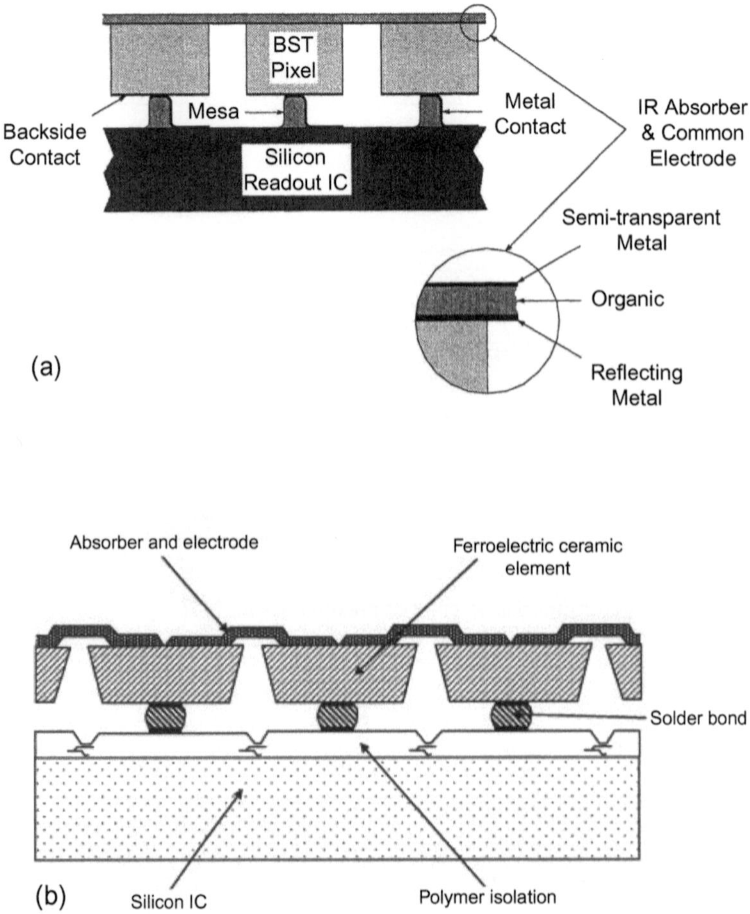

Fig. 11.30 (a), (b) Illustration of pyroelectric uncooled focal plane array [37].

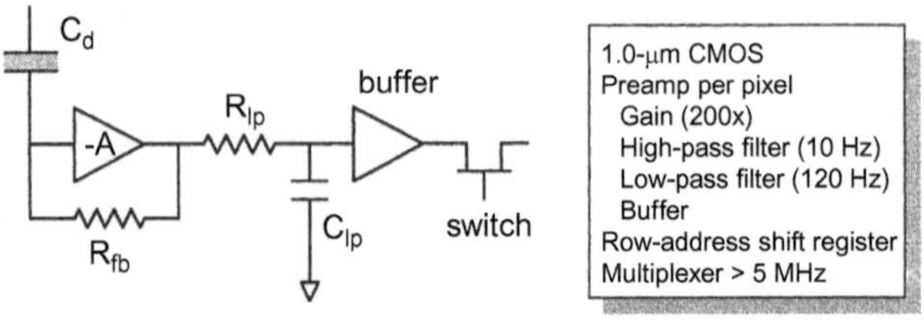

Fig. 11.31 UFPA readout IC unit cell [37].

References

[1] K.I. Schultz, M.W. Kelly, J.J. Baker, M.H. Blackwell, M.G. Brown, C.B. Colonero, C.L. David, B.M. Tyrrell, J.R. Wey, Digital-pixel focal plane array technology, MIT Lincoln Lab. J. (2014) 1–7.

[2] E.R. Fossum, B. Pain, Infrared readout electronics for space-science sensors: state of the art and future directions, SPIE Proc. 2020 (1993) 262–285.
[3] A.D. Scribner, M.R. Kruer, J.M. Killiany, Infrared focal plane array technology, Proc. IEEE 79 (1991) 66–85.
[4] A. Rogalski, Next decade in infrared detectors, in: Proceedings of SPIE 10433, Electro-Optical and Infrared Systems: Technology and Applications XIV, 104330L, SPIE Security+Defense, Warsaw, Poland, 2017, pp. 104330L-1–104330L-25.
[5] A. Rogalski, Progress in focal plane array technologies, Prog. Quantum Electron. 36 (2012) 342–473.
[6] Committee on Developments in Detector Technologies, National Research Council, 2010. http://www.nap.edu/catalog/12896.html.
[7] P. Norton, Detector focal plane array technology, in: R. Driggers (Ed.), Encyclopedia of Optical Engineering, Marcel Dekker Inc, New York, 2003, pp. 320–348.
[8] G.C. Holst, T.S. Lomheim, CMOS/CCD Sensors and Camera Systems, JCD Publishing and SPIE Press, Winter Park, 2007.
[9] M.A. Kinch, State-of-the-Art Infrared Detector Technology, SPIE Press, Bellingham, 2014.
[10] J. Robinson, M. Kinch, M. Marquis, D. Littlejohn, K. Jeppson, Case for small pixels: system perspective and FPA challenge, in: Proceedings of SPIE, 2014, p. 9100 (91000I-1-10).
[11] S.D. Gunapala, S.V. Bandara, J.K. Liu, J.M. Mumolo, D.Z. Ting, C.J. Hill, J. Nguyen, First demonstration of megapixel dual-band QWIP focal plane array, IEEE Sens. Conf. (2009) 1609–1612.
[12] P.-Y. Delaunay, B.Z. Nosho, A.R. Gurga, S. Terterian, R.D. Rajavel, Advances in III-V based dual-band MWIR/LWIR FPAs at HRL, in: B.F. Andresen, G.F. Fulop, C.M. Hanson, J.L. Miller, P.R. Norton (Eds.), Infrared Technology and Applications XLIII, Proceeding of SPIE, vol. 10177, 2017, pp. 101770T-1–101770T-12.
[13] R.K. Bhan, V. Dhar, Recent infrared detector technologies, applications, trends and development of HgCdTe based cooled infrared focal plane arrays and their characterization, Opto-Electron. Rev. 27 (2019) 174–193.
[14] S.D. Gunapala, S.V. Bandara, J.K. Liu, E.M. Luong, S.B. Rafol, J.M. Mumolo, D.Z. Ting, J.J. Bock, M.E. Ressler, M.W. Werner, P.D. LeVan, R. Chehayeb, C.A. Kukkonen, M. Ley, P. LeVan, M.A. Fauci, Recent developments and applications of quantum well infrared photodetector focal plane arrays, in: J. Rutkowski, J. Wenus, L. Kubiak (Eds.), Proceedings of SPIE International. Conference on Solid State Crystals 2000: Epilayers and Heterostructures in Optoelectronics and Semiconductor Technology, vol. 4413, 2001, pp. 323–338, https://doi.org/10.1117/12.425451.
[15] A.M. Hoang, G. Chen, A. Haddadi, M. Razeghi, Demonstration of high performance bias-selectable dual-band short−/mid-wavelength infrared photodetectors based on type-II InAs/GaSb/AlSb superlattices, Appl. Phys. Lett. 102 (2013), 011108.
[16] G. Destéfanis, P. Tribolet, M. Vuillermet, D.B. Lanfrey, MCT IR detectors in France, in: Proceedings of SPIE Infrared Technology and Applications XXXVII, Vol. 8012, 2011, p. 801235, https://doi.org/10.1117/12.886904. 1 − 12.
[17] A. Hoffman, Semiconductor processing technology improves resolution of infrared arrays, Laser Focus World 42 (2) (2006) 81–84. https://www.laserfocusworld.com/articles/print/volume-42/issue-2/features/irdetectors-semiconductor-processing-technology-improves-resolution-of-infrared-arrays.html.
[18] http://www.aeriousphotonics.com. Accessed 29 August 2019.
[19] http://www.XenICs.com. Accessed 29 August 2019.
[20] A. Rogalski, Optical detectors for focal plane arrays, Opto-Electron. Rev. 12 (2004) 221–245.

[21] S.D. Gunapala, D.Z. Ting, C.J. Hill, J. Nguyen, A. Soibel, S.B. Rafol, S.A. Keo, J.M. Mumolo, M.C. Lee, J.K. Liu, B. Yang, A. Liao, Large area III–V infrared focal planes, Infrared Phys. Technol. 54 (2011) 155–163.

[22] T. Asar, S. Özçelik, E. Özbay, Structural and electrical characterizations of $In_xGa_{1-x}As$/InP structures for infrared photodetector applications, J. Appl. Phys. 115 (2014) 104502-1–104502-7.

[23] A. Rogalski, HgCdTe infrared detector material: history, status and outlook, Rep. Prog. Phys. 68 (2005) 2267–2336.

[24] P. Norton, HgCdTe infrared detectors, Opto-Electron. Rev. 10 (3) (2002) 159–174.

[25] J.A. Wilson, E.A. Patten, G.R. Chapman, K. Kosai, B. Baumgratz, P. Goetz, S. Tighe, R. Risser, R. Herald, W.A. Radford, T. Tung, W.A. Terre, Integrated two-color detection for advanced focal plane array (FPA) applications, Proc. SPIE 2274 (1994) 117–125.

[26] R.D. Rajavel, D.M. Jamba, J.E. Jensen, O.K. Wu, P.D. Brewer, J.A. Wilson, J.L. Johnson, E.A. Patten, K. Kosai, J.T. Caulfield, P.M. Goetz, Molecular beam epitaxial growth and performance of HgCdTe-based simultaneous-mode two-color detectors, J. Electron. Mater. 27 (1998) 747–751.

[27] M.B. Reine, A. Hairston, P. O'Dette, S.P. Tobin, F.T.J. Smith, B.L. Musicant, P. Mitra, F.C. Case, Simultaneous MW/LW dual-band MOVPE HgCdTe 64x64 FPAs, Proc. SPIE 3379 (1998) 200–212.

[28] J.P. Zanatta, P. Ferret, R. Loyer, G. Petroz, S. Cremer, J.P. Chamonal, P. Bouchut, A. Million, G. Destefanis, Single- and two-color infrared focal plane arrays made by MBE in HgCdTe, Proc. SPIE 4130 (2000) 441–451.

[29] W.E. Tennant, M. Thomas, L.J. Kozlowski, W.V. McLevige, D.D. Edwall, M. Zandian, K. Spariosu, G. Hildebrandt, V. Gil, P. Ely, M. Muzilla, A. Stoltz, J.H. Dinan, A novel simultaneous unipolar multispectral integrated technology for HgCdTe IR detectors and focal plane arrays, J. Electron. Mater. 30 (2001) 590–594.

[30] G. Destefanis, J. Baylet, P. Ballet, P. Castelein, F. Rothan, O. Gravrand, J. Rothman, J.P. Chamonal, Status of HgCdTe bicolor and dual-band infrared focal arrays at LETI, J. Electron. Mater. 36 (2007) 1031–1044.

[31] F. Li, J. Deng, J. Zhou, Z. Chu, Y. Yu, X. Dai, H. Guo, L. Chen, S. Guo, M. Lan, X. Chen, HgCdTe mid-infrared photo response enhanced by monolithically integrated meta-lenses, Sci. Rep. 10 (2020) 6372, https://doi.org/10.1038/s41598-020-62433-w.

[32] A. Rogalski, Infrared detectors: an overview, Infrared Phys. Technol. 43 (2002) 187–210.

[33] S.D. Gunapala, S.V. Bandara, A. Sigh, J.K. Liu, S.B. Rafol, E.M. Luong, J.M. Mumolo, N. Q. Tran, J.D. Vincent, C.A. Shott, J. Long, P.D. Le Van, 8–9 and 14–15 μm two-color 640 x 486 quantum well infrared photo-detector (QWIP) focal plane array camera, Proc. SPIE 3698 (1999) 687–697.

[34] O. Celik, E. Inceturkmen, B. Kaplan, B. Barutcu, O. Aydin, I.E. Gonenli, R. Kepenek, C. Tunca, M. Akbulut, O. Nuzumlali, S. Keskin, K. Aydemir, C. Yildizak, 640 x 480 17 μm microbolometer uncooled detector development at ASELSAN, Inc, in: Proceediong of SPIE., 11407, Infrared Technology and Applications, XLVI, 2020, p. 1140719, https://doi.org/10.1117/12.2557669.

[35] M. Kimata, Uncooled infrared focal plane arrays, IEEJ Trans. Electr. Electron. Eng. 13 (2018) 4–12.

[36] L. Yu, Y. Guo, H. Zhu, M. Luo, P. Han, X. Ji, Low-cost microbolometer type infrared detectors, Micromachines 11 (2020) 800. 19 pages and references therein.

[37] C.M. Hanson, Uncooled thermal imaging at Texas Instruments, in: Proceeding of SPIE 2020, Infrared Technology XIX, 1993, https://doi.org/10.1117/12.160554.

A case study of an uncooled microbolometer

12

12.1 Introduction

A microbolometer is an infrared detector that is generally fabricated by micromachining technology. A microbolometer consists of a thin multilayer structure, thermally isolated and connected to a substrate, through support structures that act as heat links and electrical connectors. The two main components of the multilayer structure are an optical infrared absorber material coupled to a thermosensing material. The absorbed incident optical radiation is converted into heat by the infrared absorber material, producing changes in the resistivity of the thermosensing material that is transduced into an electrical signal.

A wide variety of materials have been used in the design and fabrication of microbolometers, and new absorbers and thermosensing materials have been considered to improve the performance of these devices in the last decade.

Historically, black metals (platinum, tungsten, titanium, and gold) have been used as absorbers of optical infrared radiation since the 1930s [1–5] and have been extensively utilized as infrared absorber coatings. Surface modification techniques such as electroplating, and chemical and mechanical etching have been traditionally used to achieve high porosity metal surfaces that are the most effective mechanism for the absorption of infrared radiation [3,6]. New surface modification techniques, such as the use of femtosecond lasers, have been implemented in the fabrication of black metals that exhibit extremely low reflectance across the infrared wavelength spectra [7].

Novel IR radiation absorber materials have emerged with extremely low reflectance exhibited across the IR spectra. Vantablack, a composite of thin film of carbon nanotubes and black silicon (BSi), a surface modification of silicon, exhibits absorptance close to unity [8–11]. Black gold films with infrared absorptance close to unity have been developed [12]; but the thin film fabrication process technology that is required for the deposition of these materials may not be compatible with all semiconductor processes and it does not allow the mass production of uncooled microbolometers [13,14]. Black gold, obtained by thermal evaporation methods, exhibits an absorptance close to unity [15–17]; but the fragile structure of this material and the degradation in absorption properties, when it is exposed to temperatures higher than 300°C, does not make it feasible for mass production [18].

Polycrystalline materials such as vanadium oxide (VO_x) and amorphous silicon (a-Si) have been widely used as IR thermosensing materials [19–21]; these materials have the advantage of high temperature coefficient of resistance (TCR), which is

Microbolometers. https://doi.org/10.1016/B978-0-08-102812-4.00008-5

required by the imaging industry in order to achieve a detector response with high sensitivity.

In the infrared imager industry, vanadium oxide (VO_x) has been the most widely used thermosensing material of choice in a variety of microbolometers [19] with temperature coefficient of resistance in the range of −2 to −3%/K [22–24] at room temperature. However, vanadium oxide is not a standard material in the integrated circuit (IC) fabrication process technology [25]. Vanadium oxide is not a single crystal material and its compositional variations have to be controlled very precisely; it has some problems such as unstable phases and poor reproducibility [26,27]. Thin films of vanadium oxide can be prepared by a variety of techniques, including reactive RF sputtering [28–30], pulsed laser deposition [31], ion beam sputtering and magnetron controlled sputtering that are high cost methods [32,33]. Today, vanadium oxide and amorphous silicon are the standard materials of choice in the manufacture of uncooled microbolometers.

In Fig. 12.1, a schematic diagram of a microbolometer fabrication process showing the various multilayers, including the IR absorbing structure, is shown [34]. In this particular study in [34], a Ti/MgF_2 absorbing layer has been considered as the IR absorber in a vanadium oxide microbolometer. The simulated and measured absorptance of Ti/MgF_2 IR absorbers are summarized in Fig. 12.2 [34]. A detailed study of several black metal candidates, as IR absorbers, for applications in microbolometers is presented in Chapter 7.

Amorphous silicon (a-Si), an alternative material that is compatible with the silicon technology, requires high values of electrical resistivity in order to achieve a high TCR [35–37]. In order to achieve a TCR of −3 to −4%/K [38] at room temperature, amorphous silicon should be exposed to high annealing temperatures resulting in a high electrical resistivity and a high pixel resistance in the order of Mega-Ohms (MΩ) [39]. This high value of resistance causes large Johnson noise and consequently low detector sensitivity as well as a mismatch with the input impedance of the CMOS read-out circuits.

The disadvantage of these materials, vanadium oxide and amorphous silicon, is that both of them exhibit high $1/f$ noise due to their noncrystalline structure. For most materials, $1/f$ noise constant is not very well documented in the literature. It is a material parameter that can vary by several orders of magnitude for different materials. Even small variations of the material composition can dramatically change the $1/f$ noise constant [40].

The complexity in the fabrication process technology, which is required for the deposition and patterning of these thermosensing materials and the high electrical resistivity needed to achieve high TCR, does not make them cost-effective for mass commercialization.

Previously, some experimental studies have been performed to develop the use of n-type silicon to optimize uncooled bolometers [41]. An uncooled microbolometer structure with n-well layer as the active element has been proposed and fabricated [42–44]. However, n-Si has not been used in the fabrication and development of commercial uncooled infrared microbolometers that is dominated by VO_x [45] and other

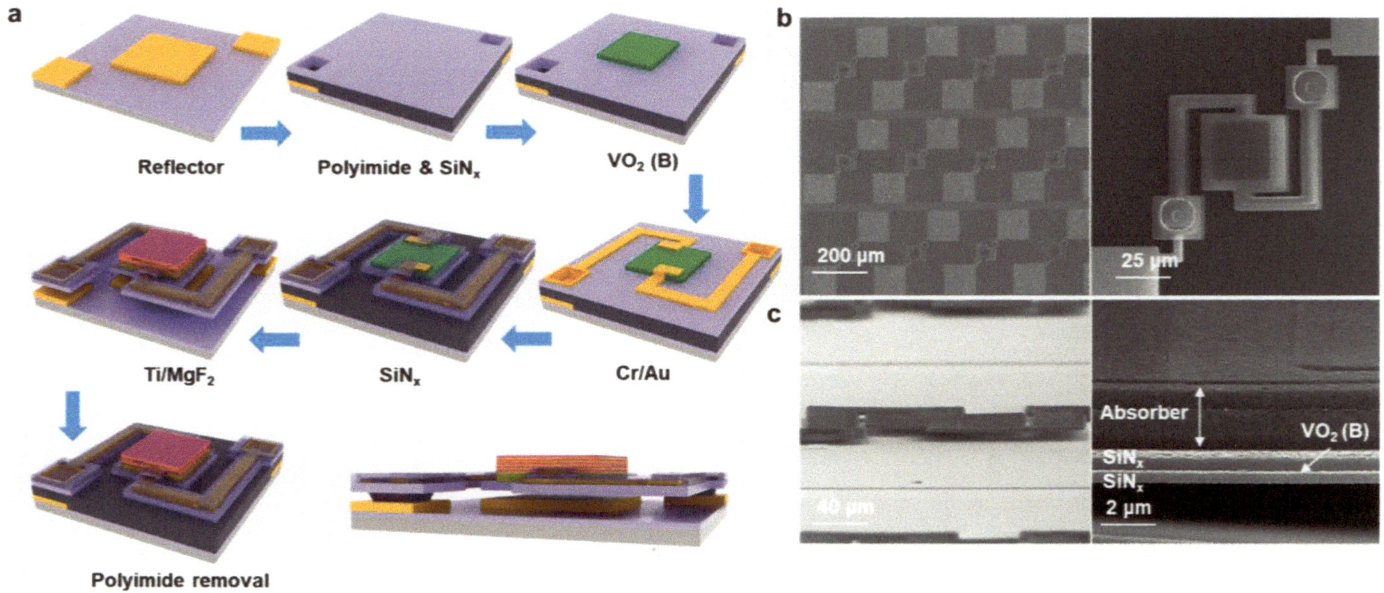

Fig. 12.1 (A) Schematic images of fabrication process of microbolometer with IR absorber. Scanning electron micrograph (B) top views and (C) cross section views of microbolometer with IR absorber [34].

From Hye Jin Lee, Dasom Wang, Tae Hyeong Kim, Dae-Han Jung, Tae-Hyeon Kil, Ki-Suk Lee, Hyung-Jin Choi, Seung-Hyub Baek, Euijoon Yoon, Won Jun Choi, Jeong Min Baik, Wide-temperature (up to 100 °C) operation of thermostable vanadium oxide based microbolometers with Ti/MgF2 infrared absorbing layer for long wavelength infrared (LWIR) detection, Appl. Surf. Sci. 547 (2021) 149142, ISSN 0169-4332, https://doi.org/10.1016/j.apsusc.2021.149142.

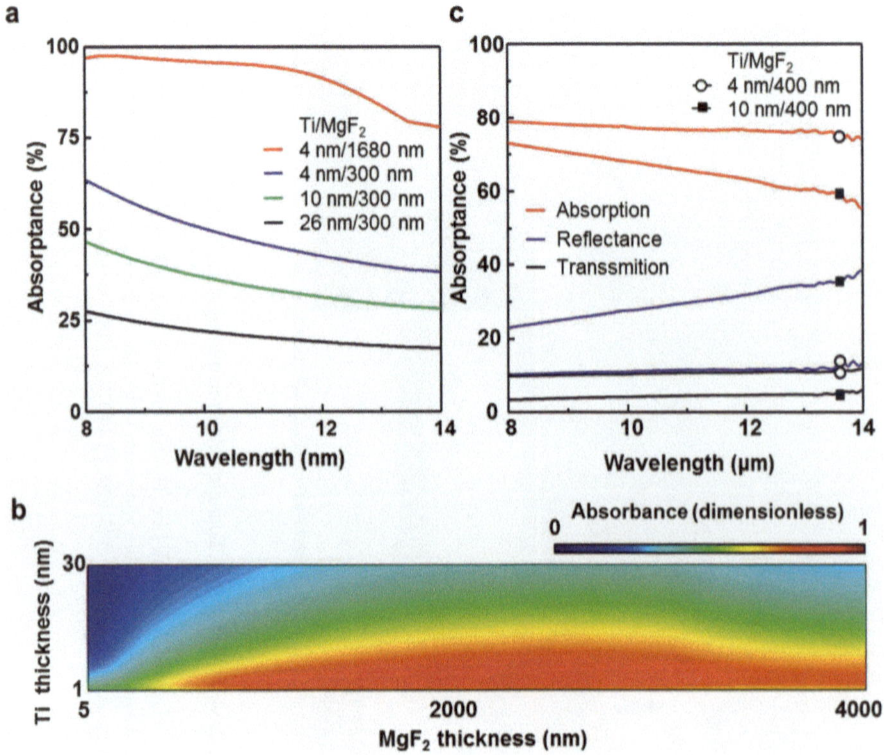

Fig. 12.2 (A) Simulated absorptance of IR absorber in the wavelength range from 8 to 14 μm. (B) Absorptance vs thickness for each layer, Ti and MgF2. (C) Measured absorptance of IR absorber in the wavelength range from 8 to 14 μm. [34].
From Hye Jin Lee, Dasom Wang, Tae Hyeong Kim, Dae-Han Jung, Tae-Hyeon Kil, Ki-Suk Lee, Hyung-Jin Choi, Seung-Hyub Baek, Euijoon Yoon, Won Jun Choi, Jeong Min Baik, Wide-temperature (up to 100 °C) operation of thermostable vanadium oxide based microbolometers with Ti/MgF2 infrared absorbing layer for long wavelength infrared (LWIR) detection, Appl. Surf. Sci. 547 (2021) 149142, ISSN 0169-4332, https://doi.org/10.1016/j.apsusc.2021.149142.

polycrystalline materials. n-Si requires an intermediate step in the form of heavily doped n-Si (n^{++}-Si) for forming an Ohmic contact with the metal.

12.2 Microbolometer: Materials selection

In this section, low doped p-type silicon Si (1.45 Ω-cm) as an active/alternative thermosensing element is introduced in combination with silicon nitride (Si_3N_4) as an infrared absorber material. The *NETD* of state-of-the-art uncooled infrared microbolometer is limited by the 1/*f* noise from the thermosensing material [46,47]. Monocrystalline materials can have a significantly lower 1/*f* noise constant as compared to amorphous or polycrystalline materials. Thus, the use of monocrystalline silicon as a thermosensing material will reduce the *NETD* of a microbolometer with a low 1/*f* noise constant [48,49]. Aluminum (Al) is considered as the interconnection material because of its electrical properties and gold (Au) is used as an infrared radiation reflector layer.

12.2.1 P-silicon as thermosensing material

The process of doping introduces impurities in the intrinsic Si semiconductor via substitutional atoms such as boron or phosphorous with lower or higher valence, respectively, via the process of diffusion. This increases the charge carrier density and the intrinsic semiconductor becomes an extrinsic semiconductor. These impurity atoms are called donors if they add electrons to the conduction band or acceptors (holes) if they add positive charges to the valence band.

Depending on the type of dominant carriers, the extrinsic semiconductor is called n-type or p-type. Silicon doped with phosphorous receives an extra free electron resulting in n-type Si. Silicon doped with boron creates holes resulting in p-type Si. The energy gap for acceptor and donor levels with respect to the corresponding energy bands is smaller than the energy gap between the valence and conduction bands [50].

As shown in Fig. 12.3, the migration of an electron from the valence band to the conduction band leaves behind a hole in the valence band. In a p-type extrinsic semiconductor, the majority carriers are holes in the valence band and they are called positive carrier type (p-type); the binding energy (activation energy E_a) is small and while the holes are the charge carriers, the electrons participate in bonding [51]. Fig. 12.3 shows the energy bands in a p-type extrinsic semiconductor, where E_a is the acceptor band energy, E_{f1} is the Fermi energy for an intrinsic semiconductor, E_c is the conduction band energy, E_f is the extrinsic Fermi energy, and E_v is the valence band energy.

When a trivalent atom such as boron is introduced substitutionally in the crystalline silicon covalent bonded lattice of silicon Fig. 12.4, one of the covalent bonding orbital is missing and a hole is created. In the presence of an external electric field, an electron carrier from another tetrahedral bond can attain sufficient energy to move to the missing bond ionizing with charge of −1 from the boron atom and creating a boron-silicon valence bond. The migration of an electron from a silicon atom to a boron atom is only 0.005 eV small compared with 1.1 eV required to transfer an electron from the valence

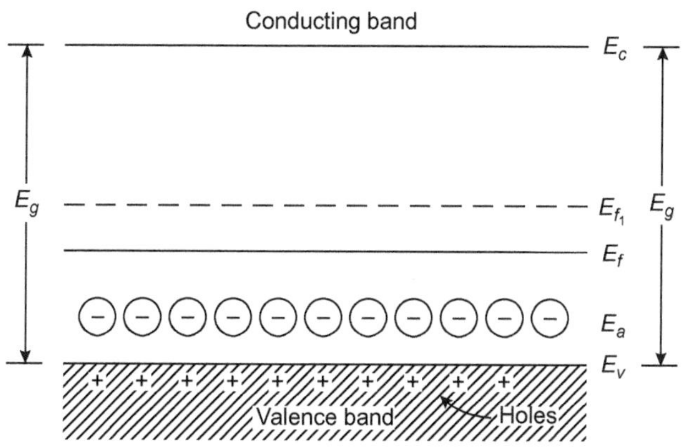

Fig. 12.3 Energy bands in a p-type extrinsic semiconductor [51].

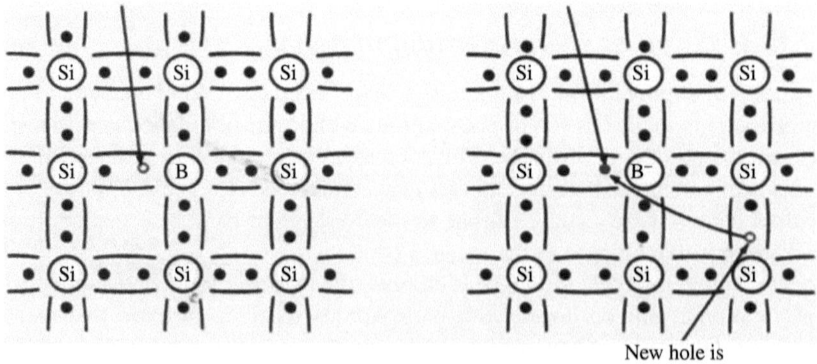

Fig. 12.4 Carrier transport in p-type extrinsic semiconductor [51].

band to the conduction band. Fig. 12.4 shows the creation of the hole by the ionization of boron; it behaves as a positive charge that moves in the silicon crystal lattice in the direction of the negative terminal [51].

Electrical conductivity in intrinsic semiconductors is an exponential function of temperature and band gap; the conductivity of extrinsic semiconductors is a function of charge carrier density and charge carrier mobility. At low temperature, the electrical conductivity increases as the charge carrier concentration increases; at intermediate temperature, the number of charge carriers is nearly independent of temperature (exhaustion range); at higher temperatures, the carrier mobility decreases as a result of increase in phonon scattering; at still higher temperatures, once the charge carriers are completely ionized, the extrinsic semiconductor behaves as an intrinsic semiconductor [52].

The thermosensing p-type silicon layer is considered with a doping concentration of 10^{16} cm^{-3} with a TCR of 0.74%/K [53,54]. Fig. 12.5 shows the results of measurements by the National Bureau of Standards (now, NIST—the National Institute of Standards and Technology) of the temperature coefficients of resistance at a temperature of 23°C for n-type and p-type silicon [54]. The TCR of silicon is determined by the type of carrier, carrier concentration and carrier mobility. Although the TCR value of p-Si is low in comparison with the standard thermosensing materials used by the infrared detector industry, the use of p-doped silicon has some significant advantages. These include low resistivity that reduces pixel resistance and Johnson noise, low resistance values that are compatible with readout circuits, negligible $1/f$ noise due to its single crystal nature and it is a cost-effective manufacturing technology (all silicon MEMS technology).

12.2.2 Radiative properties of thin films of common dielectric materials

For the manufacture of microbolometers, for excellent absorber candidates, there is a need for low-cost standard semiconductor materials with high emissivity in the IR spectral range. These materials also need to be compatible with the actual

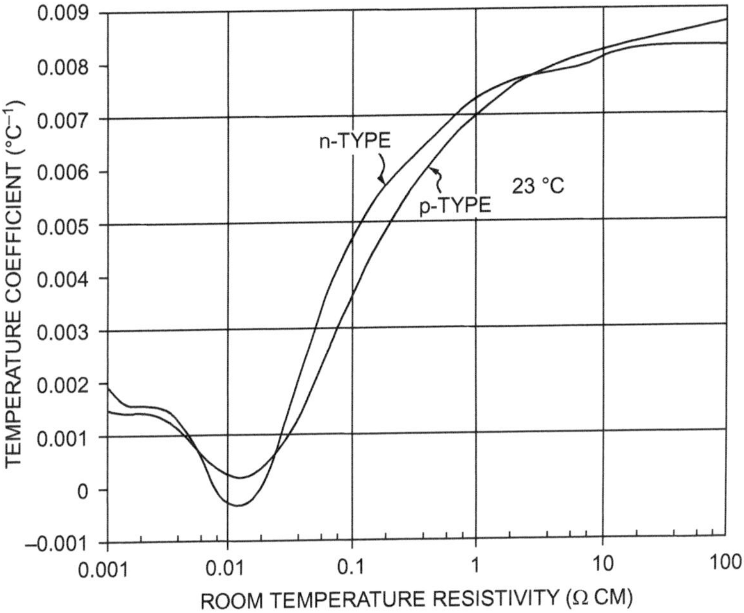

Fig. 12.5 TCR as a function of electrical resistivity for n-type and p-type silicon (23°C) [54].

manufacturing process as well as be able to integrate with the required optoelectronic/read-out circuits.

In this section, radiative properties of common dielectric materials such as Al_2O_3, SiO_2, AlN, and Si_3N_4 are simulated in the infrared spectral range of 1.5–14.2 μm. These common dielectric materials are routinely used as electrical insulators, anti-reflection coatings, optical filters, waveguides, lenses as well as passivation layers in a variety of applications including microbolometers, light emitting diodes, displays, sensors, solar cells, and other photonic devices. In particular, the design of optical coatings (highly reflective and antireflective) requires accurate knowledge of the radiative properties of the dielectric material.

With the rapid growth of application of materials in the infrared range of wavelengths and advancements in related technology, manufacturers have begun to utilize infrared materials in the design and manufacture of micro-optic components such as band-pass filters, beam splitters, lenses, mirrors, and polarizers in the IR. These IR materials vary in their optical, mechanical, thermal, and electrical properties, and these properties determine their applications in the IR.

In this study, the radiative properties of dielectric materials for varying thicknesses at room temperature, are simulated using a multilayer model, Multi-Rad. The thin films (dielectric materials) are considered to be suspended (free standing) in vacuum. A standard blackbody is considered as the source of infrared radiation with a normal angle of incidence to the thin film surface (blackbody temperature = 30°C; zero transmittance losses between source and sample). The smaller increment in

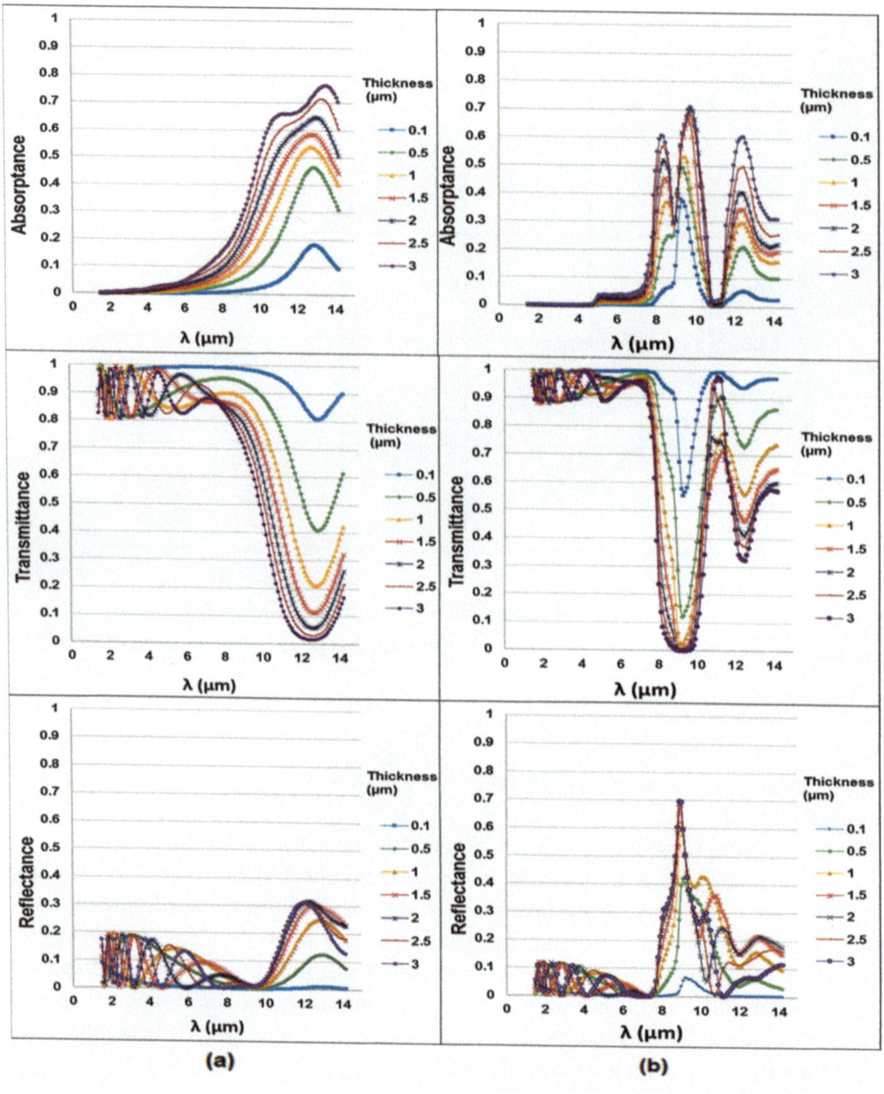

Fig. 12.6 Absorptance, transmittance, and reflectance for varying thicknesses of (A) Al_2O_3 and (B) SiO_2, at room temperature (30°C) [55].

spectral wavelength is 0.1 μm, and the range of wavelengths in which the radiative properties are simulated is 1.5–14.2 μm.

In Figs. 12.6 and 12.7, plots of emissivity, reflectance, and transmittance in arbitrary units, for varying thickness (0.1–3 μm) of the dielectric materials (Al_2O_3, SiO_2, AlN, and Si_3N_4), at room temperature (30°C), are presented for the infrared wavelength range of 1.5–14.2 μm [55]. The range of thickness of the dielectric films, considered in this study, is consistent with the industry standards.

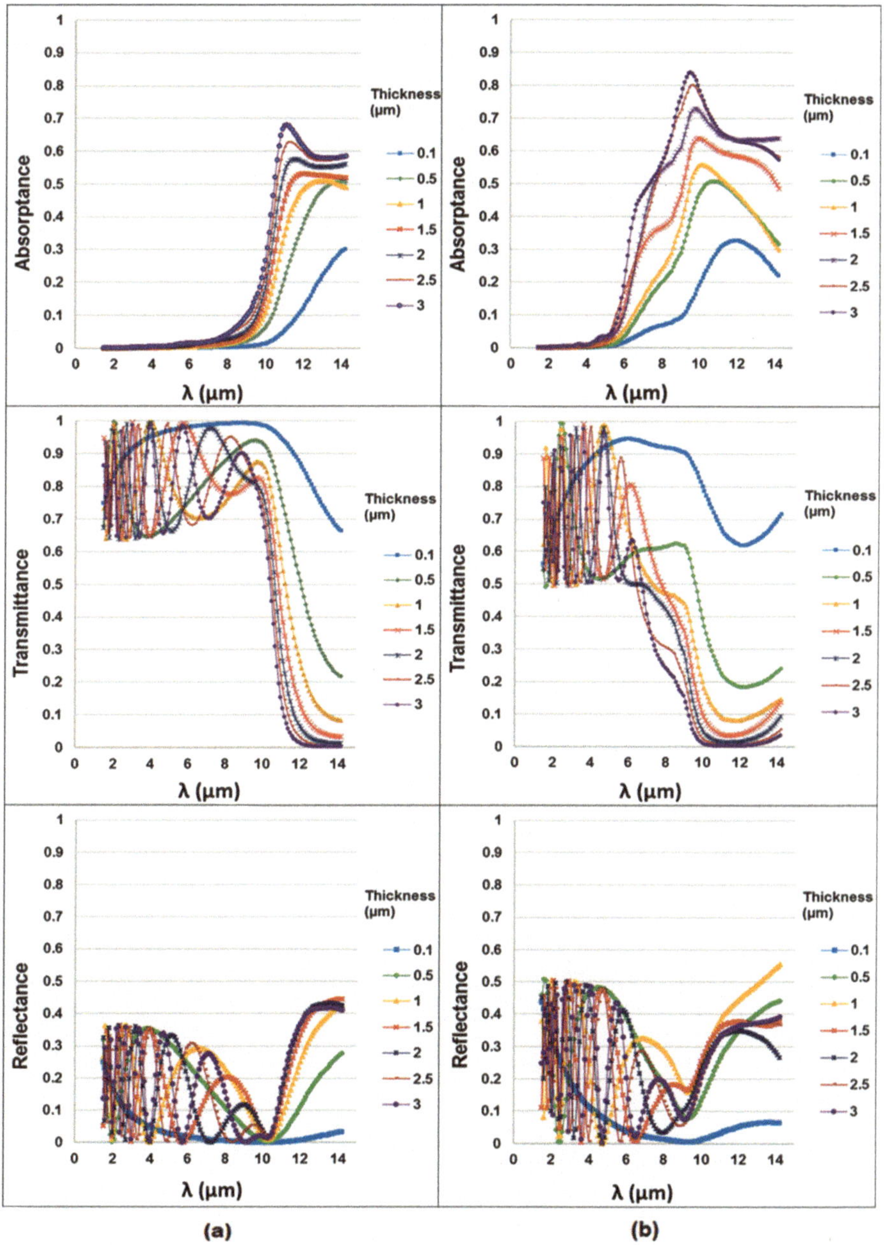

Fig. 12.7 Absorptance, transmittance, and reflectance for varying thicknesses of (A) AlN and (B) Si_3N_4, at room temperature (30°C) [55].

As can be seen in Fig. 12.6a, the absorptance increases with increase in thickness of Al_2O_3 in the wavelength range of 4–14.2 μm, reaching its peak value of 0.76 at a wavelength of 13.6 μm. It may be observed that the transmittance, with varying thickness, is higher than 0.8 in the wavelength range of 1.5–8.5 μm; it shows a decreasing trend with increasing thickness in the wavelength range of 8.5–14.2 μm. The corresponding reflectance is less than 0.2 in the wavelength range of 1.5–10.9 μm; as can be seen in the wavelength range of 9.6–14 μm, the reflectance increases as the thickness is increased, reaching a peak value of 0.31 at the wavelength of 12.3 μm.

Fig. 12.6b shows the increase in absorptance with increasing SiO_2 thickness. In the wavelength range of 4.8–14.4 μm, three peak values are observed at wavelengths of 8.4, 9.9, and 12.4 μm, with respective emissivity values of 0.60, 0.70, and 0.60. As can be noted, the transmittance is higher than 0.89 in the wavelength range of 1.5–7.5 μm, with a decreasing trend in the wavelength range of 7.5–14.2 μm as the thickness is increased. The lowest values of transmittance, with varying thickness, occur at 9.3 μm and 12.5 μm. It is observed that the reflectance increases with thickness, in the wavelength range of 7.4–14.2 μm, reaching its highest value of 0.7 at 9 μm.

As can be seen in Fig. 12.7a, the absorptance increases with increasing AlN thickness in the wavelength range of 7–14.2 μm, reaching a peak value of 0.68 at 11.2 μm. It is found that the transmittance is higher than 0.64 in the wavelength range of 1.5–10.2 μm; it shows a decreasing trend with increase in thickness for wavelengths longer than 10.2 μm. The corresponding reflectance is increasing with increase in thickness, in the wavelength range of 10.2–14.2 μm, reaching the highest value of 0.45 at 14.2 μm.

It is noted from Fig. 12.7b that for Si_3N_4, the absorptance increases, as the thickness is increased, in the wavelength range of 5–14.2 μm, reaching its highest peak value of 0.84 at 9.4 μm. An increase in the Si_3N_4 membrane thickness causes an increase in the peak amplitude and it shifts to lower wavelengths. It is found that the transmittance is higher than 0.5 in the wavelength range of 1.5–6 μm, with a trend of decrease in transmittance with increase in thickness for wavelengths longer than 6 μm.

In addition to the wavelength-dependent fundamental optical constants, n and k, the peaks in emissivity at different wavelengths (8–14 μm), for these materials, is due to the associated vibrational modes of molecules. The implication of these peaks and the corresponding wavelengths is that these materials are useful candidates in the manufacture of microbolometers.

In Tables 12.1–12.4, a summary of the total infrared radiative properties of the dielectric materials, studied in the spectral range considered (1.5–14.2 μm), is presented [55]. From these tables, it is observed that Si_3N_4 exhibits the highest total absorptance values throughout the varying thickness range. The results of the evolution of radiative properties, as a function of dielectric material thickness, suggest the

Table 12.1 Total value of radiative properties of Al_2O_3 (spectral range: 1.5–14.2 μm) [55].

Thickness (μm)	0.1	0.5	1	1.5	2	2.5	3
Absorptance	0.052	0.168	0.233	0.280	0.328	0.375	0.415
Transmittance	0.945	0.782	0.660	0.594	0.548	0.505	0.468
Reflectance	0.030	0.050	0.106	0.126	0.124	0.120	0.117

Table 12.2 Total value of radiative properties of SiO_2 (spectral range: 1.5–14.2 μm) [55].

Thickness (μm)	0.1	0.5	1	1.5	2	2.5	3
Absorptance	0.061	0.141	0.182	0.227	0.266	0.297	0.329
Transmittance	0.927	0.744	0.616	0.558	0.540	0.529	0.506
Reflectance	0.011	0.115	0.202	0.215	0.194	0.174	0.164

Table 12.3 Total value of radiative properties of AlN (spectral range: 1.5–14.2 μm) [55].

Thickness (μm)	0.1	0.5	1	1.5	2	2.5	3
Absorptance	0.065	0.167	0.206	0.235	0.264	0.289	0.311
Transmittance	0.925	0.702	0.588	0.566	0.549	0.519	0.493
Reflectance	0.010	0.124	0.205	0.199	0.188	0.191	0.197

Table 12.4 Total value of radiative properties of Si_3N_4 (spectral range: 1.5–14.2 μm) [55].

Thickness (μm)	0.1	0.5	1	1.5	2	2.5	3
Absorptance	0.170	0.316	0.351	0.462	0.545	0.567	0.589
Transmittance	0.796	0.413	0.319	0284	0.224	0.189	0.167
Reflectance	0.034	0.270	0.330	0.254	0.231	0.244	0.245

possibility to consider Al_2O and Si_3N_4 as viable candidates to be used as IR radiation absorbers in the fabrication of uncooled microbolometers in the infrared spectral range of 8–14 μm (atmospheric window).

These results demonstrate that Si_3N_4 exhibits the highest total emissivity values in the spectral wavelength range of 8–14 μm, throughout the varying thickness range. Therefore, Si_3N_4 is considered in this study as the IR radiation absorber material of choice in the design of the proposed uncooled microbolometer in the infrared spectral range of 8–14 μm (atmospheric window).

12.2.3 Dielectric materials as infrared filters

IR optical filters are used in many systems for military and commercial applications. These IR filters are essential components in the development of optical sensing systems that can be optimized by using materials better suited to the application. Materials such as calcium fluoride, germanium, optical glass N-BK7, and potassium bromide are the standard IR filter materials used by the IR industry, but at the microfabrication level, there is a need to replace them with new materials that are more compatible with the actual silicon-based microfabrication process.

Table 12.5 provides a detailed comparison of the cut-off wavelengths and IR transmittance of the dielectric materials that have been considered, with varying thickness, in the spectral range of 1.5–14.2 μm. The results show that these dielectric materials

Table 12.5 Band-pass range and transmittance for various dielectric materials considered as IR filters (spectral range: 1.5–14.2 μm; 30°C) [55].

Dielectric materials	Al_2O_3	SiO_2	AlN	Si_3N_4
Band-pass range (μm)	1.5–8.5	1.5–7.5	1.5–10.2	1.5–6.0
Transmittance	≥0.80	≥0.89	≥0.64	≥0.50

can be considered as IR band-pass filters to transmit certain wavelength band at wavelengths in the IR spectral range and to block others (by reflection and absorption) at longer wavelengths. From Figs. 12.6 and 12.7, a trend in suppressing the IR radiation at longer wavelengths, as the thickness of the dielectric materials is increased, is observed.

The results of the evolution of the radiative properties, as a function of dielectric material thickness, show the possibility to consider these dielectric materials as band-pass optical filters and absorber materials in the IR spectral range considered.

12.2.4 Silicon nitride as IR radiation absorber

Silicon nitride (Si_3N_4) is commonly used in the semiconductor industry as a passivation layer in integrated circuit (IC) fabrication, because of its ability to protect the electronics from the diffusion of water and alkali ions [56]. Silicon nitride is an excellent dielectric material [57] with a dielectric strength of the order of 10^7 V/cm and a reasonable thermal conductivity that is required to transfer heat from the microbolometer multilayer structure to the substrate [58]. Thin silicon nitride membranes are almost transparent in the infrared range except for a wide absorption peak exhibited around 12 μm [59]; this absorptance peak is due to the vibrations of the diatomic basis of silicon nitride and it grows in amplitude as the nitrogen content is raised [60]. An increase in the silicon nitride membrane thickness causes an increase in the peak amplitude and it shifts to lower wavelengths [61].

12.2.5 Microbolometer: Electrical interconnection and IR reflective layer

In this case study, aluminum (Al) is selected as an electrical interconnection material to the read-out integrated circuit (ROIC) because it provides an excellent Ohmic contact with the low doped p-Si thermosensing element [62]. When the device is heated to about 550°C, after the aluminum interconnection layer has been deposited on p-type silicon, aluminum diffuses into silicon yielding a highly conductive heavily doped p^+-region. Aluminum is commonly used as interconnect due to its low electrical resistivity ($\rho = 2.65$ μΩ-cm) that introduces a low noise contribution to the system [63]. Also, aluminum is easy to deposit, does not contaminate Si, and exhibits excellent adhesion to dielectrics.

The proposed microbolometer structure is suspended over a cavity with a gold (Au) layer deposited under it and on top of the substrate surface. Gold has a high reflectivity close to unity in the infrared region. The gold layer, in combination with the cavity thickness, creates a resonant cavity, enhancing the absorptance of the multilayer stack.

12.2.6 Optical constants of microbolometer materials

The microbolometer materials have been chosen based on their electrical, thermal, mechanical, and optical properties. The interaction of radiation with matter is determined by the wavelength of the incident radiation, the nature of the surface, and the optical properties of the material. The wavelength dependence of the index of refraction n and the extinction coefficient k describe the material optical properties and determine their radiative properties.

Silicon is almost transparent in the infrared wavelength range. The optical constants of the thermosensing p-type silicon layer, considered with a low doping concentration of 10^{16} cm^{-3}, are similar to that of undoped silicon. Fig. 12.8 shows the optical constants of silicon, in the infrared spectral range of 0–25 μm [64]. As can be seen in this figure, the extinction coefficient is negligible throughout the infrared spectral range. An increase in the silicon membrane thickness causes an increase in the absorptance of the incident radiation.

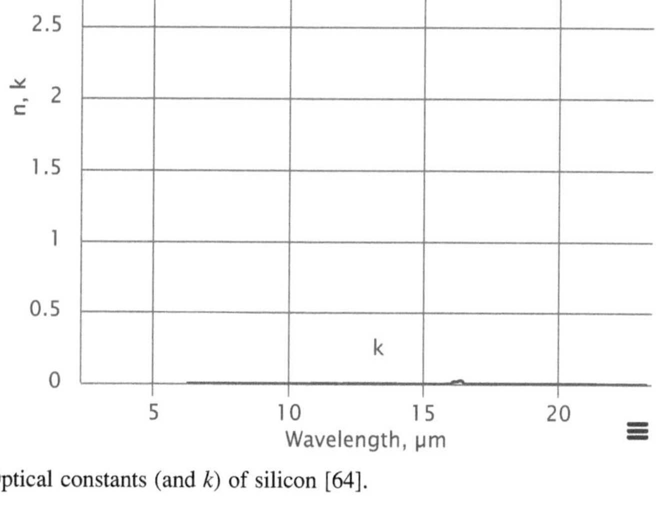

Fig. 12.8 Optical constants (and k) of silicon [64].

Thin silicon nitride membranes are almost transparent in the infrared range except for a wide absorption peak that is exhibited around 12 μm. Fig. 12.9 shows the optical constants of silicon nitride in the infrared spectral range of 0–15 μm. It is evident from this figure that the peak of the extinction coefficient is at 12 μm. This peak is due to the vibrations of the diatomic basis of silicon nitride [60]. An increase in the silicon nitride membrane thickness causes an increase in the absorptance of the incident radiation.

Metals with large electrical conductivities are characterized by a large reflectivity in the infrared spectral range. Gold and aluminum have a high reflectivity close to unity in the infrared region considered, which is suggested by large extinction coefficient throughout the infrared spectral range. Figs. 12.10 and 12.11 show optical constants of aluminum and gold in the infrared spectral range of 0–200 μm and 0–25 μm, respectively [66,67]. Gold is utilized as the infrared radiation reflective layer in order to enhance the radiation absorptance of the multilayer stack chosen in this work, because aluminum can be oxidized as a result of the microfabrication process, reducing its reflective properties in the infrared spectral range.

The total reflectivity values of aluminum and gold are predicted by Multi-Rad, as function of thin film thickness, in the spectral range considered. When the thin film thickness is larger than 0.1 μm, the total reflectivity is constant; the transmittance is negligible and the absorptance is less than 2%, as can be seen in Fig. 12.12.

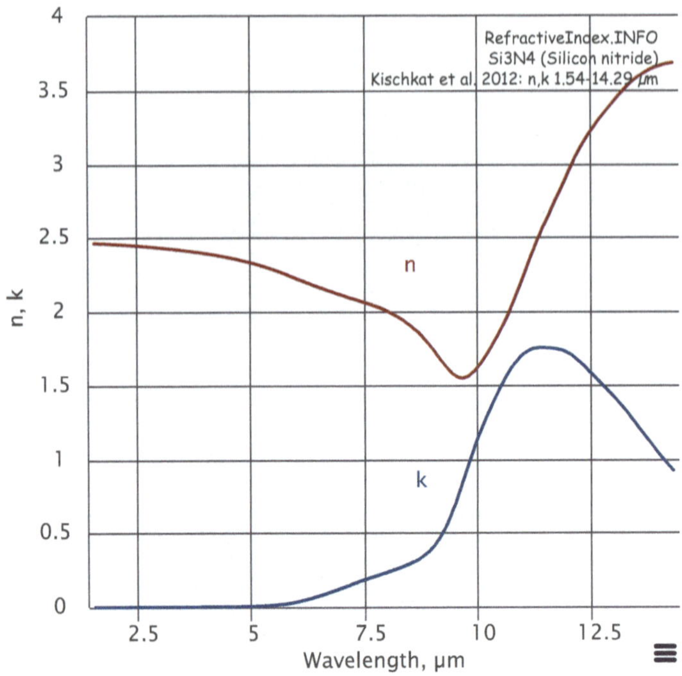

Fig. 12.9 Optical constants (and *k*) of silicon nitride [65].

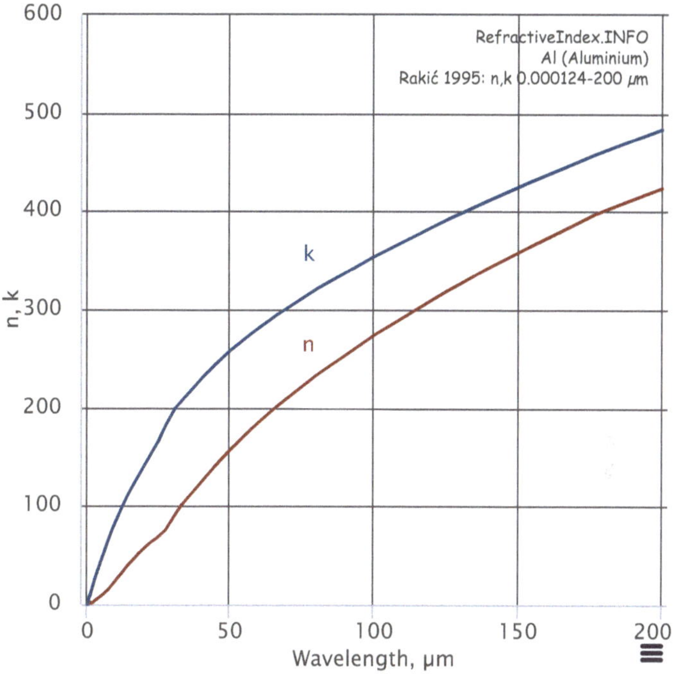

Fig. 12.10 Optical constants (and k) of aluminum [66].

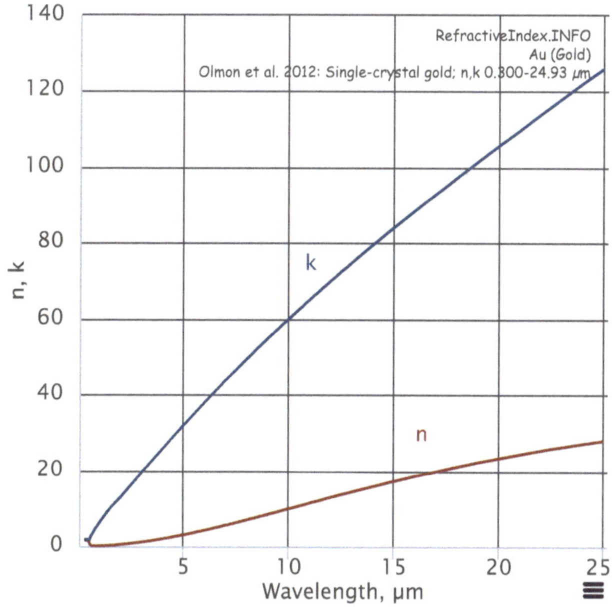

Fig. 12.11 Optical constants (and k) of gold [67].

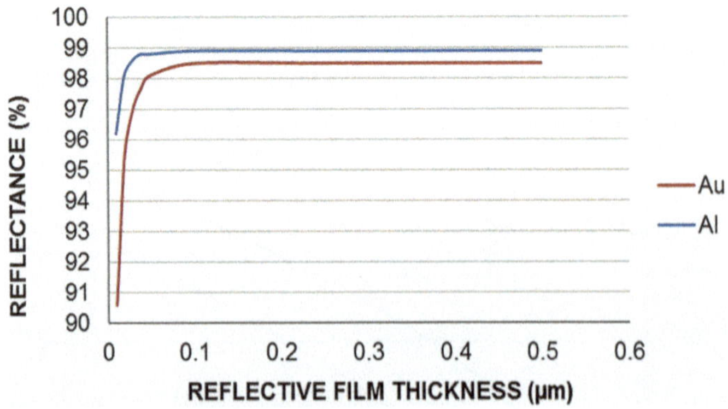

Fig. 12.12 Total reflectance of aluminum and gold.

12.3 Microbolometer structural design

In a standard infrared imaging system, single IR detector elements (pixels) are laid out in a two-dimensional array, called focal plane array (FPA). The microbolomer, proposed in this work, has the potential for applications in focal plane array imaging systems. The thermal isolation and efficient radiation absorption of single detector elements (microbolometers) are the key factors for high performance infrared focal plane array systems [68].

The uncooled microbolometer two-level structure, considered in this study, was originally developed and patented by Honeywell Inc., and licensed to several companies for the development and manufacture of focal plane arrays of uncooled microbolometers. The details of the Honeywell approach and its related intellectual property regarding uncooled microbolometers are described in this chapter. The use of pyroelectric materials such as barium strontium titanate (BST-$BaSrTiO_3$), and subsequently amorphous silicon, in the development of uncooled microbolometers, by Texas Instruments and its related technology are described in Chapter 13.

The performance of a microbolometer is determined by the material characteristics and its structural design parameters (structure geometry and dimensions) [69]. In order to obtain a high performance microbolometer, some conditions are required to be met in the design of the multilayer structure. These include the following:

- high temperature coefficient of resistance, TCR, of thermosensing material;
- efficient absorption of IR radiation by the absorber layer;
- low thermal conductance with surroundings;
- low thermal capacitance of pixel;
- fast response;
- low noise.

Efficient IR radiation absorption by the absorber layer, in combination with a thermosensing material with high TCR, provides a high responsivity of the microbolometer. A high TCR of the thermosensing material is key to the performance of the detector. In order to obtain a high TCR, a low doping concentration that leads to a high electrical resistivity is required. This increases the Johnson noise and is undesirable for the device performance. Consequently, a trade-off must be found between the TCR and the electrical resistivity of the detector.

One of the most important microbolometer design parameters is a low thermal conductance between the bolometer and its surroundings (important at low frequencies); this is the reason for conventional bolometers to be operated in a vacuum package to reduce the thermal conductance between the bolometers and their surroundings through the surrounding gas. The vacuum environment makes possible that the principal heat loss occurs by the conduction path through the support structure of the suspended membrane to the substrate. Microbolometers need to be packaged in vacuum with vacuum levels <0.01 mbar [70]. The support structure provides three functions: mechanical support, a thermal conducting path, and an electrical conducting path. The thermal conductance through the support arms can be minimal, reducing the cross-sectional area, increasing the length and using materials with low thermal conductivity. The required optimum thermal isolation of the microbolometer structure can be obtained in devices with the microbolometer multilayer structure suspended over a cavity [71].

In order to minimize the thermal capacitance (important at high frequencies) per unit of area of the microbolometer, the thermal mass of the multilayer structure must be reduced while maintaining a high absorption of the incident infrared radiation. The lower the thermal capacitance of the suspended structure, the lower is the thermal time constant response required by the imaging system [72].

The most important parameters of a thermal infrared sensor, for purposes of being utilized in an imaging system, are its high responsivity and short thermal time constant response. High responsivity is achieved by efficient absorption and a short thermal time response requires small heat capacity of the structure and a high thermal conductance to the substrate. When the thermal conductance decreases, the thermal time constant response increases and the responsivity decreases. When the thermal capacity decreases, the thermal time constant response decreases and the responsivity decreases. A desired high responsivity and fast thermal time constant response in a microbolometer cannot be achieved without compromise; a trade-off between these two critical performance parameters is needed depending on the application requirements of the microbolometer [73].

In the proposed two-level structure, in Fig. 12.13, an upper square-shaped multilayer stack is thermally isolated in vacuum and connected to the substrate through two legs, that act as structural support (Si_3N_4), thermal path and interconnection (Al) to the readout integrated circuit (ROIC) on the substrate. The thermally isolated structure consists of a suspended membrane (detector element) that is composed of three layers: Si_3N_4 layer as a passivation and antireflective element, low doped p-type silicon layer acting as a thermosensing active element and Si_3N_4 as an IR radiation absorber layer. The lower level (silicon substrate) consists of a thin layer of Au acting as an IR radiation reflection layer, deposited on a silicon dioxide (SiO_2) layer on top of the substrate (Si). The two multilayers are separated by a cavity (gap) that, in combination with the thin film of Au, creates a resonant cavity and increases the infrared radiation absorption in

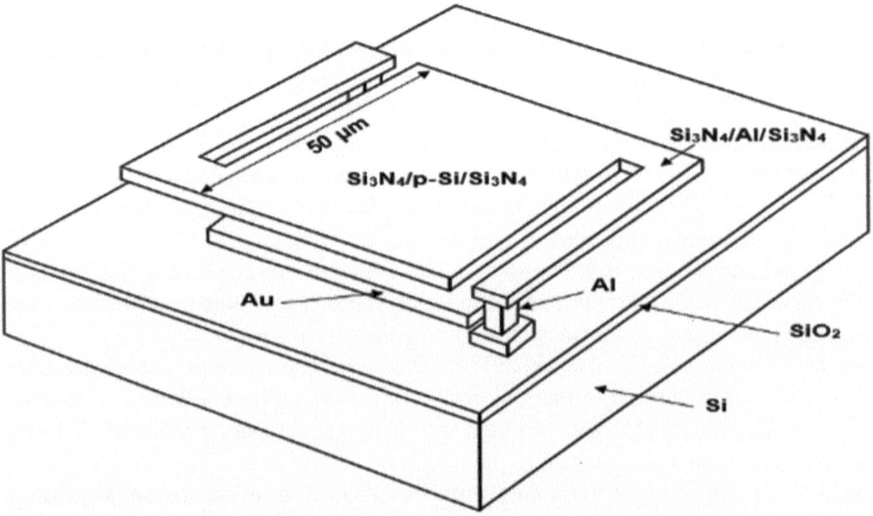

Fig. 12.13 Schematic of proposed microbolometer (two-level 50 × 50 μm^2).

the infrared wavelength range of interest. In order to enhance the IR radiation absorption in the microbolometer upper level membrane, a resonant optical cavity (Fabry-Perot) structure that is optimized for the targeted wavelength interval is required [74]. The two-level microbolometer thin film layer structure is shown in Fig. 12.13.

12.4 Optimization of radiative properties of layers in the microbolometer

Optical modeling methods for simulation of radiative properties of materials are essential tools that provide the optical performance characteristics of individual layers/multilayers that are required for consideration in the optimization and structural design approach of the microbolometer. In the proposed microbolometer, the radiative properties of the microbolometer multilayer structure are predicted using the matrix method of multilayers (Multi-Rad). The simulated results help to understand the material properties and integration process and, at the same time, reduce the fabrication costs.

The stacked layer structure of the proposed microbolometer is considered to be at room temperature (30°C), thermally isolated and suspended in vacuum, i.e., "free standing," as shown in Fig. 12.13. A standard blackbody is considered as the source of infrared radiation (30°C, zero transmission loss). The incident radiation angle is considered to be perpendicular to the surface of the stacked layer; the smaller increment in spectral wavelength is 0.1 μm, and the range of wavelengths in which the spectral radiative properties are calculated is 8–14 μm (atmospheric window).

The optical constants of the materials and their spectral dependence, considered in the fabrication (Al, Au, and Si_3N_4) of the multilayer stack, are from the literature

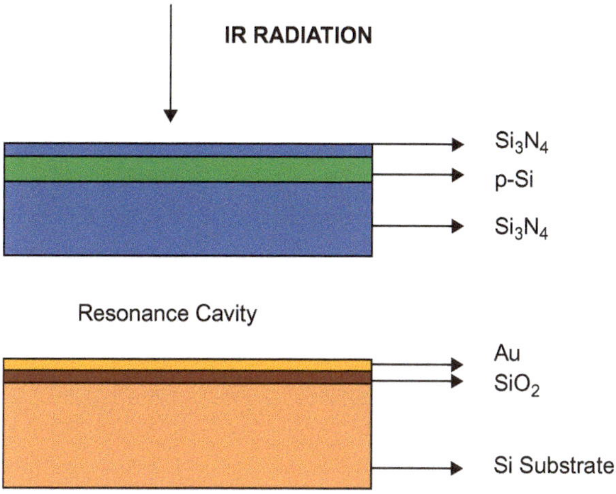

Fig. 12.14 Cross-sectional schematic of proposed microbolometer multilayer structure.

sources [60,75,76], and optical constants of Si and low doped p-Si are from the built-in library in Multi-Rad. All the material optical properties were compared and validated with experimental data available in the literature.

In the top-level structure, in Fig. 12.14, the upper Si_3N_4 layer, i.e., the passivation layer, is fixed at a thickness of 0.005 μm. The p-silicon layer is modeled with thickness of 0.097, 0.145, and 0.290 μm that correspond to a resistance value of 150, 100, and 50 k-Ohm (KΩ), respectively. The top level (pixel) is a square-shaped multilayer; as a result, the resistance of the pixel is a function of the thickness of p-Si. The thickness of the lower Si_3N_4 absorber layer and the vacuum cavity are optimized in order to achieve the highest infrared absorptance in the spectral range considered. In the bottom level structure, the thickness of the Au reflective layer is fixed at 0.1 μm to obtain the maximum possible reflectivity, the thickness of the SiO_2 layer is considered to be 0.005 μm, and the silicon substrate thickness is 525 μm. The values of the thicknesses of the various layers, considered in this study, are consistent with those of the commercially available bolometer structures.

In order to understand and interpret the optical properties of the proposed microbolometer structure, the optical properties of each component layer has been considered in detail. This analysis will further facilitate to understand the required enhancement of the IR absorptance throughout the infrared spectral range (8–14 μm).

In the simulation studies presented here, the absorptance optimization of the proposed structure is divided in two parts. First, the study of the optical properties of the top-level multilayer structure, i.e., "free standing" in vacuum, without the reflective effects of the bottom layers, as a function of the (lower) Si_3N_4 infrared absorber layer, is simulated as function of thickness. Second, the simulation of the optical properties of the entire multilayer structure, that consists of the top layers-vacuum cavity-bottom layers, as a function of the thickness of vacuum cavity, is performed. The optimization procedure has been implemented for each value of the p-Si thickness considered, in order to obtain the highest absorptance possible in each case.

Based on the above considerations, the total values of the optical properties, as a function of wavelength (8–14 μm), were simulated by Multi-Rad. In Figs. 12.15–12.17, the results of the simulated reflectance, transmittance, and absorptance, as function of Si_3N_4 thickness for the top layer (free standing) of the proposed bolometer structure, are presented.

One of the important parameters that determine the performance of a microbolometer is a low heat capacity. In order to minimize the heat capacity, the thermal mass of the pixel (top-level structure) must be reduced. In each case, the thickness of the Si_3N_4 IR radiation absorber layer is chosen to achieve the maximum absorptance, considering the effect of the Au reflective layer and the resonant cavity on top of the lower structure level, while keeping the absorber layer as thin as possible.

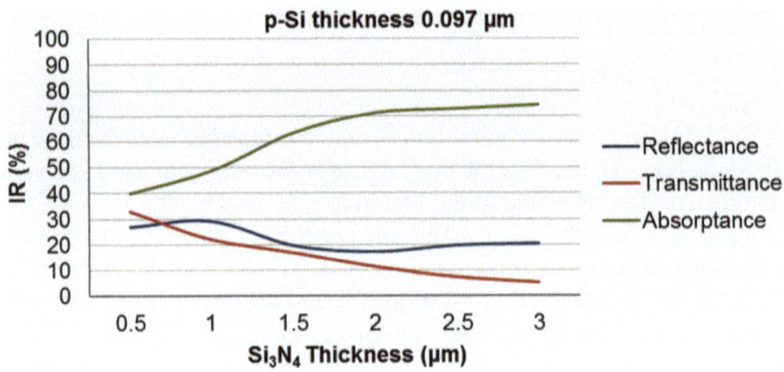

Fig. 12.15 Simulated total reflectance, transmittance, and absorptance as function of variable Si_3N_4 thickness for the top layer (free standing) of the proposed microbolometer structure (p-Si thickness = 0.097 μm).

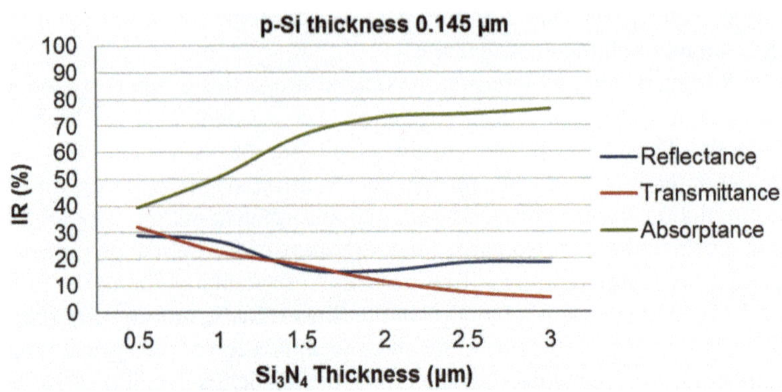

Fig. 12.16 Simulated total reflectance, transmittance, and absorptance as function of variable Si_3N_4 thickness for the top layer (free standing) of the proposed microbolometer structure (p-Si thickness = 0.145 μm).

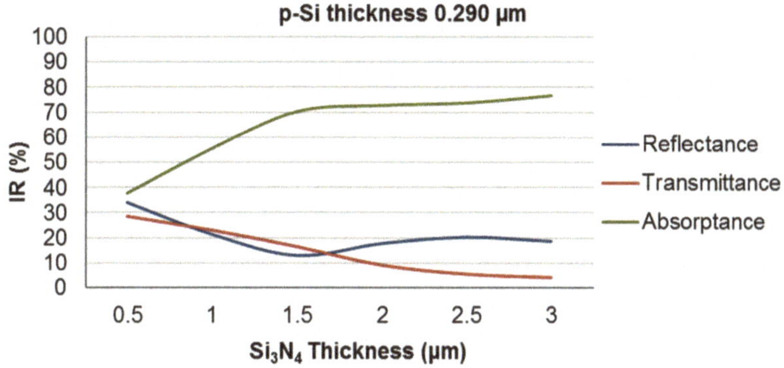

Fig. 12.17 Simulated total reflectance, transmittance, and absorptance as function of variable Si_3N_4 thickness for the top layer (free standing) of the proposed microbolometer structure (p-Si thickness = 0.290 μm).

As can be seen in Figs. 12.15 and 12.16, the optimum combination of optical properties corresponds to Si_3N_4 thickness of 2 μm and for the p-Si thickness of 0.097 μm and 0.145 μm. For the case with p-Si thickness of 0.290 μm (Fig. 12.17), the optimum combination of optical properties corresponds to Si_3N_4 thickness of 1.5 μm.

After the optimization of the thickness of the top layers, the vacuum cavity depth is optimized to establish optical resonance and maximize the IR radiation absorption. The simulation of the optical properties of the two-level multilayer structure, that consists of the top layers-vacuum cavity-bottom layers, as a function of thickness of vacuum cavity, is performed and the results are presented in Figs. 12.18–12.20.

In Figs. 12.18–12.20, the results of the simulated reflectance and absorptance, as function of cavity thickness for the proposed bolometer structure (two-level structure—Fig. 12.14), are presented. The gold layer reflects the radiation transmitted by the top-level multilayer structure with a negligible radiation transmitted through the lower multilayer structure. As can be seen in Figs. 12.18–12.20, the maximum

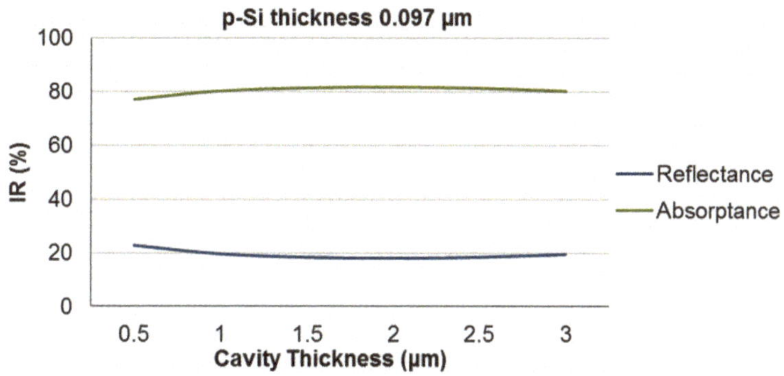

Fig. 12.18 Simulated total reflectance and absorptance as function of cavity thickness for the proposed microbolometer structure (two-level structure) (p-Si thickness = 0.097 μm).

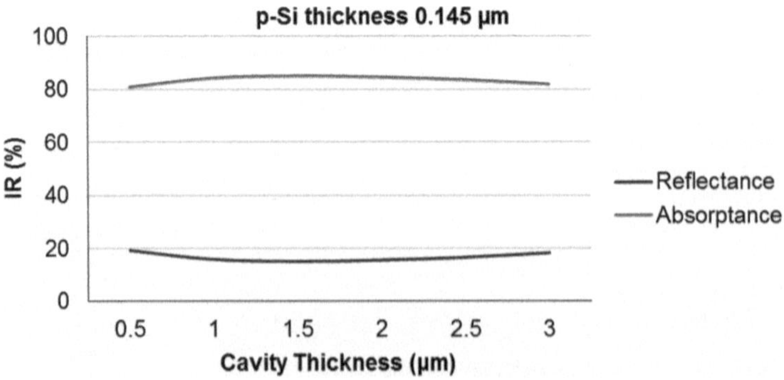

Fig. 12.19 Simulated total reflectance and absorptance as function of cavity thickness for the proposed microbolometer structure (two-level structure) (p-Si thickness = 0.145 μm).

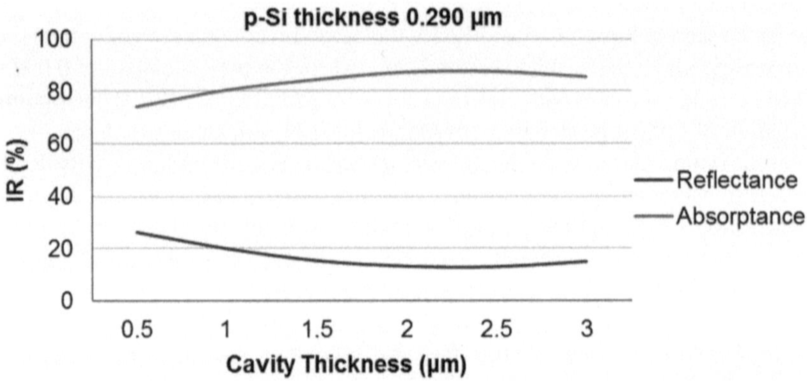

Fig. 12.20 Simulated total reflectance and absorptance as function of cavity thickness for the proposed microbolometer structure (two-level structure) (p-Si thickness = 0.290 μm).

absorptance corresponds to cavity thickness of 2.0, 1.5, and 2.5 μm for p-Si thickness of 0.097, 0.145, and 0.290 μm, respectively.

As can be expected, the optimum resonant cavity thickness is not equal to $\lambda/4$ at the center of the operating wavelength range, because the two-level microbolometer structure consists of stacked multilayers with different thickness and complex refractive indices. These multilayers act together as a complex structure of cascaded Fabry-Perot (FP) optical resonators, and thus radiative modeling is the most accurate way to determine the optimum resonant cavity thickness. Each stacked multilayer has a specific transfer matrix function, which depends on the complex refractive indices, extinction coefficients, absorptions, reflections, and transmissions of the different thin film layers.

Table 12.6 shows the optimized total IR radiation absorbed and reflected in the spectral range considered by the two-level multilayer structure for three fixed thicknesses of p-Si. The total absorptance is generally greater than 80% for the three different p-silicon thicknesses considered and the transmittance is negligible.

Table 12.6 Optical properties of proposed two-level multilayer structure (reflectance and absorptance total values).

p-Si thickness (μm)	Si_3N_4 thickness (μm)	Cavity thickness (μm)	Reflectance (%)	Absorptance (%)
0.097	2.0	2.0	18.0	82.0
0.145	2.0	1.5	15.1	84.9
0.290	1.5	2.5	12.8	87.2

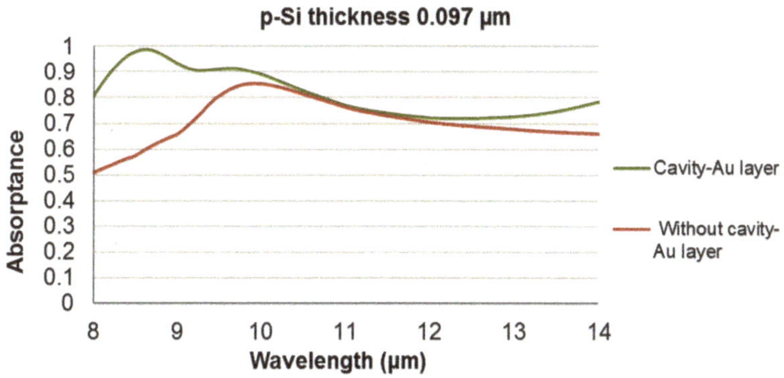

Fig. 12.21 Evolution of simulated optical spectra of the proposed two-level multilayer with/without cavity-Au layer (p-Si thickness 0.097 μm).

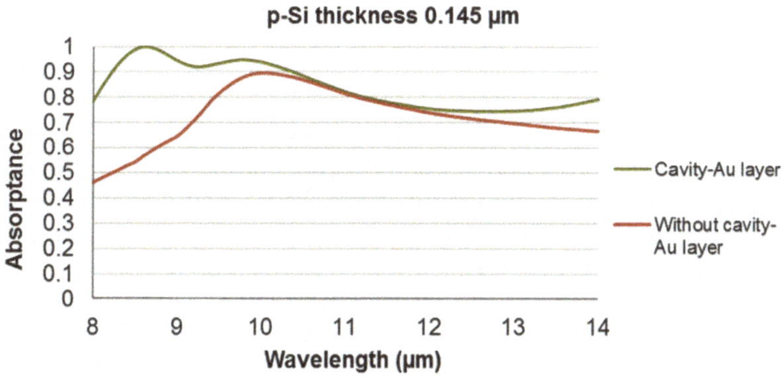

Fig. 12.22 Evolution of simulated optical spectra of the proposed two-level multilayer with/without cavity-Au layer (p-Si thickness 0.145 μm).

Figs. 12.21–12.23 show the evolution of the radiative properties of the proposed two level multilayer structure for p-Si thickness (0.097, 0.145, and 0.290 μm)—the top-level free-standing structure without the bottom level structure and with the optimized resonant cavity combined with the reflective gold layer deposited on top of the bottom level (substrate). As can be seen in Figs. 12.21–12.23, the resonant cavity in

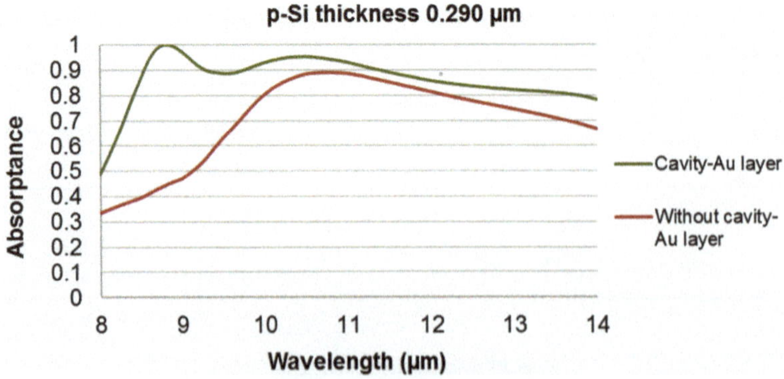

Fig. 12.23 Evolution of simulated optical spectra of the proposed two-level multilayer with/without cavity-Au layer (p-Si thickness 0.290 μm).

combination with the reflective gold layer enhances the infrared radiation absorptance throughout the spectral range considered, reaching the highest absorptance peaks at 8.8 μm and 10.5 μm. It is in this atmospheric spectral region where objects register maximum infrared radiation emissivity (λ_{max}=9.66 μm) at room-temperature (300*K*).

The optimized resonant cavity, in combination with the IR reflective gold layer, enhances the total absorptance of the free-standing multilayer by 15.28%, 15.83%, and 19.72% for p-Si thickness of 0.097, 0.145, and 0.290 μm, respectively, as can be seen in Table 12.7.

Simulations of the optical properties for the two-level stacked structure of Si_3N_4/p-Si/Si_3N_4/vacuum/Au/Si, in the considered spectral range, are presented in Figs. 12.24–12.26, considering different fixed thicknesses of p-Si (as in Table 12.7). In general, it can be observed that the optical spectra of the multilayer structure are in accord with the properties of Si_3N_4 in the long wavelength infrared, i.e., 8–14 μm; the absorptance is enhanced by the optimization of the resonant vacuum cavity thickness in combination with the Au reflective layer. The transmittance is negligible as a result of the high reflectivity of the Au layer on top of the Si substrate layer. In summary, in each case, the peak in the absorptance for the proposed bolometer structure, in the spectral range of 9–11 μm, is due to the vibrations of the diatomic basis of silicon nitride, and the peak at 8.5 μm is a result of the optimization of the vacuum resonant cavity. The maximum absorptance is achieved for p-Si layer with thickness of 0.290 μm that corresponds to the lowest resistance value (50 kΩ).

Table 12.7 Enhancement in absorptance with cavity/Au reflectance layer.

p-Si thickness (μm)	Absorptance without cavity/Au layer (%)	Absorptance with cavity/Au layer (%)	Increment in absorptance (%)
0.097	71.13	82.0	15.28
0.145	73.30	84.9	15.83
0.290	72.90	87.2	19.72

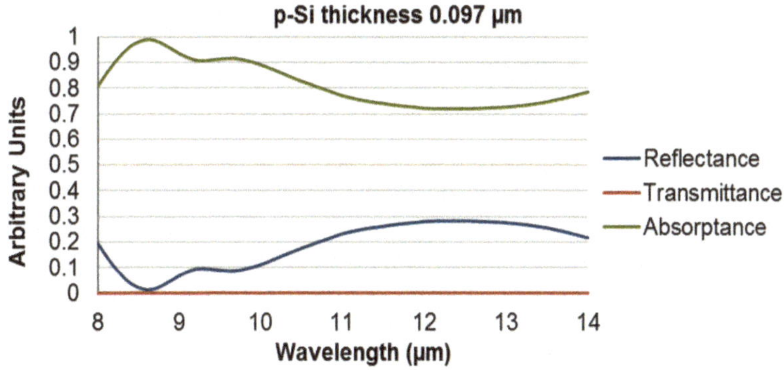

Fig. 12.24 Simulated optical spectra of Si_3N_4/p-Si/Si_3N_4/vacuum/Au/SiO_2/Si (p-Si thickness 0.097 µm).

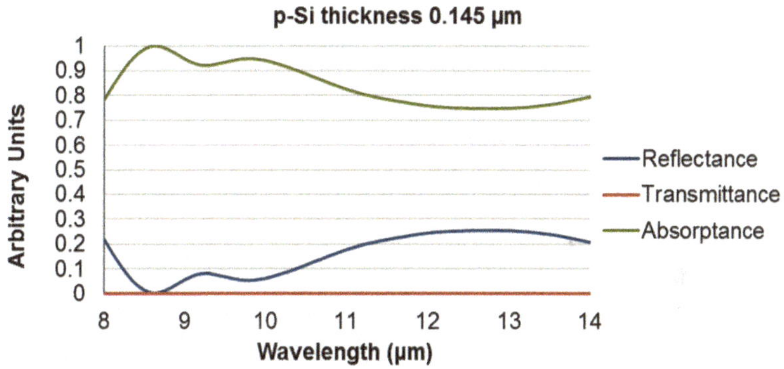

Fig. 12.25 Simulated optical spectra of Si_3N_4/p-Si/Si_3N_4/vacuum/Au/SiO_2/Si (p-Si thickness 0.145 µm).

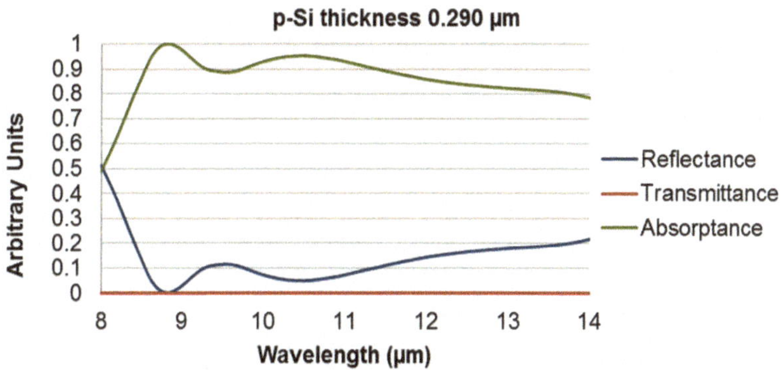

Fig. 12.26 Simulated optical spectra of Si_3N_4/p-Si/Si_3N_4/vacuum/Au/SiO_2/Si (p-Si thickness 0.290 µm).

12.5 Numerical simulations of performance of microbolometer

Numerical simulations provide predictions of a device performance based on the relation between material properties and geometric design. The performance of microbolometers is determined by the properties (optical, electrical, mechanical, and thermal) of the material used and the geometry and dimension of the structure.

From a device manufacturing perspective, the microbolometer IR absorption area (pixel) should be maximized in relation to the total area of the microbolometer. This parameter is quantified by the fill factor. The fill factor is the ratio of the IR radiation absorber area to the total area that occupy a single microbolometer (pixel). A high fill factor corresponds to a high efficiency of IR radiation energy absorption. Standard infrared microbolometer arrays typically have a fill factor between 60% and 70% [46,77].

The infrared imaging industry requires detectors with high responsivity, low noise equivalent temperature difference, and the capability to function at a high frame rate. These conditions can be met by optimizing the thermal conductance and thermal capacitance. The thermal conductance and the thermal capacitance are critical parameters and a trade-off is needed in the optimization of the microbolometer. Thermal capacitance is determined by the suspended multilayer structure and thermal conductance by the support arms to the substrate.

The performance of the microbolometer is dependent on thermal isolation. In order to maximize the thermal isolation, the thermal conductance through the support arms (heat link) between the pixel and the substrate should be minimized. Thermal conductance depends on the thermal conductivity and geometry (dimensions) of the support arms. Reducing the cross sectional area and increasing the length of the arms decreases the thermal conductance but the stability of the thermally isolated support structure must also be considered. At low frequencies, the detector responsivity is determined by the thermal conductance of the support arms. The thermal conductance is inversely proportional to the thermal time constant and to the responsivity of the detector. Therefore, an optimum geometry and dimensions of the support arms will allow to achieve the maximum possible responsivity with the thermal time constant required for applications in imaging. In the proposed microbolometer structure, the thermal conductance is determined by conduction through the considered heat links (arms) to the heat sink—in this case study, it is the combination of three layers (Si_3N_4/Al/Si_3N_4).

The thermal capacitance of the microbolometer is a function of its thermal mass. In order to minimize the thermal capacitance, its thermal mass must be reduced. At high frequencies, the responsivity is determined by the thermal conductance and the modulation frequency of the radiation. The thermal capacitance of the top-level multilayer is due to the combination of the three layers (Si_3N_4/p-Si/Si_3N_4) as shown in Fig. 12.13.

The optical modeling of the radiative properties of the proposed multilayer structure shows that the multilayer with p-type Si thickness of 0.290 μm achieves the highest total radiation absorptance (87.2%) with smallest thermal mass. This multilayer two-level structure is considered as the optimum choice in this modeling and numerical simulations.

The dimensions and materials of the proposed microbolometer structure have been determined in order to optimize the detector response for applications in infrared imaging. The proposed microbolometer (Fig. 12.13) square-shaped multilayer (pixel) dimensions are chosen to be 50 × 50 μm. This is consistent with the dimensions of commercial high-resolution imagers. The p-type Si thermosensing layer is encapsulated between two Si_3N_4 electrical insulation layers. Aluminum is used as interconnection between the p-type Si element and the ROICs on the substrate, due to its excellent Ohmic contact, low electrical resistivity and low noise contribution. The aluminum interconnection has a width of 4 μm and thickness of 0.1 μm; it is deposited over the lower layer of Si_3N_4 and the thermosensing element of p-Si to ensure the Ohmic contact, as shown in Fig. 12.27. Each support arm has a length of 55 μm with an area of 8.025 μm^2 (width = 5 μm and thickness = 1.605 μm). The fill factor of the proposed structure is 64.1%. The ROIC circuit is placed under the microbolometer membrane to increase the absorbing area of the array of microbolometer elements to conform to a focal plane array for imaging systems, maximizing thereby, the fill factor and the efficiency of the incident radiation absorption. Fig. 12.13 shows a schematic of the proposed microbolometer. Physical and thermal properties of the materials considered in the numerical calculations at 25°C [78] are summarized in Table 12.8.

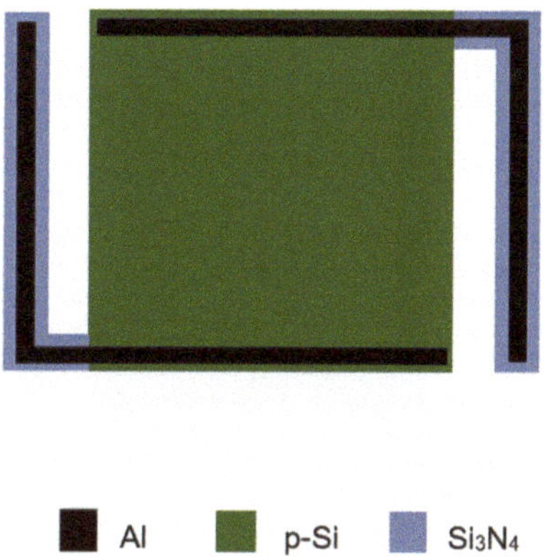

Fig. 12.27 Top-view schematic of the proposed microbolometer multilayer structure.

Table 12.8 Physical and thermal properties of materials used in the proposed microbolometer [78].

Material	Al	Si_3N_4	p-Si
Specific heat capacity ($Jg^{-1} K^{-1}$)	0.900	0.733	0.704
Density ($g\ cm^{-3}$)	2.7	3.1	2.3
Thermal conductivity ($Wm^{-1} K^{-1}$)	236	18.5	149

The figures of merit of the proposed microbolometer were calculated considering a blackbody radiation source in the infrared wavelength range of 0.8–14 μm, with an angular modulation frequency in the range: 0–10,000 Hz.

As discussed in earlier chapters, the operation of a bolometer is based on the change in electrical resistance of a thermosensing material as a function of temperature. The optical radiation absorber and the thermosensing material are connected thermally and electrically through a support structure to the substrate that acts as a heat sink. The basic representation of the thermal detector operation is shown in Fig. 12.28. The thermal detector with thermal capacitance C_{th} is connected by a thermal link of thermal conductance G_{th} to a heat sink, assuming that there is no other heat loss path [79].

In order to measure changes in its resistance, the bolometer needs to be connected to a DC circuit biased with a constant current source. The electrical circuit, commonly used to interface the bolometer into the reading system, is a voltage divider circuit configuration [80]. Fig. 12.29 shows the schematic of a voltage divider circuit. A bias voltage is applied across a load resistor and the bolometer, where V_{bias} (V) is the bias voltage applied to the circuit, R_b (Ω) is the bolometer resistance, and R_L (Ω) is the load resistance. The measurement readout electric circuit introduces Joule heating $I_b^2 R_b$ (W) as a source of heat to be considered, where I_b is constant current (A).

The bias voltage applied to the voltage divider circuit is consider to be 2, 4, and 6 V, and the load resistor is chosen to match the microbolometer resistance. Therefore, the voltage applied to the microbolometer is half the voltage bias applied to the voltage divider circuit (1, 2, and 3 V). The microbolometer is considered to operate in a vacuum package to minimize the thermal conduction between the microbolometer and the surroundings. Consequently, no transmission losses are considered by convection

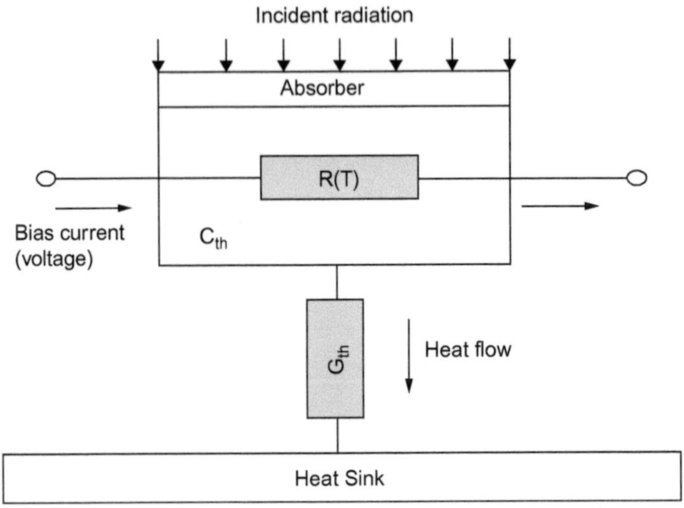

Fig. 12.28 Schematic diagram of a thermal detector operation [79].

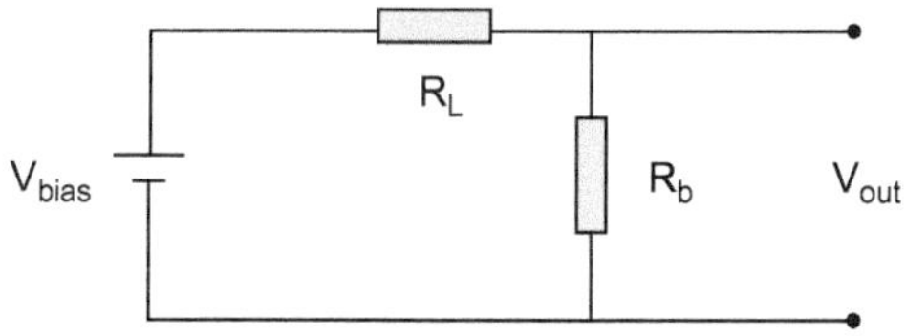

Fig. 12.29 Schematic of electrical biasing circuit for bolometer [80]. From A sub-circuit model of a microbolometer IR detector and its experimental validation; Raghvendra Sahai Saxena, Arun Panwar, Sukhvinder Singh, Lamba R.K. Bhan; Sensors Actuators A: Phys. 171(2), 2011, 138–145.

from the microbolometer. Johnson noise is considered as the predominant source of noise because the thermosensing element is single crystal silicon.

Microbolometer parameters such as thermal conductance, thermal conductivity, thermal time constant, detector noise bandwidth, and electrical noise depend on emissivity and absorptance of the multilayer stack. These parameters also depend on the geometrical and thermal properties of the materials and are independent of the angular frequency of modulation of radiation. Table 12.9 presents a summary of the numerical results of these parameters for p-Si with varying thickness. The figures of merit (responsivity, normalized detectivity, and noise equivalent power) of the microbolometer depend on various parameters, as detailed in Chapter 3, and are function of the radiation modulation frequency and the applied bias voltage. The values of responsivity and noise determine the performance of the microbolometer, calculated in terms of *NEP* and D^*. Johnson noise is a function of noise bandwidth and the detector resistance and is summarized in Table 12.9.

12.5.1 Figures of merit

Figures of merit have been calculated for the three two-level multilayer structures proposed, and the structure with p-Si thickness 0.290 μm, results in the highest values of R_v and D^*, and with the lowest values of *NEP* throughout the IR source modulation frequency considered. Figs. 12.30–12.32 show these figures of merit (p-Si thickness 0.290 μm) as function of modulation frequency of radiation source (0–1000 Hz) with a bias voltage applied to the microbolometer of 1, 2, and 3 V.

Table 12.9 Parameters of the proposed microbolometer.

p-Si thickness (μm)	0.097	0.145	0.290
Resistance R_{bol} (kΩ)	150	100	50
Thermal conductance G_e (W K^{-1})	1.001×10^{-5}	1.001×10^{-5}	8.357×10^{-6}
Thermal capacitance C_{th} (J K^{-1})	2.054×10^{-8}	2.054×10^{-8}	1.779×10^{-8}
Thermal time constant τ_e (ms)	2.052	2.052	2.129
Cut-off frequency f_c (Hz)	487	487	469
Noise bandwidth Δf (Hz)	77.543	77.546	74.754
Johnson noise V_j (V)	1.255×10^{-8}	1.300×10^{-8}	2.489×10^{-7}

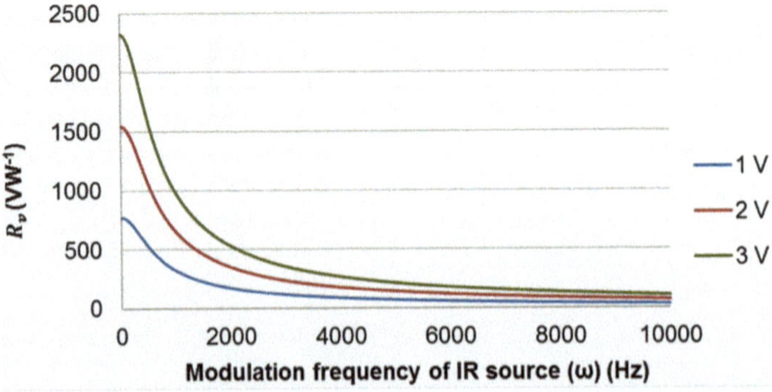

Fig. 12.30 Calculated R_v as function of applied bias voltage and modulation frequency ω of IR source (p-Si thickness 0.290 μm).

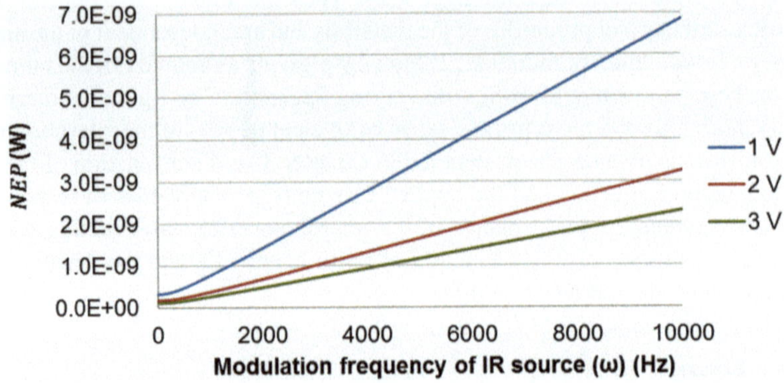

Fig. 12.31 Calculated *NEP* as function of applied bias voltage and modulation frequency ω of IR source (p-Si thickness 0.290 μm).

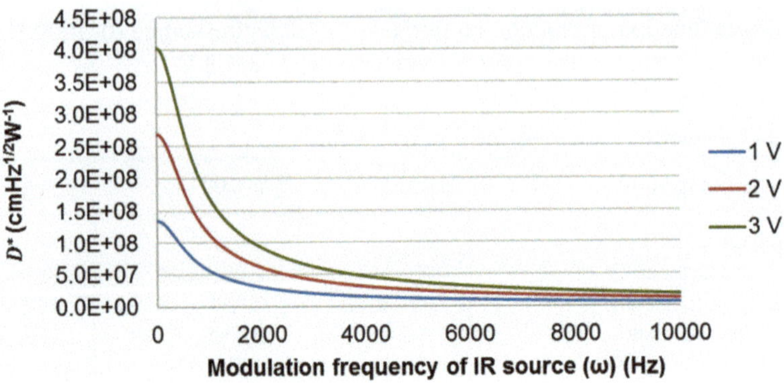

Fig. 12.32 Calculated D^* as function of applied bias voltage and modulation frequency ω of IR source (p-Si thickness 0.290 μm).

As can be seen in Fig. 12.30 and 12.32, the values of R_v and D^* show a decreasing trend as function of modulation frequency of the IR source radiation. At the cut-off frequency f_c (−3 dB frequency) of 469 Hz, the responsivity and normalized detectivity reach a value of 70% of their value when the IR source radiation modulation frequency is 0 Hz. It is evident that, at low frequencies, the detector responsivity is inversely proportional to the thermal conductance of the heat link; and at high frequencies, the detector responsivity is inversely proportional to the thermal capacitance and decreases inversely with the frequency of the optical signal.

The responsivity and normalized detectivity can be enhanced by increasing the bias voltage, but it will also increase Joule heating. The normalized detectivity attains values of 10^8 which is comparable to many commercial devices.

12.5.2 NETD *of microbolometer*

In the numerical calculation of *NETD*, the IR source is considered to be a blackbody source. The transmittance of the optics system τ_0 is considered to be 0.9 (Ge lens), and the atmospheric transmittance τ_a is considered to be lossless. Germanium is commonly used as IR filter, with a radiation transmittance of 0.9 over the wavelength range considered of 8–14 μm. The focal number of the optical detection system $f/_\#$ is considered to be 1, 2, and 3 that correspond to optics lens diameters of equal, half, and one-third of the focal length, respectively. The imaging system is shown in Fig. 12.33 (schematic diagram of optical system).

Figs. 12.34–12.36 show the results of the simulated *NETD* (p-Si thickness 0.290 μm), as function of source radiation modulation frequency and applied bias voltage for focal number equal to 1, 2, and 3 of the optical detection imaging system.

The results of *NETD* show that, for practical applications, the proposed two-level structures with p-silicon thickness 0.097 and 0.145 μm could be used as IR radiation sensing materials in the proposed microbolometer. Nevertheless, they do not conform to the requirements of the actual imaging industry that requires *NETD* lower than 100 mK.

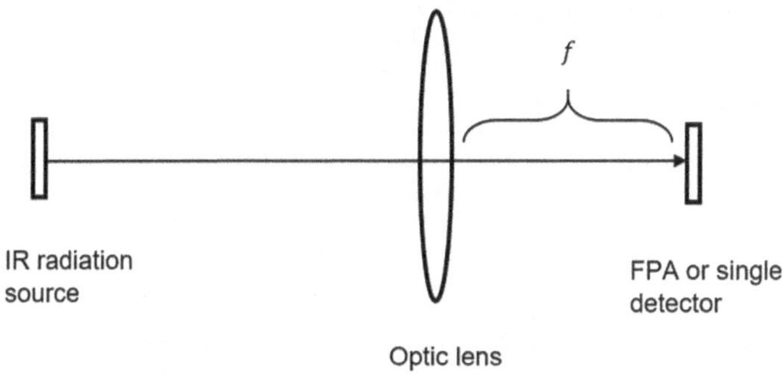

Fig. 12.33 Schematic diagram of optical system.

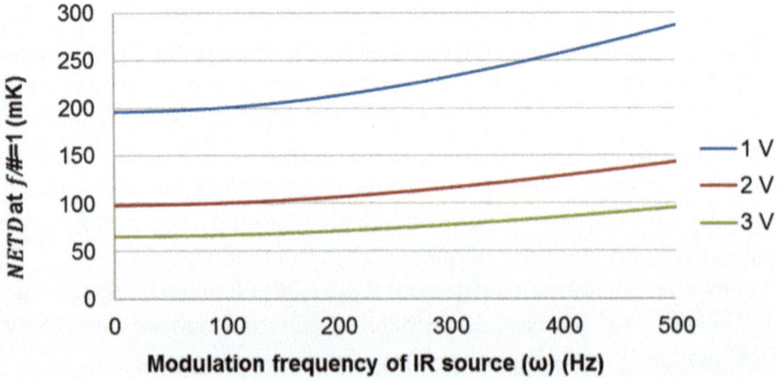

Fig. 12.34 Calculated *NETD* as function of applied bias voltage and modulation frequency ω of IR source at $^f/_{\#}$=1 (p-Si thickness 0.290 μm).

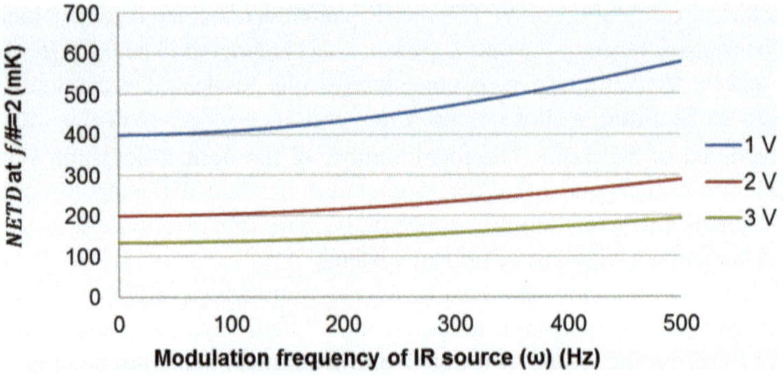

Fig. 12.35 Calculated *NETD* as function of applied bias voltage and modulation frequency ω of IR source at $^f/_{\#}$=2 (p-Si thickness 0.290 μm).

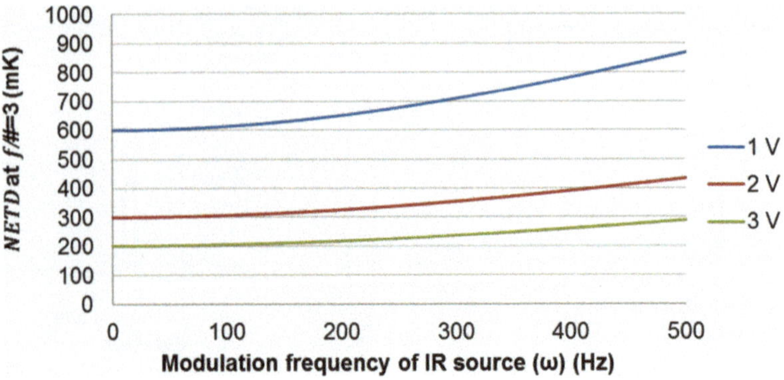

Fig. 12.36 Calculated *NETD* as function of applied bias voltage and modulation frequency ω of IR source at $^f/_{\#}$=3 (p-Si thickness 0.290 μm).

As is evident in Figs. 12.34–12.36, the microbolometer with thickness of p-Si of 0.290 μm exhibits an *NETD* of less than 100 mK, for a voltage bias of 3 V, $^{f}/_{\#}=1$, and for a source modulation frequency of 469 Hz. The thermal time constant of 2.129 ms allows the imaging system to operate at a standard frame rate in the range of 10–157 Hz.

12.6 Numerical calculations of improved microbolometer

In order to achieve *NETD* values lower than 100 mK, some of the parameters of the proposed two-level structure must be modified. Equation (12.1) defines *TD*, and to reduce its value, D^* of the detector must be increased.

A significant figure of merit of the infrared imaging system is the noise equivalent temperature difference *NETD* (mK) [81]; *NETD* is defined as

$$NETD = \frac{4\left(^{f}/_{\#}\right)^2\sqrt{\Delta f}}{\tau_a \tau_0 \sqrt{A_d} D^* \left(^{\Delta L}/_{\Delta T}\right)_{\lambda 1-\lambda 2}}, \tag{12.1}$$

where $^{f}/_{\#}$ is the focal number of the optical detection imaging system, A_d is the detector area, τ_0 is the transmittance of the optics system, τ_a is the atmospheric transmittance, and $\left(^{\Delta L}/_{\Delta T}\right)_{\lambda 1-\lambda 2}$ is the temperature contrast of total spectral radiance emittance at 300 K and represents the change in IR radiation power detected on the detector element per unit change in temperature of the source object within the spectral wavelength band under consideration.

D^* is proportional to R_v, as described in Eqs. (12.2) and (12.3). The thermal capacitance C_{th} is optimized and fixed; therefore, G_{cond} is the parameter that needs to be modified in order to reduce the *NETD*.

The detector output signal must be above the detector noise level that determines the minimum absorbed optical radiation level the detector can discern; the noise equivalent power *NEP* (W) is the signal power needed to achieve a signal-to-noise ratio of 1, and it can be expressed in terms of responsivity as [82],

$$NEP = \frac{V_n}{R_v}. \tag{12.2}$$

The normalized detectivity D^* (cmHz$^{1/2}$ W^{-1}) is the reciprocal of the noise equivalent power normalized to 1-cm^2 area and 1-Hz noise equivalent bandwidth and is defined as [82],

$$D^* = \frac{\sqrt{A_d \Delta f}}{NEP}, \tag{12.3}$$

where A_d (cm^2) is the area of the detector (area of radiation absorption) and Δf (Hz) is the detector noise equivalent bandwidth related to the modulation frequency of the radiation source. The normalized detectivity defines the resolving power of the detector and allows to compare detector performance with different sizes and technology.

By increasing the length of the support arms to the heat sink, the thermal conductance will decrease as can be seen from Equation (12.4); this increases the responsivity and as a final result, *NETD* decreases.

The thermal conductance by conduction G_{cond} is the heat conductance through the link to the heat sink and is given by

$$G_{cond} = \sum_i \frac{K_i A_i}{L_i}, \tag{12.4}$$

where K ($\mathrm{Wm^{-1}\,K^{-1}}$) is the thermal conductivity of the link, A ($\mathrm{m^2}$) is the area of the link, L (m) is the respective length of the link to the heat sink, the subscript i denotes the thermal conductance, and the dimensions of each thin film layer of the heat link.

In order to reduce the *NETD*, the proposed two-level structure has been modified by doubling the lengths of the arms to 110 μm, while the rest of the geometrical parameters are kept unchanged. The fill factor of this modified structure is 51.0%. Fig. 12.37 shows the schematic of the improved two-level microbolometer thin film layer structure. The new microbolometer parameters for each p-Si thickness considered are numerically calculated and summarized in Table 12.10. The thermal time response is less than 5 ms for the three cases of p-Si with varying thickness, allowing them

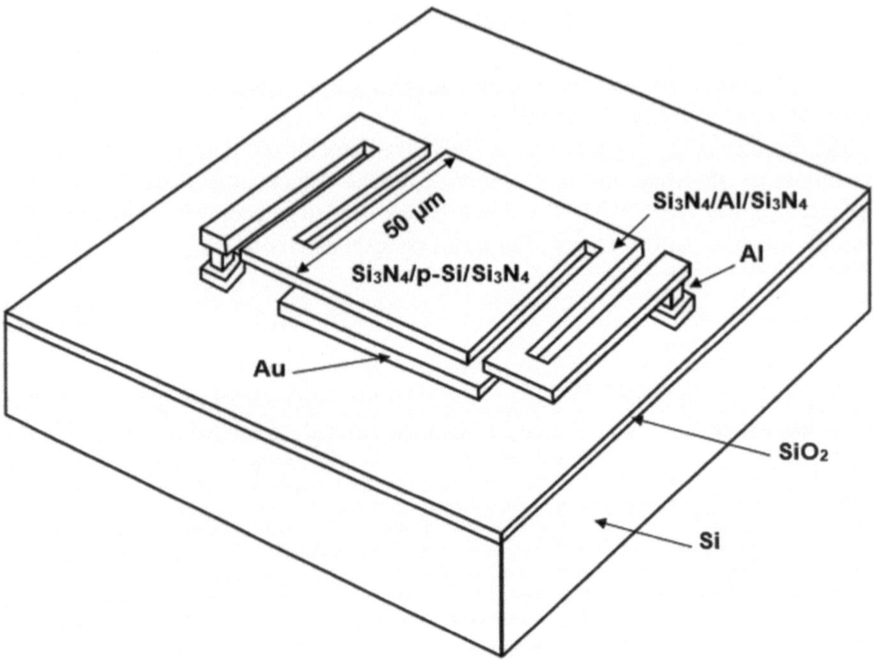

Fig. 12.37 Schematic of improved proposed microbolometer (two-level 50 × 50 μm^2).

Table 12.10 Parameters of modified (proposed) microbolometer.

p-Si thickness (μm)	0.097	0.145	0.290
Resistance R_{bol} (kΩ)	150	100	50
Thermal conductance G_e (W K^{-1})	5.101×10^{-6}	5.101×10^{-6}	4.261×10^{-6}
Thermal capacitance C_{th} (J K^{-1})	2.054×10^{-8}	2.054×10^{-8}	1.779×10^{-8}
Thermal time constant τ_e (ms)	4.027	4.026	4.176
Cut-off frequency f_c (Hz)	248	248	239
Noise bandwidth Δf (Hz)	39.53	39.53	38.12
Johnson noise V_j (V)	3.134×10^{-7}	2.559×10^{-7}	1.777×10^{-7}

to be implemented in the imaging system that operate at frame rates in the range of 10–80 Hz. Comparing Tables 12.9 and 12.10, it may be noted that G_e and Δf decrease with the increment in the length of the arms, while τ_e increased, as may be expected.

12.6.1 Figures of merit of improved microbolometer

The calculations of figures of merit for the modified structure show an increase in R_v and D^*, with decrease in *NEP* ; this is a significant improvement over the original proposed structure. The small increase in τ_e still allows the microbolometer to operate at frame rates that are in the standard commercial range of imaging systems. The changes in the figures of merit are a direct result of the increase in G_e.

Figures of merit were calculated for the proposed three two-level multilayer structures, and the structure with p-Si thickness of 0.290 μm results in the highest values of R_v and D^*, and with the lowest values of *NEP* throughout the modulation frequency of the IR source. Figs. 12.38–12.40 show the figures of merit as function of modulation frequency of the radiation source (0–10,000 Hz) with a bias voltage of 1, 2, and 3 V to the modified microbolometer with p-Si thickness = 0.290 μm.

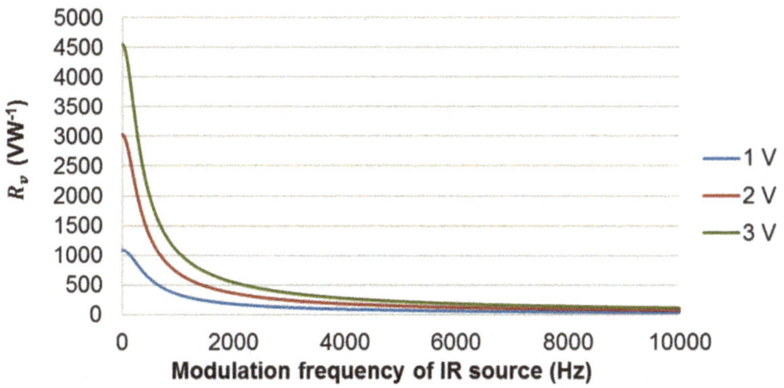

Fig. 12.38 Calculated R_v as function of applied bias voltage and modulation frequency ω of IR source (p-Si thickness = 0.290 μm).

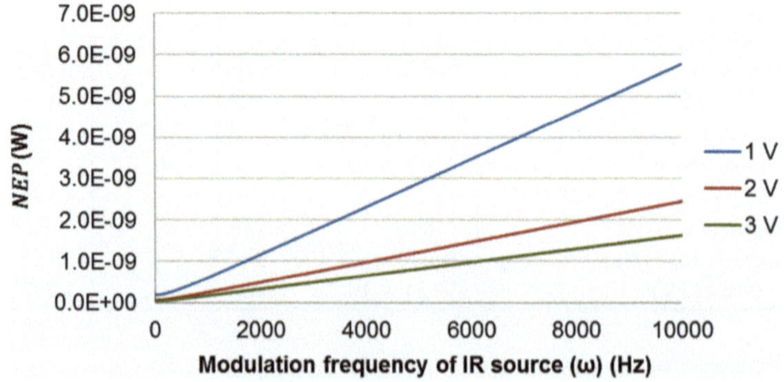

Fig. 12.39 Calculated *NEP* as function of applied bias voltage and modulation frequency ω of IR source (p-Si thickness = 0.290 μm).

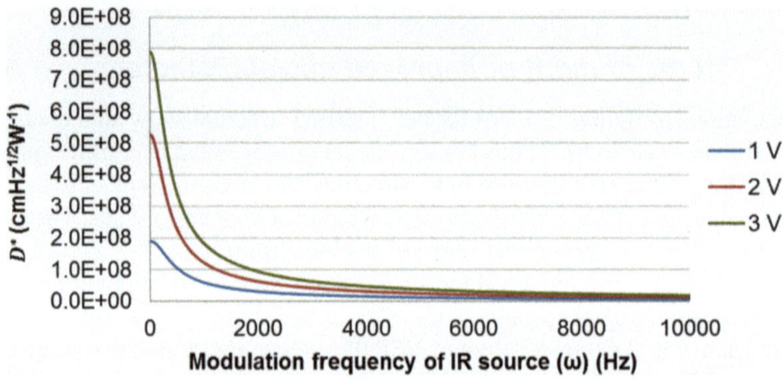

Fig. 12.40 Calculated D^* as function of applied bias voltage and modulation frequency ω of IR source (p-Si thickness = 0.290 μm).

12.6.2 NETD *of improved microbolometer*

Figs. 12.41–12.43 show the *NETD* for the improved microbolometer with p-Si thickness of 0.290 μm, as function of modulation frequency of the IR source and applied bias voltage to the microbolometer for focal number equal to 1, 2, and 3.

The *NETD* results in Fig. 12.41 show that at $f/\# = 1$ and at all bias voltages considered, the *NETD* is lower than 100 mK for sources with modulation frequencies under 239 Hz. Fig. 12.42 shows that at $f/\# = 2$ and at bias voltages of 2 and 3 V, the *NETD* is lower than 100 mK for sources with modulation frequencies under 239 Hz. Fig. 12.43 shows that at $f/\# = 3$ and at bias voltage of 3 V, the *NETD* is lower than 100 mK for sources with modulation frequencies under 239 Hz.

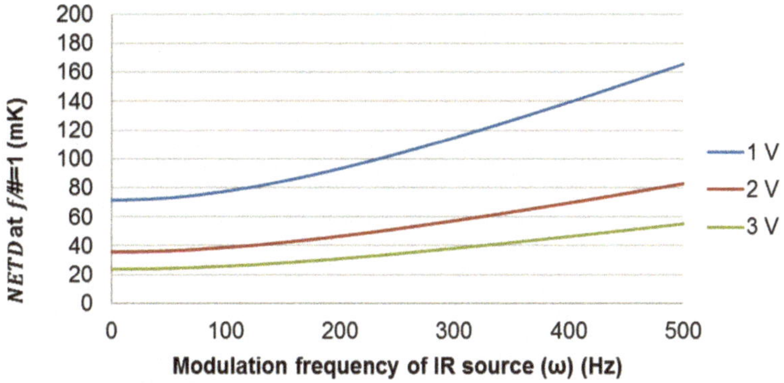

Fig. 12.41 Calculated *NETD* as function of applied bias voltage and modulation frequency ω of IR source at $f/_{\#}=1$ (p-Si thickness 0.290 μm).

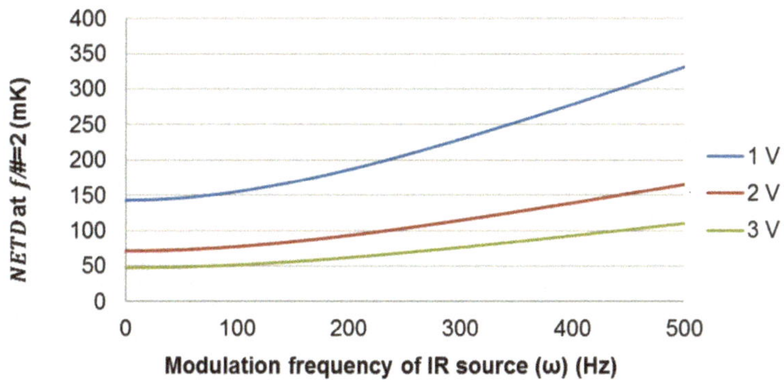

Fig. 12.42 Calculated *NETD* as function of applied bias voltage and modulation frequency ω of IR source at $f/_{\#}=2$ (p-Si thickness 0.290 μm).

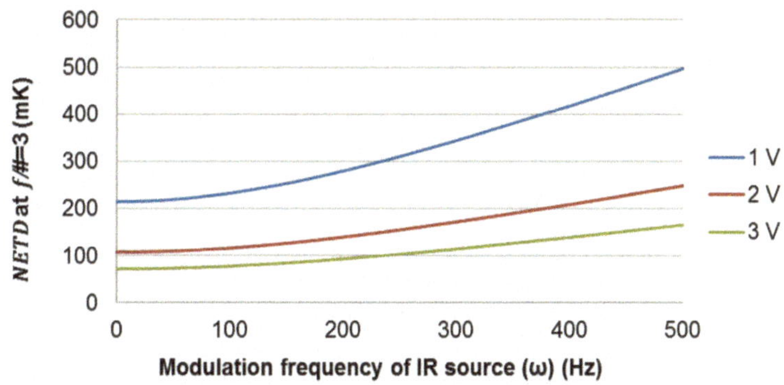

Fig. 12.43 Calculated *NETD* as function of applied bias voltage and modulation frequency ω of IR source at $f/_{\#}=3$ (p-Si thickness 0.290 μm).

12.7 Concluding remarks

In this chapter, a case study of p-Si as a thermosensing active element, in combination with Si_3N_4 layer as IR absorber material, has been proposed as potential candidates in the manufacture of uncooled IR microbolometers. The maturity of the silicon processing technology makes p-type silicon a viable material to be considered as a thermosensing element with easy integration and compatibility with standard CMOS fabrication processes.

An optical model-based simulated evolution of absorptance, transmittance, and reflectance, as function of thin film thickness, of various common dielectric materials has been presented. Results show that the optical properties of the dielectric materials considered, Si_3N_4, SiO_2, AlN, and Al_2O_3, exhibit thickness dependence throughout the spectral wavelength range (1.5–14.2 μm); the simulation further corroborates the high absorptance of Si_3N_4 in the wavelength range of 8–14 μm. This study provides wavelength-dependent radiative properties of common dielectrics to be considered and to facilitate in the design and fabrication of IR band-pass filters and lenses with applications in the infrared spectrum range of 1.5–14.2 μm.

A two-level suspended, thermally isolated, microbolometer square-shaped multilayer structure (Si_3N_4/p-Si/Si_3N_4/vacuum/Au/Si) is considered in this chapter. Low doped p-silicon is proposed as an alternative with a goal to reduce the standard high element (pixel) resistance, noise, and fabrication costs. Its radiative properties are modeled and optimized in the wavelength spectral range of 8–14 μm (atmospheric transmission window) at room temperature, utilizing a matrix method (Multi-Rad simulation package) of representing the optical properties. The proposed two-level multilayer structure is optimized as a function of thin film multilayer and resonance cavity thickness and exhibits a total infrared absorptance of more than 80% over the spectral range considered.

Results of the simulation show that the combination of a resonant cavity with a reflective gold layer results in the enhancement of the IR radiation absorptance with negligible IR radiation transmittance. Total values of the radiative properties, as a function of the wavelength, have been simulated and calculated. The performance of the proposed microbolometer is modeled by numerical calculations of figures of merit as function of applied bias voltage, modulation frequency of the radiation source (blackbody) and focal number of the optical system. Two microbolometer structures are proposed and compared obtaining, thereby significant improvements in the figures of merit for the modified microbolometer structure. Even though the low doped p-Si thermosensing layer has low TCR, the simulated normalized detectivity D^* achieved values of 10^8 are comparable with the state-of-the-art commercial devices. The *NETD* values lower than 50 mK at $f/\# = 1$ and the low thermal time constant τ_e of 4.176 ms show that the modified microbolometer structure is suitable for imaging systems with a video frame rate of 10–80 Hz. The results obtained are critical in the design and fabrication of uncooled two-level microbolometer structures. In summary, the simulated results of the radiative properties and numerical calculations of Figures of Merit suggest the possibility to consider low doped p-type Si and the proposed two-level multilayer structure in the design and fabrication of uncooled infrared imaging arrays with a reasonable performance and low cost.

References

[1] A.H. Pfund, The optical properties of metallic and crystalline powders, J. Opt. Soc. Am. 23 (1933) 375–378.

[2] L. Harris, et al., The preparation and optical properties of gold blacks, J. Opt. Soc. Am. 38 (1938) 582–589.

[3] W. Lang, et al., Absorbing layers for thermal infrared detectors, Sens. Actuators A 34 (1992) 243–248.

[4] F.J.J. Clarke, J.A. Larkin, Measurement of total reflectance, transmittance and emissivity over the thermal IR spectrum, Infrared Phys. 25 (1985) 359–367.

[5] F.J.J. Clarke, J.A. Larkin, Low reflectance coatings: platinum black for infrared technology, Platin. Met. Rev. 30 (1986) 21–22.

[6] J. Goldsmith, et al., Black aluminum: a novel anti-reflective absorbing coating, in: NAECON—IEEE National Aerospace and Electronics Conference, 2014, pp. 69–70.

[7] A.Y. Vorobyev, C. Guo, Femtosecond laser blackening of metals, in: 52nd IEEE International Midwest Symposium on Circuits and Systems, 2009, pp. 905–908.

[8] P.G. Maloney, et al., Emissivity of microstructure silicon, Appl. Opt. 49 (2010) 1065–1068.

[9] S.R. Marthi, et al., Optical properties of black silicon: an analysis, JOM 67 (2015) 2154–2159.

[10] N.M. Ravindra, et al., Modeling of optical properties of black silicon/crystalline silicon, J. Sci. Ind. Metrol. 1 (2015) 1–7.

[11] X. Liu, et al., Black silicon: fabrication methods, Properties and Solar Energy Applications, Energy Environ. Sci. 10 (2014) 1–95.

[12] R.R. Neli, et al., Development of process for far infrared sensor fabrication, Sensors Actuators A 132 (2006) 400–406.

[13] R.R. Neli, et al., Characterization of fast-response and low noise poly Si uncooled far infrared sensor, in: 2014 IEEE 5th Latin American Symposium on Circuits and Systems, 2014, pp. 1–4.

[14] C.N. Chen, W.C. Huang, A CMOS-MEMS thermopile with low thermal conductance and a near-perfect emissivity in the 8–14-μm wavelength range, IEEE Electron Device Lett. 32 (2011) 96–98.

[15] N. Nelms, J. Dowson, Goldblack coating for thermal infrared detectors, Sensors Actuators A 120 (2005) 403–407.

[16] S. Ilias, et al., Deposition and characterization of gold black coatings for thermal infrared detectors, Proc. SPIE 7750 (2010), 77501J-10.

[17] Y. Fang, et al., Preparation and characterization of ultra-low density porous gold foams, Mater. Lett. 67 (2012) 46–48.

[18] F. Forsberg, et al., Very large scale heterogeneous integration (VLSHI) and wafer-level vacuum packaging for infrared bolometer focal plane arrays, Infrared Phys. Technol. 60 (2013) 251–259.

[19] C. Lamsal, N.M. Ravindra, Simulation of spectral emissivity of vanadium oxides (VOx)-based microbolometer structures, Emerg. Mater. Res. 3 (2014) 194–202.

[20] A. Rogalski, Infrared detectors for the future, Acta Phys. Pol. A 116 (2009) 389–406.

[21] A. Rogalski, Recent progress in infrared detector technologies, Infrared Phys. Technol. 54 (2011) 136–154.

[22] P.W. Kruse, D.D. Skatrud, Uncooled Infrared Imaging Arrays and Systems, Semiconductors and Semimetals, Academic Press, 1997, ISBN: 978-0127521558.

[23] C. Chen, et al., Micromachined uncooled IR bolometer linear array using VO2 thin films, Int. J. Infrared Millimeter Waves 22 (2001) 53–58.

[24] C. Chang-hong, et al., Infrared responsivity of uncooled VO_2-based thin films bolometer, Acta Phys. Pol. A 50 (2001) 450–452.
[25] P. Wang, et al., High sensitivity 17 μm pixel pitch 640 × 512 uncooled infrared focal plane arrays based on amorphous vanadium oxide thin films, IEEE Electron Device Lett. 36 (2015) 923–925.
[26] S. Chen, et al., Characterization of nanostructured VO2 thin films grown by magnetron controlled sputtering deposition and post annealing method, Opt. Express 17 (2009) 24153–24161.
[27] S. Moon, Novel infrared absorbing material coupled uncooled microbolometer, Proc. IEEE 2 (2004) 658–660.
[28] P. Eriksson, et al., Thermal characterization of surface-micromachined silicon nitride membranes for thermal infrared detectors, J. Microelectromech. Syst. 6 (1997) 55–61.
[29] S. Chen, et al., Fabrication and performance of microbolometer arrays based on nanostructured vanadium oxide thin films, Smart Mater. Struct. 16 (2007) 696–700.
[30] F. Guinnetona, et al., Comparative study between nanocrystalline powder and thin film of vanadium dioxide VO2: electrical and infrared properties, J. Phys. Chem. Solids 62 (2001) 1229–1238.
[31] R.T.R. Kumar, et al., Study of a pulsed laser deposited vanadium oxide based microbolometer array, Smart Mater. Struct. 12 (2003) 188–192.
[32] S. Chen, et al., Characterization of nanostructure VO_2 ThAn films grow by magnetron controlled sputtering deposition and post annealing method, Opt. Express 17 (2009) 24153–24161.
[33] S.V. Grayli, et al., Room temperature deposition of highly sensitive vanadium oxide films for infrared light sensing applications, IEEE Sens. (2015) 1–4.
[34] H.J. Lee, et al., Wide-temperature (up to 100 °C) operation of thermostable vanadium oxide based microbolometers with Ti/MgF_2 infrared absorbing layer for long wavelength infrared (LWIR) detection, Applied Surface Science 547 (2021) 149142.
[35] M. Almasri, et al., Amorphous silicon two-color microbolometer for uncooled IR detection, IEEE Sensors J. 6 (2006) 239–300.
[36] X. Liu, et al., A novel uncooled a-Si microbolometer for infrared detection, in: 8th International Conference on Solid-State and Integrated Circuit Technology Proceedings, 2006, pp. 626–628.
[37] F. Simoens, et al., Active imaging with THz fully-customized uncooled amorphous-silicon microbolometer focal plane arrays, in: International Conference on Infrared, Millimeter, and Terahertz Waves, 2011, pp. 1–2.
[38] A. Heredia, et al., Low temperature electrical properties of a boron-doped amorphous silicon bolometer, Proc. Electrochem. Soc. (2003). Abs.881.
[39] J.-H. Su, et al., Temperature coefficient of resistance related to amorphous silicon/metal contact design of microbolometers, in: 3rd International Conference on Devices, Circuits and Systems, 2016, pp. 63–67.
[40] V.Y. Zerov, et al., Calculational modeling of the main characteristics of an uncooled linear microbolometer array, J. Opt. Technol. 7 (2004) 153–157.
[41] M. Boninsegni, et al., low-temperature bolometer array, Rev. Sci. Instrum. 60 (1989) 661–665.
[42] S. Eminoglu, et al., Low-cost uncooled infrared detectors in CMOS process, Sensors Actuators A 109 (2003) 102–113.
[43] D.S. Tezcan, et al., A low-cost uncooled infrared microbolometer detector in standard CMOS technology, IEEE Trans. Electron Devices 50 (2003) 494–502.

[44] D.S. Tezcan, et al., An uncooled microbolometer infrared detector in any standard CMOS technology, in: The 10th International Conference on solid-State Sensors & Actuators (Transducer'99), 1999, pp. 610–613.

[45] P. Wang, et al., High sensitivity pixel pitch uncooled infrared focal plane arrays based on amorphous vanadium oxide thin films, IEEE Electron Device Lett. 36 (2015) 923–925.

[46] M. Kohin, N. Buttler, Performance limits of uncooled VOx microbolometer focal plane arrays, Proc. SPIE 5406 (2004) 447–453.

[47] P.W. Kruse, Can the 300 K radiating background noise limit be attained by uncooled thermal imagers? Proc. SPIE 5406 (2004) 437–446.

[48] F. Niklaus, et al., Uncooled infrared bolometer arrays operating in a low to medium vacuum atmosphere: performance model and tradeoffs, Proc. SPIE 6542 (2007). 1M.1-1M.12.

[49] F. Niklaus, et al., Performance model for uncooled infrared bolometer arrays and performance predictions of bolometers operating at atmospheric pressure, Infrared Phys. Technol. 51 (2007) 168–177.

[50] N.W. Aschroft, N.D. Mermin, Solid State Physics, Saunders College Publishing, 1976, ISBN: 0-03-083993-9.

[51] W.F. Smith, Foundations of Materials Science and Engineering, second ed., WCB/McGraw-Hill, 1993, ISBN: 0-07-059202-0.

[52] B.S. Mitchell, An introduction to Materials Engineering and Science for Chemical and Materials Engineer, John Wiley & Sons, 2004, ISBN: 0-471-43623-2.

[53] W.M. Bullis, et al., Temperature coefficient of resistivity of silicon and germanium near room temperature, Solid State Electron. 11 (1968) 639–646.

[54] P. Norton, J. Brandt, Temperature coefficient of resistance for p- and n-type silicon, Solid State Electron. 21 (1978) 969–974.

[55] A. Bañobre, S.R. Marthi, N.M. Ravindra, Radiative properties of thin films of common dielectric materials in the IR spectral range of 1.5–14.2 μm: application to infrared imaging, JOM 70 (2018) 1267–1273.

[56] N. Maluf, An Introduction to Microelectromechanical Systems Engineering, Artech House, Inc., 2004, ISBN: 978-1580535908.

[57] K. Jhansirani, et al., Deposition of silicon nitride films using chemical vapor deposition for photovoltaic applications, Results Phys. 6 (2016) 1059–1063.

[58] S. Franssila, Introduction to Microfabrication, John Wiley & Sons, Ltd., 2010, ISBN: 9780470749838.

[59] P.J. Van Zwol, et al., Emissivity of freestanding membranes with thin metal coatings, J. Appl. Phys. 118 (2015) 1–5.

[60] J. Kischkat, et al., Mid-infrared optical properties of thin films of aluminum oxide, titanium dioxide, silicon dioxide, aluminum nitride, and silicon nitride, Appl. Opt. 51 (28) (2012) 6789–6798.

[61] S.F. Jutzi, et al., Far-infrared sensor with LPCVD-deposited low-stress Si-rich nitride absorber membrane—part 1. Optical absorptivity, Sens. Actuators A 152 (2009) 119–125.

[62] D.K. Schroder, D.L. Meier, Solar cell contact resistance—a review, IEEE Trans. Electron Devices 31 (1984) 637–647.

[63] J. Tao, et al., Electromigration reliability of tungsten and aluminum vias and improvements under AC current stress, IEEE Trans. Electron Devices 40 (1993) 1398–1405.

[64] Refractive Index Database, https://refractiveindex.info/?shelf=main&book=Si&page=ChandlerHorowitz, Accessed 12 February 2017.

[65] Refractive Index Database, https://refractiveindex.info/?shelf=main&book=Si3N4&page=Kischkat, Accessed 12 February 2017.

[66] Refractive Index Database, https://refractiveindex.info/?shelf=main&book=Al&page=Rakic, Accessed 12 February 2017.
[67] Refractive Index Database, https://refractiveindex.info/?shelf=main&book=Au&page=Olmon-sc, Accessed 12 February 2017.
[68] M. Alhussein, S.I. Haider, Simulation and analysis of uncooled microbolometer for serial readout architecture, J. Sens. 2016 (2016) 1–13.
[69] S. Gupta, et al., Design optimization of pixel structure for α-Si based uncooled infrared detector, Def. Sci. J. 63 (2013) 581–588.
[70] D. Murphy, et al., Performance improvements for VOx microbolometer FPAs, Proc. SPIE 5406 (2004) 531–540.
[71] M. Moreno, et al., Microbolometers based on amorphous silicon–germanium films with embedded nanocrystals, IEEE Trans. Electron Devices 62 (2015) 2120–2127.
[72] F. Sizov, IR region challenges: photon or thermal detectors? Outlook and means, Semicond. Phys. Quantum Electron. Optoelectron. 15 (2012) 183–199.
[73] F. Niklaus, MEMS-based uncooled infrared bolometer arrays—a review, Proc. SPIE 6836 (2007) 68360D (1–15).
[74] Q. Cheng, et al., Design of dual-band uncooled infrared microbolometer, IEEE Sensors J. 11 (2011) 167–175.
[75] A.D. Rakić, Algorithm for the determination of intrinsic optical constants of metal films: application to aluminum, Appl. Opt. 34 (1995) 4755–4767.
[76] R.L. Olmon, et al., Optical dielectric function of gold, Phys. Rev. B 86 (2012) 235147. 1–9.
[77] E. Mottin, et al., Uncooled amorphous silicon technology enhancement for 25 μm pixel pitch achievement, Proc. SPIE 4820 (2003) 200–207.
[78] G.T.A. Kovacs, Micromachined Transducer Sourcebook, McGraw-Hill, 1998, ISBN: 0-07-290722-3.
[79] A. Kosarev, et al., Thin film uncooled microbolometers based on plasma deposited materials, Can. J. Phys. 92 (2014) 570–575.
[80] J.W. Nilsson, Electric Circuits, Addison-Wesley Publishing Company, 1983, ISBN: 0-201-06238-0.
[81] J. Baars, M. Schulz, Performance characteristics, measurements procedures, and figures of merit for infrared planes arrays, Proc. SPIE 2470 (1995) 141–155.
[82] R.D. Hudson Jr., Infrared System Engineering, John Wiley & Sons Inc., 1969. ISBN: 471-41850.

Present and future trends

13

13.1 Present and future trends in infrared detectors and infrared imaging

Infrared radiation detectors and infrared imaging systems are used in a wide variety of security, surveillance, military, civilian, and commercial applications. Their ability to detect thermally emitted radiation makes them well suited for night vision, detection, tracking, search and rescue, nondestructive evaluation, gas analysis, traffic monitoring, automotive applications, biomedical imaging, surveillance, process monitoring and control, temperature measurements, 3D printing, and condition monitoring, to name just a few of the growing list of uses of this versatile technology [1]. Table 13.1 presents a summary of the various examples of applications of infrared detectors and infrared imaging technology.

There are two main types of detectors that can be classified by their infrared radiation absorption mechanisms: photon detectors and thermal detectors. Photon detectors depend on the wavelength of the incident radiation. Photon detectors are devices that detect optical radiation by direct interaction of individual photons with active charge carriers within the atomic lattice of the detector material [2]. The photodetector response is proportional to the number of photons absorbed. Thermal detectors are wavelength independent because the signal does not depend on the nature of the radiation. Photon detectors require a cooling system, while most thermal detectors operate at room temperature even for the incident far infrared wavelength radiation [3]. The cooling systems of infrared photon detectors reduce the thermal generation of charge carriers, but add to the system bulk, weight, and tend to increase the total system cost.

Typically, infrared detector systems are characteristically classified as either cryogenically cooled or uncooled [4]. Cryogenically cooled infrared radiation detectors and cameras based on photon detectors—such as indium antimonide (InSb) and mer cury cadmium telluride (HgCdTe or MCT) arrays—are many times more sensitive than uncooled microbolometer-based thermal IR imagers. On the other hand, the lower cost and reduced SWaP (size, weight, and power) of cameras, based on uncooled thermal infrared detectors, result in systems that are suitable for low-cost applications required by low video-rate surveillance and short-focal-length optics [5].

The infrared wavelength range represents an important focus of military and defense research and development because much of surveillance and targeting occurs under the cover of darkness. New developments in infrared technology, including sensors fabricated using novel materials, decreased pixel size, improved cooling systems and vacuum technology, makes imaging process possible in total darkness. Infrared imaging is used in many defense applications to enable high-resolution vision and identification in near and total darkness. Thanks to recent advancements, companies, and government laboratories, that focus on optics and optical components, are continuing to improve low-light-level vision, identification capability, power

Microbolometers. https://doi.org/10.1016/B978-0-08-102812-4.00005-X

Table 13.1 Examples of applications of infrared detectors and infrared imaging.

Application	Parameters measured/end results
Infrared industrial imaging	• Excess electrical current; electrical damage or failure; failure analysis; nondestructive evaluation
Process monitoring and control	• Fill levels (solid, liquid, gas) • Material flow, materials processing, leaks
Maintenance	• Excessive wear and tear, predictive maintenance; hot spots that lead to mechanical malfunctions
Thermography	• General thermography/machine vision/inspection; firefighting • search and rescue; medical/physiotherapy/health/cancer-related applications; • thermofluid dynamics; in-flight velocity and temperature measurement of cold spray particles; monitoring environment; turbines, windmills; solar cell panels; rechargeable batteries/power supplies/transformers, ultracapacitors; space and aeronautics; homes/buildings, bridges and structures; semiconductor devices, circuits, and systems; thermal management; real-time spatial and temporal measurement of temperature in semiconductor processing
3D Printing	• Real-time spatial and temporal measurements of temperature
Small UAV/portable applications; autonomous vehicles	• Perimeter control; law enforcement; autonomous sensors
Military/defense related applications	• Thermal weapon sight, portable vision enhancement, vehicle vision enhancers, remote weapon stations; unmanned aerial vehicles (UAV), unmanned ground vehicles (UGV), unmanned ground seekers (UGS), missile seekers

conservation, minimal weight, and system cost [6]. From a fundamental scientific interest point of view, astronomy continues to be one of the prime users of infrared imaging technology [7]. Due to its noninvasive nature, interest in early detection of cancer [8,9] using infrared imaging, as well as intraoperative near-infrared imaging [10] is on the rise.

Detection is only the first step of the military's surveillance and reconnaissance technology. Recognition is the second step; it is the ability to discriminate between a human being and another living being. The third step, identification, involves being able to discern whether a person is friend or foe, which is key in advanced infrared imaging technology today. Long-wavelength infrared imagers, which sense thermal signatures, provide excellent detection capability in low-light-level conditions. However, they do not provide enough information. Conventional long-wavelength IR imagers enable soldiers to detect targets from very far distances, but they cannot identify them. In order to perform this task, users require visible or infrared short wavelengths, which detect ambient light reflected off the object. These technologies use detector arrays to sense the reflected light and enable easier recognition and identification of distant objects from features such as clothing on humans or the structural details of a truck [6].

The next-generation infrared systems for applications in the defense/military sector are transitioning to 12 and 8 μm pitch. The 12-μm-pixel-pitch UFPA (uncooled focal plane array) provides sensor systems with SWaP savings as well as cost advantages over existing devices. In order to meet the market demand, DRS has improved its production facilities to accommodate 17-μm-pixel detector manufacturing with an integrated two-color imaging system by combining a VO_x microbolometer (for 8 to 14 μm) and InGaAs (0.7 to 1.6 μm) detectors into a single focal plane array [6]. The evolution in new semiconductor technologies, time to market and the associated impact on cost reduction is shown in Fig. 13.1 [11].

Imaging in the infrared can involve a wide range of detectors or sensors. Indium gallium arsenide (InGaAs) and germanium (Ge) are commonly used in cooled photon infrared detectors [12]. Uncooled microbolometers can be fabricated from vanadium oxide (VO_x) or amorphous silicon. However, these semiconductor materials are expensive; for example, a glass lens for visible imaging may cost $ 100. However, a Ge lens in the infrared spectrum range will cost $ 5000. Reducing the pixel pitch, while maintaining performance, is of the outmost importance. Currently, sensors with 12-μm and 8-μm pixel pitch are in the development stage at several companies, and they have even been demonstrated at SWIR, MWIR, and LWIR wavelengths using mercury cadmium telluride (HgCdTe or MCT and indium antimonide (InSb) with simple readout integrated circuits. A nonexhaustive list of companies pursuing 15-μm pitch sensors includes Raytheon (Waltham, Mass., U.S.A.), Goodrich/Sensors

Impact on Cost Reduction

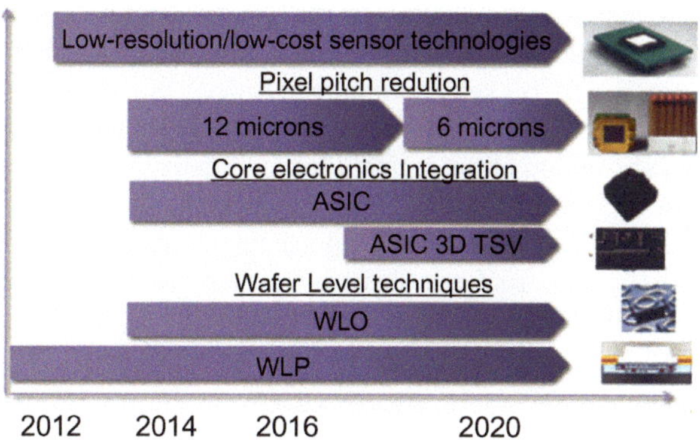

Fig. 13.1 New technologies' time-to-market and impact on cost reduction [11].

Unlimited (Princeton, N.J., U.S.A.), DRS Technologies (Parsippany, N.J., U.S.A.), AIM INFRAROT-MODULE GmbH (Heilbronn, Germany), and Sofradir (Châtenay-Malabry, France). The technology has come a long way in a short time to improve performance, reduce noise, and array size, but many barriers remain [6].

The requirement for small size, weight, and power (SWaP) sensor configurations is spurring the development of new detector technologies that promise to be less expensive to manufacture, offer increased capabilities, and operate at warmer temperatures compared to traditional infrared focal-plane arrays (FPAs) that are cooled to liquid-nitrogen temperatures [12]. Infrared detector/imaging system manufacturing companies have the challenge to come up with lower-cost solutions that perform well, and this will require innovation.

Advances in thermal infrared focal plane array (FPA) technology have resulted in a new generation of infrared detectors/imagers systems that offer increased capability while operating at higher temperatures than traditional cooled infrared systems. These developments have led to infrared cameras that are more versatile, require less power, and feature reductions in size, weight, and cost, allowing their use in an ever-widening range of critical applications [12].

13.2 Microbolometer-based infrared detector technology: Challenges and trends

Thermal infrared imaging and uncooled microbolometer thermal detectors have been the subject of intense research during the last decade. The choice of the detector material is highly application specific, depending on a myriad of factors such as intended absorption bandwidth, detector noise, detector thermal time constant, pixel size, cooling requirements, and detector uniformity. Uncooled microbolometers are receiving significant attention as they can be operated at room temperature and have lower cost, lower weight, wider spectral response, compared to their cooled photon detector counterparts that have a significant cost, power, and weight in cooling requirements.

The operation of a microbolometer IR imager starts with the absorption of IR energy that is incident on it which heats up the bolometer and causes a change in resistance in its thermistor material accordingly [13]. For higher performance imagers, the microbolometers should be kept thermally isolated from their surroundings; they should exhibit a larger change in resistance with respect to temperature as well as exhibit efficient absorption of the incident infrared radiation with low thermal capacity. The readout electronics of the imager is tasked with converting the resistance change to an electrical signal, such as a voltage or a current depending on the readout architecture. The signal from the detector is then amplified and is ready to be converted by an analog-to-digital converter for processing the digitized video information [14, 15].

An infrared thermal camera is a thermal system that converts infrared (IR) radiation into visible image. Fig. 13.2 shows the main components of an infrared camera. Camera cores are the modules that include imagers along with the electronics and

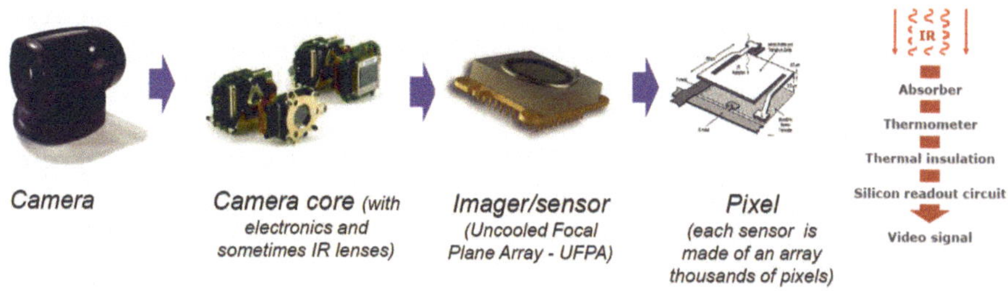

Fig. 13.2 Main components of an infrared camera [16].

sometimes lenses. Without lenses, it is also called a video engine. Imagers are also referred to as UFPA (uncooled focal plane arrays) or IR sensors. Detectors provide only a detection signal and not an actual image [16].

Crucial for the development and manufacturing of low cost uncooled microbolometer infrared detectors, but challenging is the reduction in microbolometer pixel size, the use of more efficient thermosensing elements and infrared absorber materials, and the development of wafer level packaging technology. The development of high performance and low cost uncooled infrared cameras require the reduction of the pixel size to be <12 μm, a detector noise equivalent temperature detectivity (NETD) as low as 50mK and the infrared system should operate shutterless with a high speed of operation.

Advancements in pixel technology allows for smaller pixels. Smaller pixels allow the increase in the number of devices per wafer, reducing the cost of the infrared array system and the cost of optical lens. Fig. 13.3 shows the evolution of the microbolometer pixel size [16] over the last 25 years.

Vanadium oxide (VO_x) and amorphous silicon (αSi) are the standard thermosensing materials that are used by the infrared detector industry. These materials exhibit high values of temperature coefficient of resistance (TCR), with a high sensitivity to the temperature changes. However, VO_x is not a standard material in IC fabrication process technology, has unstable phases and poor reproducibility; and αSi exhibits high values of electrical resistivity and pixel resistance in the order of MΩ (mismatch with electronic circuits) [17]. The disadvantage of these materials is that

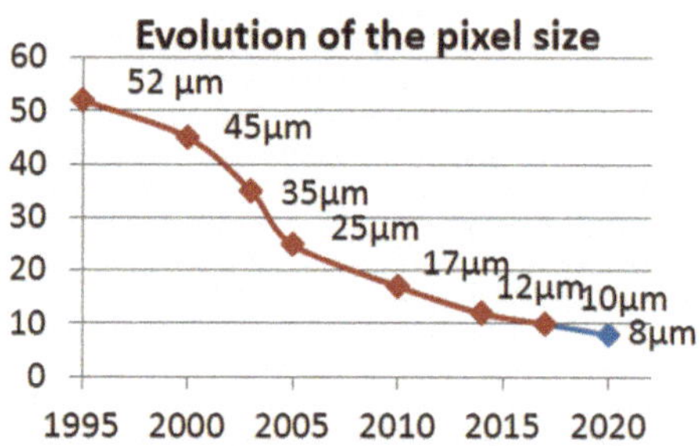

Fig. 13.3 Evolution of the microbolometer pixel size [16].
Reproduced with permission from Telydyne Scientific & Imaging.

both of them exhibit high 1/*f* noise due to their noncrystalline structure [18]. As a result of these disadvantages, new thermosensing materials should be considered in order to lower the microbolometer fabrication process costs and to increase the performance.

Thin silicon nitride membranes are almost transparent in the infrared range except for a wide absorption peak that is exhibited around 12 μm (this absorptance peak is due to the vibrations of the diatomic basis of silicon nitride) [19]. Materials such as black gold and black silicon, with absorptance close to unity, are incorporated as options to increase the infrared absorption.

New surface modification techniques such as the use of femtosecond lasers have been implemented in the fabrication of black metals that exhibit extremely low reflectance across the infrared wavelength spectra [20]. In recent years, new IR radiation absorber materials have emerged with extremely low reflectance that is exhibited across the IR range of wavelengths. Vantablack, a composite of thin film of carbon nanotubes and black silicon (BSi), a surface modification of silicon, exhibits absorptance close to unity [21,22]. Black gold with infrared absorptance close to unity has been developed [23]; but the thin film fabrication process technology that is required for the deposition of these materials may not be compatible with all semiconductor processes; further, the process technology does not allow the mass production of uncooled microbolometers. Black gold, obtained by thermal evaporation methods, exhibits an absorptance close to unity [24]; but the fragile structure of this material and the degradation in absorption properties, when it is exposed to temperatures higher than 300°C, does not make it feasible for mass production or manufacture.

Wafer level packaging technology is under development in order to replace the conventional ceramic package. Although conventional packaging costs are declining, wafer level packaging technology can lower the manufacturing costs considerably [25]. Fig. 13.4 shows a conventional ceramic package (infrared imager/detector) [16]. Fig. 13.5 [26] and Fig. 13.6 [27] show a wafer-level-package of VO_x microbolometers and a packaged imager/detector of VO_x microbolometers, respectively.

Fig. 13.4 Conventional ceramic package (infrared imager/detector) [16].
Reproduced with permission from Telydyne Scientific & Imaging.

Fig. 13.5 Wafer-level-packaged of VO_x microbolometers [26].

Fig. 13.6 Packaged imager/detector of VO_x microbolometers [27].

Teledyne DALSA is one of the leading companies involved in the development of wafer level packaging technology. By stacking a cap wafer, using vacuum sealing technology, onto a CMOS high-speed readout integrated circuit (ROIC) wafer populated with VO_x pixels, Teledyne manufactures smaller, lighter devices with leading edge performance and game-changing lower costs, enabling compact detectors with advanced integrated features that will drive the next phase in the evolution of uncooled IR imaging.

The global infrared camera imagers market is expected to witness considerable growth on account of growing adoption in previously untapped application areas. Increasing consumerization and development of advanced microbolometers has redefined the application development landscape. High investment in developing camera-based scanners is also anticipated to fuel the handheld imagers market demand over the forecast period. Camera-based solutions provide high price-performance ratio and facilitate ease of use, which is further expected to provide avenues for market growth over the next few years. Fig. 13.7 shows a study of the growth of global uncooled IR camera market from 2016 to 2022 [28] reported by Yole development.

Technological innovation has led to the miniaturization and development of lightweight handheld imagers, which is expected to positively impact market growth over the forecast period. The availability of cost effective handheld imagers owing to mass production, and a wide range of diverse applications, integrating this technology, are

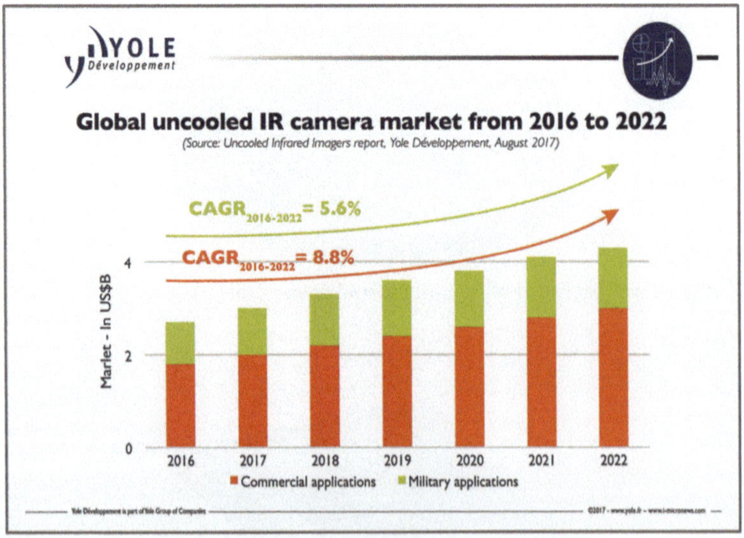

Fig. 13.7 Global uncooled IR camera market from 2016 to 2022 [28].

anticipated to be the key market drivers. Furthermore, the handheld imager market is also expected to witness growth due to declining manufacturing costs and product price [29].

The handheld camera imager product segment consists of stud finders, IR scanners, millimeter wave scanners, and microbolometers. Stud finders are widely used in the construction industry. Magnetic stud detectors and internal capacitor stud finders are the two main types of stud finders. IR scanners are expected to witness high growth through the forecast period. These scanners have the ability to read, scan, and output printed codes and items to a computer and find widespread use in the construction sector. Millimeter wave scanners make use of special electromagnetic waves for detecting objects concealed under a person's clothing. These are extensively used for security related applications in government buildings and airport security checkpoints. Uncooled microbolometers are used as detectors in thermal cameras. They do not require cooling unlike other infrared detecting equipment such as thermopiles, pyroelectrics, MCT, and InGaAs-based infrared detectors. The availability of low-cost microbolometers is also expected to fuel market demand over the next few years. The key industry participants in this sector include Zircon Corporation, Fluke Corporation, FLIR Systems, and ULIS [29].

References

[1] A. Rogalski, Infrared Detectors, CRC Press, Boca Raton, 2010.

[2] A. Rogalski, Infrared detectors: an overview, Infrared Phys. Technol. 43 (2002) 187–210.

[3] A. Rogalski, History of infrared detectors, Opt. Rev. 20 (3) (2012) 279–308.

[4] A. Rogalski, Infrared detectors: status and trends, Prog. Quantum Electron. 27 (2003) 59–210.

[5] A. Rogalski, Infrared detectors for the future, Acta Phys. Pol. A 3 (116) (2009) 308–406.
[6] https://www.osaopn.org/home/articles/volume_22/issue_4/features/seeing_in_the_dark_defense_applications_of_ir_ima/, Accessed 11 March 2019.
[7] D. Buscher, et al., Towards high-resolution astronomical imaging, Astron. Geophys. 60 (3) (2019) 3.22–3.27. https://doi.org/10.1093/astrogeo/atz146.
[8] J.M. Sebastien, et al., Breast cancer detection using infrared thermal imaging and a deep learning model, Sensors 18 (9) (2018) 2799. https://doi.org/10.3390/s18092799.
[9] Infrared Imaging Technology Being Developed to Better Detect Breast Cancer, https://medicalxpress.com/news/2019-06-infrared-imaging-technology-breast-cancer.html, June 26 2019. Accessed 26 January 2020.
[10] A. Newton, et al., Intraoperative near-infrared imaging can identify neoplasms and aid in real-time margin assessment during pancreatic resection, Ann. Surg. 270 (1) (2019) 12–20. https://doi.org/10.1097/SLA.0000000000003201.
[11] http://www.yole.fr/iso_upload/Samples/Yole_Uncooled_thermal_imaging_2014_sample.pdf, Accessed 11 March 2019.
[12] A. Rogalski, Infrared Photon Detectors. SPIE The International Society of Optical Engineering, SPIE Optical Engineering Press, 1995, ISBN: 978-1420076714.
[13] A. Rogalski, Progress in focal plane array technologies, Prog. Quantum Electron. 36 (2012) 342–473.
[14] R.K. Bhan, R.S. Saxena, C.R. Jalwania, S.K. Lomash, Uncooled infrared microbolometer arrays and their characterization techniques, Def. Sci. J. 59 (2009) 580–589.
[15] S. Gupta, R.K. Bhan, A. Katiyar, R. Muraldharan, Design optimization of pixel structure for α-Si based uncooled infrared detector, Def. Sci. J. 63 (2013) 581–588.
[16] https://www.stemmer-imaging.com/media/uploads/dalsa/TE/TELEDYNE-DALSA-EN-Challenges-and-trends-in-microbolometer-design.pdf, Accessed 11 March 2019.
[17] D. Murphy, J. Wyles, M. Rey, J. Asbrock, Performance improvements for VOx microbolometer FPAs, Proc. SPIE 5406 (2004) 531–540.
[18] Q. Cheng, S. Paradis, T. Bui, M. Almasri, Design of dual-band uncooled infrared microbolometer, IEEE Sensors J. 11 (1) (2011) 67–175.
[19] A. Kosarev, A. Torres, M. Moreno, R. Ambrosio, Thin film uncooled microbolometers based on plasma deposited materials, Can. J. Phys. 92 (2014) 570–575.
[20] A.Y. Vorobyev, C. Guo, Femtosecond laser blackening of metals, in: 52nd IEEE International Midwest Symposium on Circuits and Systems, 2009, pp. 905–908.
[21] P.G. Maloney, P. Smith, V. King, C. Billman, M. Winkler, E. Mazure, Emissivity of microstructure silicon, Appl. Opt. 49 (2010) 1065–1068.
[22] S.R. Marthi, N. Ravindra, Optical properties of black silicon: an analysis, The Journal of The Minerals, Metals & Materials Society (TMS) 67 (2015) 2154–2159.
[23] S. Ilias, P. Topart, C. Larouche, P. Beaupré, D. Gay, C. Proulx, T. Pope, C. Alain, Deposition and characterization of gold black coatings for thermal infrared detectors, Proc. SPIE (2010) 77501J–10.
[24] Y. Fang, X. Luo, Q. Zhang, H. Ren, Y. Wang, Preparation and characterization of ultra-low density porous gold foams, Mater. Lett. 67 (2012) 46–48.
[25] F. Forsberg, N. Roxhed, A.C. Fischer, B. Samel, P. Ericsson, N. Hoivik, A. Lapadatu, M. Bring, G. Kittilsland, G. Stemme, F. Niklaus, Very large scale heterogeneous integration (VLSHI) and wafer-level vacuum packaging for infrared bolometer focal plane arrays, Infrared Phys. Technol. 60 (2013) 251–259.
[26] https://www.teledynedalsa.com/en/products/imaging/infrared-detectors/, Accessed 11 March 2019.

[27] https://www.teledynedalsa.com/en/products/imaging/infrared detectors/microbolometers/, Accessed 11 March 2019.
[28] http://www.yole.fr/Uncooled_IR_Imagers_MarketStatus_IRForum.aspx#.Xb3-p-hKiM9, Accessed 11 March 2019.
[29] https://www.grandviewresearch.com/industry-analysis/hand-held-imagers-market, Accessed 11 March 2019.

Applications

14

14.1 Infrared detector applications

An infrared detector is an optoelectronic device that converts the incident infrared radiation into a readable electrical signal. Electromagnetic radiation interacts with matter in various ways and depending on the mechanism involved in the detection process, the interaction is considered to be a photon effect or thermal effect. The radiation-matter interaction determines the infrared radiation absorber material and the infrared radiation spectral range [1].

Night vision represents an area of research that is of significant interest to applications in astronomy as well as in military/defense sector. Infrared detectors are broadly classified into photon detectors and thermal detectors depending on the nature of the radiation-matter interaction. Infrared photon detectors detect infrared optical radiation by direct interaction of photon radiation with active charge carriers within the atomic lattice of the sensing element of the detector. Two major types of photon detectors are photoconductive and photovoltaic devices. Infrared thermal detectors absorb infrared radiation and, as a result, cause changes in the detector electrical properties. Bolometers, thermopiles, and pyroelectric detectors represent the most used infrared thermal detectors in the industry today [2].

A comparison of infrared thermal and photon detector characteristics is presented in Table 14.1. Thermal detectors are slower than photon detectors, because the infrared radiation absorption and conversion process, in a thermal detector, is a two-step process, while in the photon detector, the absorption of a photon and the generation of an electron-hole pair is faster [3]. Thermal detectors are less sensitive than photon detectors, but they can operate in a broad range of wavelengths in contrast to the photon detector that operates at a specific and narrow wavelength range [4]. Thermal detectors operate at room temperature, in contrast to photon detectors that need to be cooled. In infrared photon detectors, the thermal generation-recombination of charge carriers increases the noise effect, and in order to reduce the noise, the temperature of the detector must be maintained at cryogenic levels [4]. The required use of cooling systems, such as Stirling cooling and Peltier cooling, increases the subsequent weight and cost of photon detectors [5].

Infrared photon detectors and thermal detectors have been developed by the infrared detection/imaging industry since the 1960s. These two approaches to detect and measure infrared radiation have advantages and disadvantages that are determined by the specific field of application. The device cost, the range of infrared spectrum wavelength to detect, the response time, and the size are factors to be considered at the time to decide which detector is the most appropriate device for a specific application [6].

Microbolometers. https://doi.org/10.1016/B978-0-08-102812-4.00007-3

Table 14.1 Infrared thermal and photon detector characteristics.

IR detector	Time response	Wavelength range	Operation temperature	Cost
Photon detector (photoconductive, photovoltaic)	Fast (ps)	Specific and narrow	Cryogenic (77 K) liquid nitrogen (thermally induced generation of charge carriers at room temperature	High
Thermal detector (bolometer, pyroelectric, thermopile)	Two-step (conservative ms)	Broadband	Room temperature (300 K)	Low

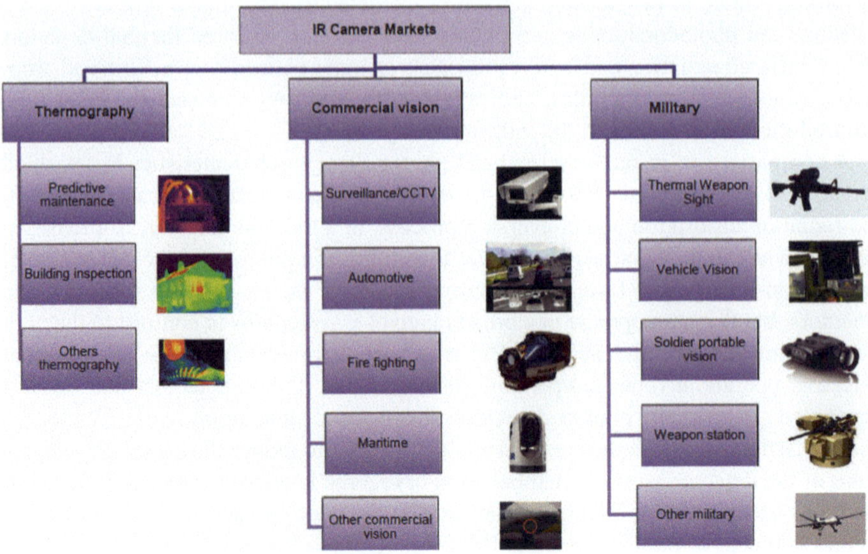

Fig. 14.1 Infrared thermal and photon detector markets [7].

Fig. 14.1 presents a summary of the infrared thermal and photon detector markets [7]. In addition to applications in the military sector, some of the major applications of infrared detectors and imagers are as follows:

- thermal cameras
- particle detectors
- detection of concealed weapons
- fingerprint scanners

- air surveillance
- detection of forest fire
- applications in astronomy.

Each component of the infrared spectrum provides different information that is specific to the wavelength and, therefore, targets different markets. There are five major categories of range of wavelengths for the infrared spectrum [7]:

- NIR (near IR) (0–7 μm—1.4 μm): active vision enhancement, high temperature thermography and material analysis.
- SWIR (short Wave IR) (1.4–3 μm): active vision enhancement, high temperature thermography and material analysis.
- MWIR (medium Wave IR) (3–5 μm): thermography, passive vision enhancement, material analysis.
- LWIR (long wave IR) (8–15 μm): thermography and passive vision enhancement.
- FWIR (far wave IR) (15–1000 μm): thermography, screening, and communication.

Fig. 14.2 shows the infrared electromagnetic spectrum for four main atmospheric absorption windows used by the infrared detection and imaging industry [7].

In order to operate at temperatures between −70°C and −150°C, cooled photon detectors require cooling system equipment. These cooling systems consume a significant amount of power and are costly and bulky. They also require a cooling-down time, which delays the time to first image anywhere from 10 to 20 min. For some applications, the military requires a performance technology that is superior to thermal detectors that can deliver at long distances [8]. Therefore, in certain high-end applications, such as the military, these disadvantages are acceptable [9]. Other military applications require more portable infrared detectors with low power consumption, lightweight, and fast response to access immediate to critical information, such as imaging systems mounted on helmets or weapons.

In the construction and building industry, infrared thermal detectors are the detection systems of choice. Thermal infrared devices that are compact, portable, and

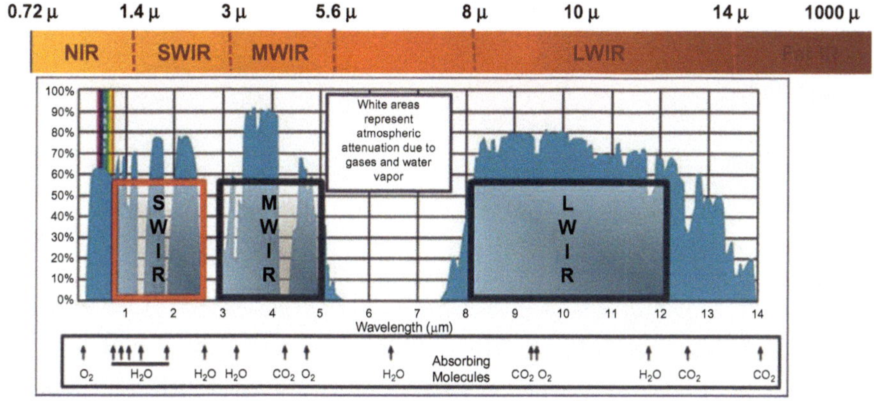

Fig. 14.2 Infrared radiation transmission in earth's atmosphere, over 6000 ft. horizontal path at sea level, containing 17 mm of water precipitate [7].

inexpensive are used by building inspectors and energy auditors that rely on handheld infrared thermography devices to detect issues such as electrical and plumbing problems, water leaks through roofs and foundations, structural defects, moisture accumulation and mold growth, heat loss, and missing insulation.

The automotive industry is incorporating infrared imaging as part of a safety system that can identify objects, animals, or humans on the road and warn drivers before they encounter these potential hazards. In the automotive identification industry, the use of thermal detectors is considered adequate, since human vision sees motion at 60 Hz. Although thermal imaging cameras are still expensive, it is projected that their cost reduction in the near future will be significant and it will be mandatory for their use on all vehicles [10].

Security and video surveillance is another important market that offers potential for the implementation of infrared technology detection. As the cost of uncooled infrared thermal technology continues to decrease, the thermal infrared imaging cameras are the top choice for night-vision applications, such as detecting the movement of individuals, animals, and objects [9]. FLIR [11] is positioned as one of the global leaders in the development and manufacture of infrared photon and infrared thermal detectors, manufacture of infrared detector/imaging devices and offers infrared detector/imaging cores.

Fig. 14.3 shows examples of hyperspectral cameras for Unmanned Aircraft Systems (UAS) [12,13]. These cameras use uncooled VOx detectors that are coupled with absorber technology to capture infrared images in the long-wave infrared spectral range of 8 to 14 μm [13].

Fig. 14.3 Examples of hyperspectral cameras for UAS: (A) Photonfocus MV1-D2048x1088-HS05-G2; (B) RESONON Pika L; (C) SENOP HSC-2; (D) SPECIM FX10; (E) XIMEA xiSpec Series [12,13].
From Adamopoulos, E.; Rinaudo, F. UAS-Based Archaeological Remote Sensing: Review, Meta-Analysis and State-of-the-Art. Drones 2020, 4, 46. https://doi.org/10.3390/drones4030046.

Fig. 14.4 Compact long wavelength infrared (LWIR) thermal core camera (Model Boson 640, 95 (HFOV) 4,9 MM) [14].

The Boson long-wave infrared (LWIR) thermal camera module sets a new standard for size, weight, power, and performance (SWaP). It utilizes FLIR infrared video processing architecture to enable advanced image processing and several industry-standard communication interfaces while keeping the power consumption low. The 12 μm pitch Vanadium Oxide (VO_x) uncooled detector comes in two resolutions: 640 × 512 or 320 × 256. It is available with multiple lens configurations, adding flexibility to integration programs [11]. Fig. 14.4 shows the thermal camera core Boson 640, 95 (horizontal field of view, HFOV) 4,9 MM) [11, 14].

Where SWaP are design constraints, and when cost and performance is desired, Neutrino is the ideal solution. The Neutrino offers FLIR's common image processing architecture and the resulting image from the 640 × 512 InSb array is excellent. Weighing in at less than 16 oz. (<450 g) and consuming less than 5 watts at steady-state, it is FLIR's balanced MWIR core optimizing performance, cost, and ease of integration [14]. Fig. 14.5 shows the high performance cooled infrared camera mid-wavelength infrared (MWIR) core (Model Neutrino) [14].

14.2 Applications of microbolometers

A microbolometer is a specific type of bolometer used to detect infrared radiation and it is integrated into a thermal detector or camera. A microbolometer absorbs infrared radiation, transducing the heat due to the incident radiation into a readable output, which can be represented graphically. The infrared sensing area is a grid of microbolometers.

The infrared radiation is absorbed by infrared absorbers such as silicon nitride and silicon oxide. The heat, as the result of the radiation absorption, changes the electrical properties of sensing materials such as vanadium oxide or amorphous silicon

Fig. 14.5 High performance cooled infrared camera mid-wavelength infrared (MWIR) core (Model Neutrino) [14].
Reproduced with permission from Neutrino SWaP+C Series | Teledyne FLIR.

(electrical resistivity). Microbolometers work in the long-wavelength infrared band from 7.5 to 15 μm. Microbolometers offer several advantages over traditional cooled detectors, including smaller size, lighter weight, lower cost, immediate power-up capabilities, and decreased power consumption [15].

The low cost and portability are the main reasons why uncooled thermal detectors are considered as an alternative to replace cooled photon detectors. Microbolometer approach, with its low weight, complexity, size, and lower power consumption, is the ideal candidate to be able to manufacture low cost and portable thermal/imaging systems.

The key benefits of bolometers include the following:

- Operate at room temperature and hence do not require cooling.
- Extremely efficient in terms of sensitivity and energy resolution when compared to other infrared detectors.
- Low cost with a reasonable performance.

Microbolometer technology has been accepted extensively for applications in infrared detection and imaging, where it is impractical to use cooled infrared detectors (for example, military systems, space applications, and portable electronics). Advantages of microbolometers include operation at the desired temperature including room temperature, flat response over the infrared spectrum, and conservative time response; this allows to consider microbolometer infrared detectors as the ideal choice in the development of uncooled infrared detectors and imaging systems [14].

Uncooled thermal detectors have certain disadvantages when compared with their cooled counterparts. They are less sensitive and are intrinsically slow (the typical time constant is 10 ms). These limitations do not affect their suitability for an ever-increasing number of cost-sensitive commercial applications that simply need acceptable image quality, not high performance or speed. The high cost of cooled infrared detectors, in fact, makes them unattainable for the vast majority of commercial applications [15].

When high-end infrared detectors are required, cooled photon infrared detectors are needed in order to achieve high sensitivities and distances of the target [16]. These specific applications will continue to use less portable, more expensive cooled photon detectors/imaging systems that do not provide instant-on image acquisition. Other applications, including military and commercial sectors that need immediate access to critical information, require the use of thermal systems that are portable, lightweight with low battery power consumption [17].

There are several infrared detector manufacturing companies, each of them has developed their own technologies, using different materials and sensing element configurations that position them as leaders in specific application areas. Fig. 14.6 presents a summary of the main infrared uncooled camera manufacturers positioning in the industry [7].

FLIR Systems is the world's largest commercial company specializing in the design and production of thermal imaging cameras, components and imaging detectors. The main characteristics of these thermal imaging cameras are as follows:

- Detector type: long-life, uncooled vanadium oxides (VO_x) microbolometer
- Pixel pitch: 17 μ
- Sensitivity: <50 μmK @ 25°C F# 1.0

FLIR manufactures photon and thermal infrared cameras and components for a wide variety of commercial and military applications. Figs. 14.7-14.11 show some examples of thermal imaging cameras manufactured by FLIR that use cooled microbolometer as infrared radiation detection elements [18–22].

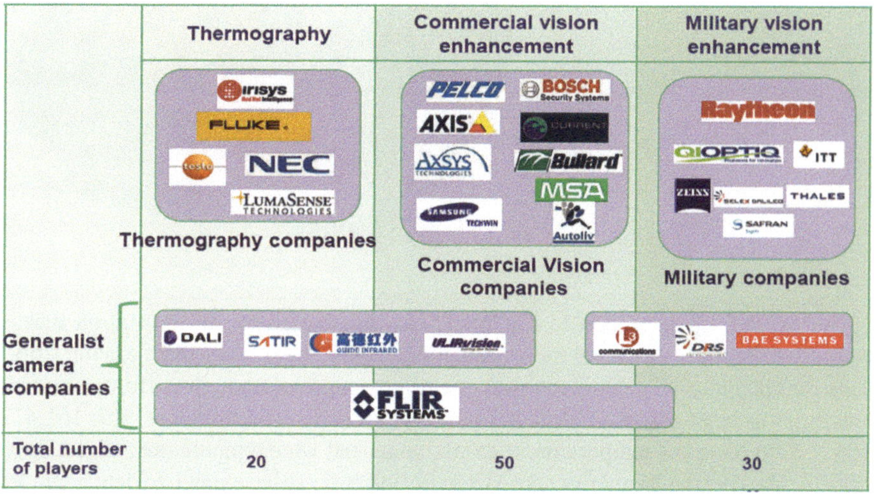

Fig. 14.6 Summary of infrared uncooled camera manufacturers positioning in the industry [7].

Fig. 14.7 Industrial advanced thermal camera FLIR E95 model: FLIR E95 24 [18].

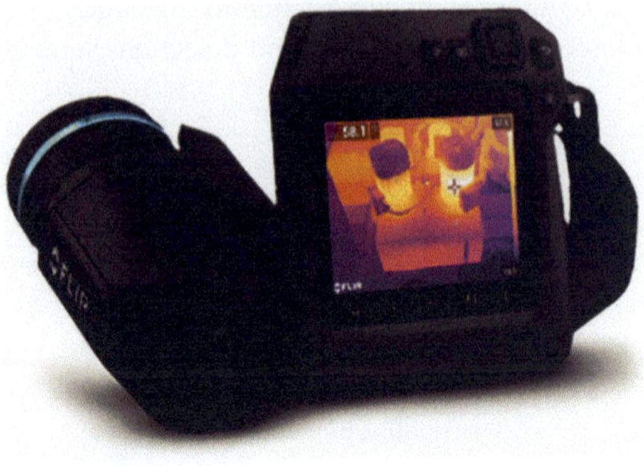

Fig. 14.8 Professional thermal camera FLIR T530 model: FLIR T530, 24 [19].

The FLIR E95, presented in Fig. 14.7, has the sensitivity and resolution that is required to diagnose electrical faults, find hidden deficiencies, and maintain the workplace running smoothly. This thermal imaging camera offers high-performance features, such as laser-assisted autofocus and on-screen area measurement, plus 161,472 (464×348) points of temperature measurements and wide temperature ranges, up to 1500°C. The E95 has a superior spot-size ratio and is sensitive enough to detect minute temperature differences [18].

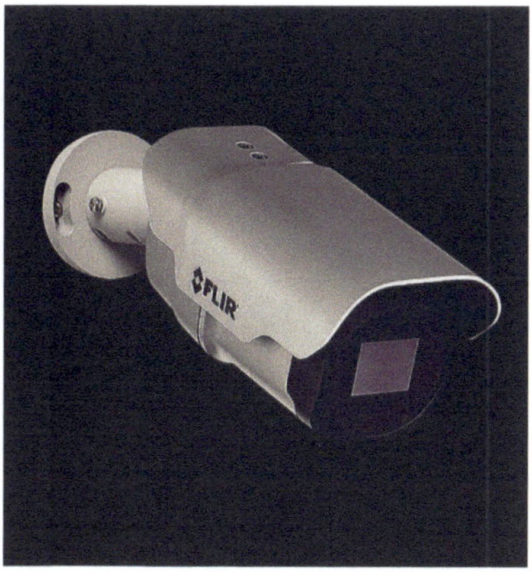

Fig. 14.9 Thermal analytics security camera Elara FB-series ID Model: FB-695 ID, 95 degree FOV, 17 MM [20].

Fig. 14.10 Mobile accessories pro-grade thermal camera for camera for smartphone FLIR one Pro model: FLIR ONE PRO—IOS [21].

The FLIR T530, shown in Fig. 14.8, has the features that professionals need to accurately troubleshoot hot spots and potential faults. With the 180 degree rotating lens platform and a bright 4" LCD, the FLIR T530 is engineered to help users diagnose hard-to-reach components in any environment. Advanced on-camera measurement

Fig. 14.11 Military soldier thermal binocular with EO, laser rangefinder, and laser pointer FLIR Recon [21].

tools, laser-assisted autofocus, and FLIR's industry-leading image quality ensure that the user will find and diagnose problems quickly [19].

The FLIR Elara FB-Series ID thermal security camera, displayed in Fig. 14.9, uses onboard analytics to classify human or vehicular intrusions, making it ideal for perimeter intrusion detection and sterile-zone monitoring. The Elara FB-Series ID can also hand off classified intrusions to FLIR PTZ cameras for autonomous tracking of intruders. Easy to set up, Elara FB-Series cameras provide reliable detection with low false alarm rates in challenging environments, bad weather, and complete darkness [20].

The FLIR ONE Pro helps the user to find invisible problems faster than ever, whether the application involves inspecting electrical panels, troubleshooting mechanical systems, investigating HVAC related problems, or finding water damage. The FLIR ONE Pro-Series camera offers 4 × the native resolution of the FLIR ONE Pro LT, for sharper image clarity that is further enhanced by the revolutionary FLIR Vivid IR image processing. The camera permits measurement of temperatures that is more than 3 × higher than any FLIR ONE model—up to 400°C (752°F)—with a sensitivity that detects temperature differences down to 70 mK. Packed with powerful measurement tools, the FLIR ONE Pro is reported to be very user-friendly [21].

The Recon V, shown in Fig. 14.11, is a compact and light yet rugged multisensor thermal binocular designed for 24/7 day and night field operations that require enhanced imagery and long standoff range. Its internal GPS, DMC, and long-range laser pointer provide accurate range to target as well as precise location. The 10 × continuous zoom optics, MEMS-based electronic stabilization, and high-definition color video display combine to yield unmatched image quality and flexibility [21].

BAE Systems [23], with its 12-μm uncooled pixel technology, facilitates the imaging system manufacturers to reduce lens size by 50%, thereby reducing the costs of the optics by 20% without compromising image quality. This results in reduced weight and reduced power requirements of the infrared camera. The TWV640 camera core, shown in Fig. 14.12, has been reported to capture images through fog, smoke, dust, and haze at 60 Hz.

Fig. 14.12 TWV 640 thermal camera core [23,24].
From https://www.flir.in/products/boson/.

14.3 Examples of applications of microbolometers

Historically, while the defense and military sectors of the global economy have been the primary drivers of much of the research in the development and manufacture of microbolometers, in recent years, interest in civilian applications of microbolometers is on the rise. Some of these applications will be illustrated in this section of the chapter.

Military/Defense Sector: The year 2023 marks the sixtieth year since the inception of the forward-looking infrared (FLIR) technology by Texas Instruments in 1963 [25]. Today, Raytheon [26] continues to advance the technology by addressing the industry needs such as long range, improved sensor integration, reduced power requirements, and reduced weight. In addition to providing advanced targeting capabilities such as

Fig. 14.13 Long-range advanced scout surveillance system [26].

Fig. 14.14 Heads-up display [28].
From https://en.wikipedia.org/wiki/Head-up_display#/media/File:PZL_TS-11F_Iskra_(HUD).jpg.

seeing through smoke, rain, snow, and fog, today's defense technology offers products such as the Long-Range Advanced Scout Surveillance System, shown in Fig. 14.13. Another example of the functionality of the forward sensor is to display the image to the pilot through the aircraft's heads-up display (HUD) [27]—Fig. 14.14 [28]. HUD

Fig. 14.15 StormCaster payload family—SkyRaider [29].

facilitates eyes-forward capability by improving safety and enhancing operations in real-time [28].

The integration of infrared imaging with Unmanned Aerial Systems (UAS) represents another example of recent advancements in the applications of microbolometers. FLIR has recently introduced StormCaster Payload Family for its SkyRaider and SkyRanger UAS [29] that feature a FLIR Boson thermal camera (Fig. 14.15). The camera facilitates in the delivery of sharp, clear images, day or night, at maximum range. The objective is to be able to detect objects, recognize them, and acquire targets. These multisensor, multifunction UAS should help in the supervision, search, and rescue as well as accomplish missions in a timely manner.

Some of the military applications include motion detection, object/target detection, object/target recognition, target acquisition, search and rescue, soldier vision, thermal weapon sight, and weapon station. Automatic target recognition (ATR), Perimeter Monitoring Systems (PMS), and Moving Target Identification (MTI) are critical for the detection, tracking, and classification of targets from video and still images. Thermal imaging has unique potential due to its ability to operate under all weather conditions [27]. Surveillance is key to the success of each and every mission and is one of the critical applications of infrared imaging. Extended Air Defense (EAD) and Ballistic Missile Defense (BMD) require dual mode sensors. The concept of integration of various components in the missile seeker is shown in Fig. 14.16 [30].

Civilian Sector: Spatial and temporal measurements of temperature, remotely, is critical for a variety of applications including food [31], medicine [32], skin [33], water deposits in porous materials [34], sweat retention in porous materials [35], as well as in nondestructive testing of materials (Fig. 14.17) [36]. Allied Vision [37] manufactures short-wavelength infrared cameras for use in infrared microscopy to detect defects during processing of semiconductors. TECHIMAGING [38] offers infrared thermography for a variety of applications, including aerial thermography, nondestructive material testing, etc. Since abnormal body temperature is a sign of illness [39], infrared thermography offers a quick, passive, noncontact, and noninvasive approach to monitor body temperature. While mammography is the traditional method for screening breast cancer, thermography is being adopted as an alternative [40].

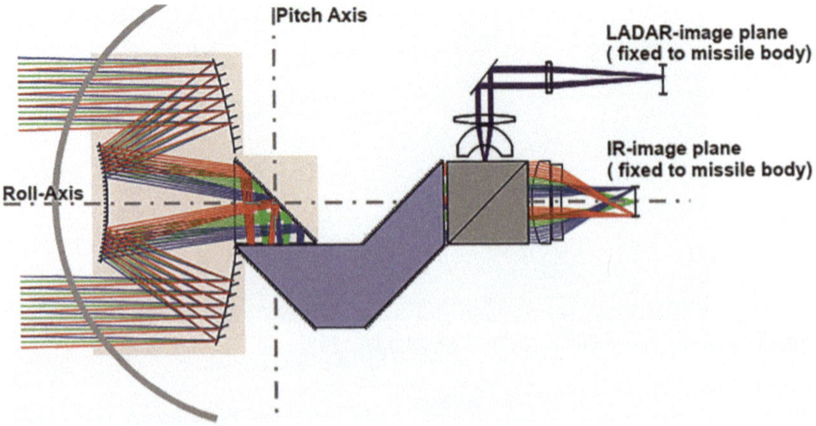

Fig. 14.16 Integration concept of the missile seeker [30].

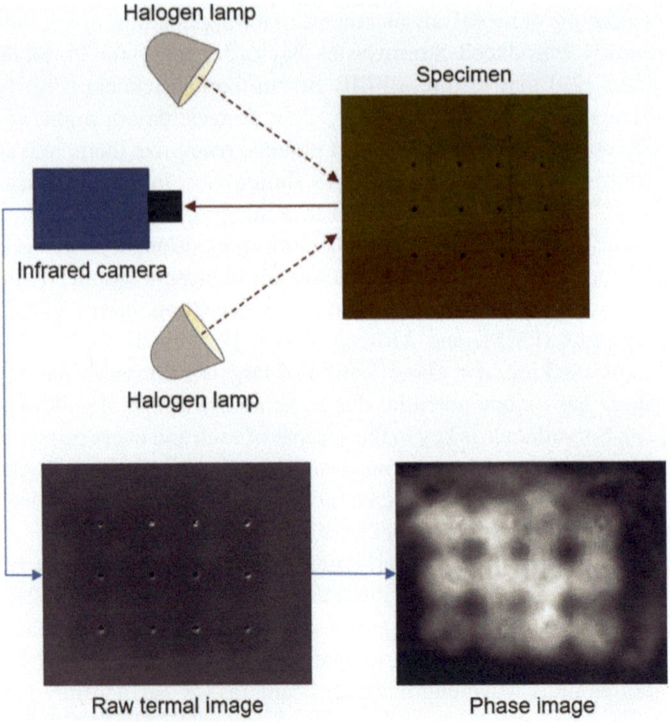

Fig. 14.17 Schematic setup of a possible active thermographic inspection [36].

In addition to causing billions of dollars of damage, forest fires are known to increase the carbon dioxide levels in the atmosphere contributing, thereby, to the greenhouse effect and climate change [41]. Subsequently, the resulting ashes destroy much of the nutrients. This leads to soil erosion causing flooding and landslides.

In Fig. 14.18, a Seek Thermal [42] infrared camera attached to a smart phone is shown. While this product is intended for experts and contractors in industrial and commercial building trades [42], Seek Thermal [43] has been very actively involved in the manufacture and supply of thermal imaging cameras for fire-fighting applications (Fig. 14.18).

With the rapid growth in the autonomous vehicle industry, by addressing the necessity for safety, thermal imaging has become a key component in self-driving cars. An image as seen through the eyes of a thermal infrared camera is shown in Fig. 14.19 [44].

The ability to customize the shape, geometry, design, size, and choice of materials, while at the same time, minimizing waste and risk [45] are some of the benefits of additive manufacturing (AM) or commonly known as 3D printing. The spatial and

Fig. 14.18 Seek CompactPRO thermal imaging camera [42]. From https://www.flir.com/products/flir-one-pro/.

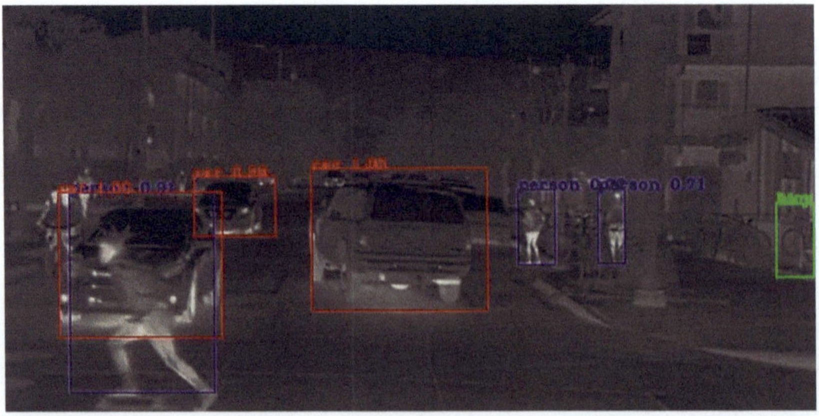

Fig. 14.19 Through the "eyes" of a thermal infrared camera [44].

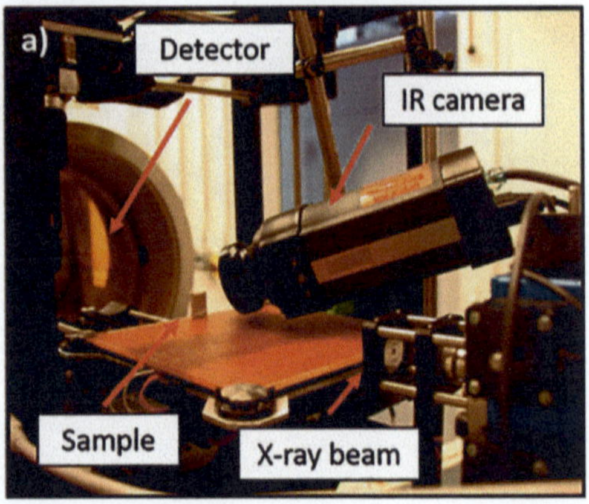

Fig. 14.20 Enlarged image showing placement of the IR thermal imaging camera and the X-ray detection system [47].

temporal measurement of temperatures of the extruded material, the substrate, and the decay temperature [46] are critical to the understanding of the growth and adhesion of layers as well as their mechanical properties. In Fig. 14.20, a photograph of the simultaneous in situ X-ray scattering and infrared imaging systems for monitoring polymer extrusion in additive manufacturing [47] is shown.

Astronomy will continue to benefit from advancements in infrared imaging. One such example is graphene-based bolometers [48]. Two-dimensional materials are expected to lead to several advantages in the manufacture of bolometers. These include light weight, low-power, high mobility, high-frequency, improved radiation hardness, and superior mechanical properties.

The ability to incorporate microsensors, along with functional coatings, on surfaces for the fabrication of smart structures [49,50], by methods such as cold spray technology, will lead to new technologies. In recent years, there has been renewed activity in understanding the fundamental mechanisms of cold spray technology through

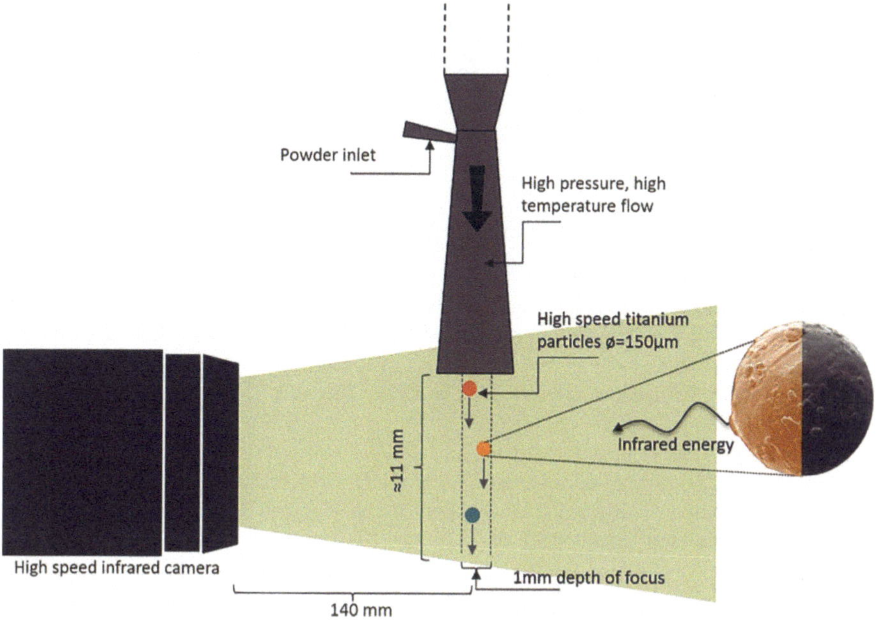

Fig. 14.21 Cold spray of titanium particles and infrared imaging processes setup for particle surface temperature readings. The camera is positioned to frame both the nozzle exit area and in-flight particles [50].

computational fluid dynamics as well as particle in-flight temperature measurements by utilizing high-speed infrared imaging—Fig. 14.21 [50,51]. Telops [52] is one of the major designers and manufacturers of high speed infrared cameras for such applications in cold spray technology.

Integration of microbolometers as infrared imagers with drones, 3D printing, and autonomous vehicles are just a few of the examples of developments that will continue to evolve. In addition, spatial and temporal temperature measurements of semiconductor devices, circuits and systems, rechargeable batteries, ultracapacitors, transformers, power supplies, solar cell panels, turbines, and windmills, in real-time, for a variety of energy-related and thermal management applications will continue to catalyze the growth of the infrared imaging industry. The applications, utilization, and advantages of deploying infrared imaging to diagnose human health is still in its infancy. There are a significant number of health-related opportunities; these include in situ during surgery or as a post-operation tool, that need to be explored.

References

[1] R.K. Bhan, R.S. Saxena, C.R. Jalwania, S.K. Lomash, Uncooled infrared microbolometer arrays and their characterization techniques, Def. Sci. J. 59 (2009) 580–589.

[2] A. Kosarev, A. Torres, M. Moreno, R. Ambrosio, Thin film uncooled microbolometers based on plasma deposited materials, Can. J. Phys. 92 (2014) 570–575.

[3] S. Gupta, R.K. Bhan, A. Katiyar, R. Muraldharan, Design optimization of pixel structure for α-Si based uncooled infrared detector, Def. Sci. J. 63 (2013) 581–588.
[4] M. Alhussein, S.I. Haider, Simulation and analysis of uncooled microbolometer for serial readout architecture, J. Sens. (2016) 1–13.
[5] P.W. Kruse, The Photon Detection Process in Optical and Infrared Detectors, Springer Verlag, New York, Heidelberg, Berlin, 1980 (Rogalski A. Infrared Detectors. CRC Press, Boca Raton; 2010).
[6] A. Rogalski, Infrared Photon Detectors. SPIE The International Society of Optical Engineering, SPIE Optical Engineering Press, 1995, ISBN: 978-1420076714.
[7] http://www.sensorsportal.com/DOWNLOADS/Uncooled_Infrared_Cameras_and_Detectors_Report_sample.pdf, Accessed 31 October 2019.
[8] C. Corsi, History highlights and future trends of infrared sensors, J. Mod. Opt. 57 (18) (2010) 1663–1686.
[9] A. Rogalski, Infrared detectors: an overview, Infrared Phys. Technol. 43 (2002) 187–210.
[10] A. Rogalski, History of infrared detectors, Opto-Electron. Rev. 20 (3) (2012) 279–308.
[11] https://www.flir.com/products/boson/, Accessed 31 October 2019.
[12] E. Adamopoulos, F. Rinaudo, UAS-Based Archaeological Remote Sensing: Review, Meta-Analysis and State-of-the-Art. Drones 2020, 4, 46. https://doi.org/10.3390/drones4030046.
[13] https://resonon.com/?gclid=CjwKCAiAm7OMBhAQEiwArvGi3KOMujGVD4rqwwy5cJE1PivvKfJXPlREkoJJTpeJjgCRz9XFYY2tCRoClEYQAvD_BwE. Accessed 11 November 2021.
[14] https://www.flir.com/products/neutrino/, Accessed 31 October 19.
[15] Q. Cheng, S. Paradis, T. Bui, M. Almasri, Design of dual-band uncooled infrared microbolometer, IEEE Sensors J. 11 (2011) 167–175.
[16] F. Sizov, IR region challenges: photon or thermal detectors? Outlook and means, Semicond. Phys., Quantum Electron. Optoelectron. 15 (2012) 183–199.
[17] N. Shen, J. Yu, Z. Tang, An uncooled infrared microbolometer array for low-cost applications, IEEE Photon. Technol. Lett. 27 (12) (2015) 1247–1249.
[18] https://www.flir.com/products/t530/, Accessed 31 October 2019.
[19] https://www.flir.com/products/fb-series-id/, Accessed 31 October 2019.
[20] https://www.flir.com/products/flir-one-pro/, Accessed 31 October 2019.
[21] https://www.flir.com/products/recon-v/, Accessed 31 October 2019.
[22] https://www.flir.com/products/e95/, Accessed 31 October 2019.
[23] Inspiring a New Class of Infrared Imaging Systems, https://www.baesystems.com/en-us/article/inspiring-a-new-class-of-infrared-imaging-systems, Accessed 2 December2019.
[24] TWV 640 Thermal Core, https://www.fairchildimaging.com/products/thermal-cores/twv640-thermal-camera-core, Accessed 2 December 2019.
[25] Fifty Years of Owning the Night: The History of Infrared Imaging, https://www.raytheon.com/news/feature/rtn13_flir01, Accessed 2 December 2019.
[26] https://www.raytheon.com/capabilities/products/lras3, Accessed 2 December 2019.
[27] A. Akula, R. Ghosh, H.K. Sardana, Thermal imaging and its application in defense systems, optics: phenomena, materials, devices and characterization, AIP Conf. Proc. 1391 (2011) 333–335, https://doi.org/10.1063/1.3643540.
[28] Head-up Display (HUD), https://en.wikipedia.org/wiki/Head-up_display#/media/File:PZL_TS-11F_Iskra_(HUD).jpg. Accessed 2 December 2019.
[29] FLIR Introduces StormCaster Payload Family for its SkyRaider and SkyRanger Unmanned Aerial Systems, in Business Wire, November 19th 2019.

[30] J. Barenz, R. Baumann, H.D. Tholl, Eyesafe imaging LADAR/infrared seeker technologies, in: Proceeding of SPIE 5791, Laser Radar Technology and Applications X, 2005, https://doi.org/10.1117/12.603329 (Event: Defense and Security, 2005, Orlando, Florida).
[31] A.A. Gowen, B.K. Tiwari, P.J. Cullen, K. McDonnell, C.P. O'Donnell, Applications of thermal imaging in food quality and safety assessment, Trends Food Sci. Technol. 21 (4) (April, 2010) 190–200.
[32] J.R. Whitney, A. Rodgers, E. Harvie, W.F. Carswell, S. Torti, A.A. Puretzky, C.M. Rouleau, D.B. Geohegan, C.G. Rylander, M.N. Rylander, Spatial and temporal measurements of temperature and cell viability in response to nanoparticle-mediated photothermal therapy, Nanomedicine 7 (11) (2012), https://doi.org/10.2217/nnm.12.66.
[33] L. Gao, Y. Zhang, V. Malyarchuk, L. Jia, K.I. Jang, R.C. Webb, H. Fu, Y. Shi, G. Zhou, L. Shi, H.X. Shah, B. Xu, C. Yu, Y. Huang, J.A. Rogers, Epidermal photonic devices for quantitative imaging of temperature and thermal transport characteristics of the skin, Nat. Commun. 5 (2014) 4938, https://doi.org/10.1038/ncomms5938.
[34] N.P. Avdelidis, A. Moropoulou, P. Theoulakis, Detection of water deposits and movement in porous materials by infrared imaging, Infrared Phys. Technol. 44 (2003) 183–190.
[35] M. Raccuglia, C. Heyde, A. Lloyd, S. Hodder, G. Havenith, The use of infrared thermal imaging to measure spatial and temporal sweat retention in clothing, Int. J. Biometeorol. 63 (2019) 885–894.
[36] R. Usamentiaga, P. Venegas, J. Guerediaga, L. Vega, J. Molleda, F.G. Bulnes, Infrared thermography for temperature measurements and non-destructive testing, Sensors 14 (2014) 12305–12348, https://doi.org/10.3390/s140712305.
[37] Imaging through semiconductor material, https://www.alliedvision.com/en/news/detail/news/imaging-through-semiconductor-material.html, Accessed 3 December 2019.
[38] https://www.techimaging.com/applications/infrared-thermal-imaging-applications, Accessed 2 December 2019.
[39] B.B. Lahiri, S. Bagavathiappan, T. Jayakumar, J. Philip, Medical applications of infrared thermography: a review, Infrared Phys. Technol. 55 (2012) 221–235.
[40] Comparing mammography and thermography, https://www.medicalnewstoday.com/articles/316632.php, July 22 2019.
[41] https://www.ompe.org/en/causes-et-consequences-of-forest-fires/, Accessed 2 December 2019.
[42] https://www.thermal.com/press-releases/seek-thermal-announces-its-first-affordable-professional-grade-infrared-thermal-imaging-camera-for-the-smartphone, Accessed 3 December 2019.
[43] https://www.thermal.com/seek-thermal-supplies-the-entire-los-angeles-fire-department-with-lifesaving-thermal-imaging-cameras.html, Accessed 3 December 2019.
[44] Autonomous Vehicle Industry Remains Cool to Thermal Imaging, https://spie.org/news/thermal-infrared-for-autonomous-vehicles?SSO=1, 25 September 2018.
[45] A. Pearson, 10 Advantages of 3D Printing, 2018. https://3dinsider.com/3d-printing-advantages/.
[46] R.B. Dinwiddie, V. Kunc, J.M. Lindal, B. Post, R.J. Smith, L. Love, C.E. Duty, Infrared imaging of the polymer 3D-printing process, in: Proceeding of SPIE 9105, Thermosense: Thermal Infrared Applications XXXVI, 910502, 2014, https://doi.org/10.1117/12.2053425 (Event: SPIE Sensing Technology+Applications, 2014, Baltimore, Maryland).
[47] Y. Shmueli, J. Jiang, Y. Zhou, Y. Xue, C.-C. Chang, G. Yuan, S.K. Satija, S. Lee, C.-Y. Nam, T. Kim, G. Marom, D. Gersappe, M.H. Rafailovich, Simultaneous in situ X-ray scattering and infrared imaging of polymer extrusion in additive manufacturing, ACS Appl. Polym. Mater. 1 (6) (2019) 1559–1567.

[48] D.K. Efetov, R. Shiue, Y. Gao, B. Skinner, E.D. Walsh, H. Choi, J. Zheng, C. Tan, G. Grosso, C. Peng, J. Hone, K.C. Fong, D. Englund, Fast thermal relaxation in cavity-coupled graphene bolometers with a Johnson noise read-out, Nat. Nanotechnol. 13 (2018) 797–801, https://doi.org/10.1038/s41565-018-0169-0.

[49] J. Villafuerte, Current trends in cold spray technology: looking at the future, Mater. Today (2010).

[50] A. Nastic, B. Joroin, Evaluation of heat transfer transport Coefficient for cold spray through computational fluid dynamics and particle in-flight temperature measurement using a high-speed IR camera, J. Therm. Spray Technol. 27 (2018) 1491–1517.

[51] A. Sova, M. Doubenskaia, Petrovskiy, I. Smurov, Visualization of particle jet in cold spray by infrared camera: feasibility tests, Int. J. Adv. Manuf. Technol. 95 (2018) 3057–3063.

[52] https://www.telops.com/about-us, Accessed 4 December2019.

Index

Note: Page numbers followed by *f* indicate figures and *t* indicate tables.

FSC
www.fsc.org
MIX
Papier | Fördert gute Waldnutzung
FSC® C083411

Kolibri 360 GmbH
Ferdinand-Jühlke-Straße 7
99095 Erfurt, Deutschland
produktsicherheit@kolibri360.de